移动互联网开发技术丛书

"十三五"江苏省高等学校重点教材
编号：2019-2-177

Android
案例开发项目实战

施冬梅 孙翠改 主编

清华大学出版社
北京

内 容 简 介

本书以 Android Studio 为开发环境，对 Android 基础编程和 Android 网络编程中最重要的基础内容分任务进行了讲解，知识的学习与任务的实施得到了很好的结合。全书共包含 11 个项目：初识 Android、猜猜我的星座、智能计算器、打地鼠小游戏、记忆的仓库——备忘录、多彩水果店、唱歌跳舞小管家——媒体播放器、我的第一桶金——理财通 App 的设计与实现、我的第一道菜——菜谱 App 的设计与实现、实战案例——移动互联网软件开发竞赛1、实战案例——移动互联网软件开发竞赛2。内容涵盖了 Android 基本理论、Activity、基础 UI 编程、高级 UI 编程、Intent、BroadcastReceiver、SQLite 数据存储、ContentProvider 数据共享、Service 服务及网络编程等。

本书重点突出，强调动手操作能力，以项目驱动的方式，引导读者在完成一个又一个项目的同时，轻松掌握每个项目支撑的知识点。另外，还通过微课详细讲述了每个知识点和操作演示，方便读者自学，使得读者能够快速理解掌握 Android 的基础知识及基本应用。本书使用 Android 11（API 30）版本并向下兼容。

本书适用面广，可作为高校、培训机构的 Android 教材，也可作为计算机科学与技术、软件外包、计算机软件、计算机网络、电子商务等专业的程序设计课程的教材。

本书封面贴有清华大学出版社防伪标签，无标签者不得销售。
版权所有，侵权必究。举报：010-62782989，beiqinquan@tup.tsinghua.edu.cn。

图书在版编目（CIP）数据

Android 案例开发项目实战/施冬梅，孙翠改主编．—北京：清华大学出版社，2021.10（2024.8重印）
（移动互联网开发技术丛书）
ISBN 978-7-302-58287-8

Ⅰ．①A… Ⅱ．①施… ②孙… Ⅲ．①移动终端—应用程序—程序设计 Ⅳ．①TN929.53

中国版本图书馆 CIP 数据核字（2021）第 105894 号

策划编辑：魏江江
责任编辑：王冰飞　薛　阳
封面设计：刘　键
责任校对：焦丽丽
责任印制：刘海龙

出版发行：清华大学出版社
　　　　网　　址：https://www.tup.com.cn，https://www.wqxuetang.com
　　　　地　　址：北京清华大学学研大厦A座　　邮　编：100084
　　　　社 总 机：010-83470000　　邮　购：010-62786544
　　　　投稿与读者服务：010-62776969，c-service@tup.tsinghua.edu.cn
　　　　质量反馈：010-62772015，zhiliang@tup.tsinghua.edu.cn
　　　　课件下载：https://www.tup.com.cn，010-83470236
印 装 者：涿州汇美亿浓印刷有限公司
经　　销：全国新华书店
开　　本：185mm×260mm　　印　张：30.5　　字　数：744 千字
版　　次：2021 年 11 月第 1 版　　印　次：2024 年 8 月第 6 次印刷
印　　数：10001～12000
定　　价：89.80 元

产品编号：086807-02

前 言
FOREWORD

Android 是 Google 公司开发的基于 Linux 的开源操作系统，主要应用于智能手机、平板电脑等移动设备。随着 Android 系统在全球的大规模推广，还可应用于穿戴设备、智能家居等领域。据不完全统计，Android 系统已经占据全球智能手机操作系统的 80% 以上的份额，我国市场的占有率更是高达 90% 以上。由于 Android 的迅速发展，导致市场对 Android 开发人才需求猛增，因此越来越多的人开始学习 Android 技术，以适应市场需求寻求更广阔的发展空间。

本书将 Android 开发中的基础知识整理出来，分布到项目的实施过程中。项目的规模和难度呈阶梯形递增，符合编程开发的学习规律，结合多位任课教师多年的教学经验进行教材内容的设计，力争教材结构合理、实用性强。

本书既适合作为各类院校的 Android 课程教材，也适合想快速入门并提高 Android 编程技能的初、中级用户自学使用。

主要内容

本书以 Android Studio 为开发环境，对 Android 基础编程和 Android 网络编程中最重要的基础内容分任务进行了讲解，知识的学习与任务的实施得到了很好的结合。全书包含 11 个项目，各项目内容如下。

项目 1　初识 Android。主要介绍手机操作系统、Android 的发展历史和版本、Android 特征、Android 平台架构、Android 开发环境、创建 Android 工程的方法、创建 Android 模拟器的方法、运行与调试 Android 程序的方法。

项目 2　猜猜我的生肖。主要介绍 TextView 控件的属性和方法、EditText 控件的属性和方法、Button 控件的监听器、ImageView 控件的属性和方法、DatePicker 控件的属性和方法、App 之间的通信方式。

项目 3　智能计算器。主要介绍 Android 的 UI 的含义、Android 的几种常见布局、Android 系统中的样式和主题。

项目 4　打地鼠小游戏。主要介绍 Toast 消息提示的方法和使用、Dialog 对话框的方法和使用、Spinner 控件的方法和使用、Android 中的适配器、Menu 菜单的使用方法。

项目 5　记忆的仓库——备忘录。主要介绍 SharedPreferences 实现简单的数据存储方法、Android 中的文件存储方法、调用摄像头和相册获取图片的方法、Android 运行时权限设置。

项目 6　多彩水果店。主要介绍 ListView 控件的使用、RecyclerView 控件的使用、Android 系统中的样式和主题、Serializable 序列化接口的应用。

项目 7　唱歌跳舞小管家——媒体播放器。主要介绍 Timer 及 TimerTask 的使用、Handler 的工作机制、RecyclerView 控件的使用、MediaPlayer 类的使用、SeekBar 控件的使用、VideoView 控件的使用、媒体控制柄 mediaController 的使用项目。

项目 8　我的第一桶金——理财通 App 的设计与实现。主要介绍 Android 系统的数据存储方式、SQLite 数据库存储数据、数据库的增删改查操作。

项目 9　我的第一道菜——菜谱 App 的设计与实现。主要介绍 ViewPager 控件的使用、HTTP、Java 中的输入/输出流操作、HttpURLConnection 访问网络、第三方网络通信库 OkHttp 访问网络、解析 JSON 格式的数据。

项目 10　实战案例——移动互联网软件开发竞赛 1，主要介绍 MVP 设计模式、UI 设计标准、MaterialDesign、四大组件、资源使用、Handler/多线程/定时器、网络请求框架、数据封装和解析、多媒体、手势识别、依赖注入、事件传递、内存泄漏管理、数据存储、业务逻辑、数据挖掘和开源图表库 MPAndroidChart API。

项目 11　实战案例——移动互联网软件开发竞赛 2。主要介绍请求框架、Handler/多线程/定时器、数据封装和解析、多媒体、依赖注入、事件传递、数据挖掘和开源图表库 MPAndroidChart API。

为了更便于教学，每章都设置了"任务实战"和"习题"两个模块，使选用本书作为 Android 课程的教师可以更加合理地安排实训教学。读者也能及时检验学习成果，并举一反三地制作出更多精彩的 App。

本书特点

（1）构建线上线下混合式教材，实现优质教学资源共享。

本书通过教研课题的积累，逐步形成"电子教材—电子课件—微课—案例—动画—课程网站"等一系列的教学资源，积极打造完整的网络信息教学资源，让枯燥的知识变得生活化、情景化、动态化、形象化，进而实现优质教材资源共享。

（2）深化校企合作机制，在教材建设上出亮点。

本书吸收行业企业专家、高校教师等共同参与，建立高校、行业企业与出版机构共同参与的教材建设协作联动机制。将行业企业中典型案例等拆解、消化，逐步形成适合学生学习的知识点案例、综合案例。

（3）理论结合实践，充分体现专业教育特色。

本书全面对接现代产业体系、职业标准和岗位需求，在强调"理论够用"的原则下，充分体现教学实践性，编写过程中始终遵循"项目引领—任务驱动"的思路，积极调动学生的实践动手能力。

（4）以工作过程为导向，精心规划教材内容。

本书精心整合与序化教学内容，设计出由简单到复杂的基于工作过程的项目，遵循学生认知规律，促进学生知识、能力和素质的稳步提升。

(5)赛、教、学结合,注重教学资源的成果转化。

本书中的拓展案例,是由全国职业技能大赛"移动互联网应用软件开发"赛项中的智能交通模块转化而来,此模块中涵盖了Android开发中所涉及的最新技术,通过案例的训练,给未能参赛的同学提供参赛体验,开阔学生的眼界和思维。

为了使读者更加轻松地掌握Android技术,本书制作了配套微课资源、源代码及习题答案。微课均为多年从事Android教学的资深老师录制,语音讲解,真实操作演示,让读者一学就会!

本书作者

参加本书编写的作者均为从事Android教学工作多年的资深教师和Android工程师,有着丰富的教学经验和Android项目开发经验。

本书主编为施冬梅(编写第6章、第7章、第11章)、孙翠改(编写第2章、第3章、第5章、第8章、第10章),副主编为盛雪丰(编写第9章)、陈双华(编写第1章、第4章)。

敖建华参与了创作和编写工作。华为技术有限公司对本书创作给予了支持和帮助,在此一并表示感谢。

相关资源

立体出版计划,为读者建构全方位的学习环境!最先进的建构主义学习理论告诉我们,建构一个真正意义上的学习环境是学习成功的关键。学习环境中有真情实境、有协商和对话、有共享资源的支持,才能使读者高效率地学习,并且学有所成。因此,为了帮助读者建构真正意义上的学习环境,作者以图书为基础,为读者专门设置了一个图书服务网站。

网站提供相关图书资讯,以及相关资料下载和读者俱乐部。在这里读者可以得到更多、更新的共享资源;还可以交到志同道合的朋友,相互交流、共同进步。

图书服务网站

目　录
CONTENTS

随书资源

项目1　第一个 Android 程序 ·· 1

　1.1　Android 简介 ·· 1

　　1.1.1　手机操作系统 ·· 1

　　1.1.2　Android 发展史 ··· 2

　　1.1.3　Android 特征 ·· 3

　　1.1.4　Android 平台架构 ·· 4

　1.2　Android 开发环境 ·· 6

　　1.2.1　Android 开发所需的工具介绍 ·· 6

　　1.2.2　搭建 Android 开发环境 ··· 6

　　1.2.3　创建 Android 原生模拟器 ··· 17

　　1.2.4　安装第三方 Android 模拟器 ·· 22

　　1.2.5　搭建真机测试环境 ·· 25

　1.3　实战案例——第一个 Android 程序 ·· 26

　　1.3.1　界面分析 ·· 26

　　1.3.2　实现思路 ·· 27

　　1.3.3　任务实施 ·· 27

　项目小结 ·· 40

　习题 ··· 40

项目2　猜猜我的生肖 ·· 41

　2.1　Android 常用 UI 控件 ··· 41

　　2.1.1　TextView 控件 ··· 41

2.1.2　EditText 控件 ……………………………………………………… 48
　　　2.1.3　Button 控件 …………………………………………………………… 52
　　　2.1.4　ImageView 控件 ……………………………………………………… 58
　　　2.1.5　DatePicker 控件 ……………………………………………………… 61
　2.2　APP 之间的通信 ……………………………………………………………… 65
　　　2.2.1　显式 Intent …………………………………………………………… 66
　　　2.2.2　隐式 Intent …………………………………………………………… 71
　2.3　实战案例——猜猜我的生肖 ………………………………………………… 75
　　　2.3.1　界面分析 ……………………………………………………………… 75
　　　2.3.2　实现思路 ……………………………………………………………… 75
　　　2.3.3　任务实施 ……………………………………………………………… 76
项目小结 ……………………………………………………………………………… 82
习题 …………………………………………………………………………………… 82

项目3　智能计算器 …………………………………………………………………… 84

　3.1　Android UI 常用布局 ………………………………………………………… 84
　　　3.1.1　UI 简介 ………………………………………………………………… 84
　　　3.1.2　LinearLayout 布局 …………………………………………………… 86
　　　3.1.3　RelativeLayout 布局 ………………………………………………… 93
　　　3.1.4　FrameLayout 布局 …………………………………………………… 100
　3.2　Android 开发中的样式设计 ………………………………………………… 102
　　　3.2.1　自定义控件样式 ……………………………………………………… 103
　　　3.2.2　自定义背景样式 ……………………………………………………… 104
　3.3　Android 开发中的主题设计 ………………………………………………… 108
　3.4　实战案例——智能计算器 …………………………………………………… 110
　　　3.4.1　界面分析 ……………………………………………………………… 110
　　　3.4.2　实现思路 ……………………………………………………………… 110
　　　3.4.3　任务实施 ……………………………………………………………… 111
项目小结 ……………………………………………………………………………… 118
习题 …………………………………………………………………………………… 118

项目4　打地鼠小游戏 ………………………………………………………………… 119

　4.1　Android 的事件处理与交互实现程序设计 ………………………………… 119
　　　4.1.1　提示 …………………………………………………………………… 119
　　　4.1.2　对话框 ………………………………………………………………… 122
　　　4.1.3　菜单 …………………………………………………………………… 127
　　　4.1.4　下拉列表框 …………………………………………………………… 132
　4.2　实战案例——打地鼠小游戏 ………………………………………………… 136
　　　4.2.1　界面分析 ……………………………………………………………… 136

4.2.2　实现思路 136
　　　4.2.3　任务实施 136
　项目小结 143
　习题 143

项目5　记忆的仓库——备忘录 144

　5.1　SharedPreferences 存储 144
　　　5.1.1　SharedPreferences 简介 144
　　　5.1.2　SharedPreferences 的使用 146
　5.2　Android 的文件存储 151
　　　5.2.1　内部存储 151
　　　5.2.2　外部存储 157
　5.3　调用摄像头和相册 163
　　　5.3.1　调用摄像头拍照 163
　　　5.3.2　从相册中选择照片 167
　5.4　Android 运行时权限设置 171
　　　5.4.1　Android 权限机制介绍 171
　　　5.4.2　在程序运行时申请权限 173
　5.5　实战案例——记忆的仓库：备忘录 178
　　　5.5.1　界面分析 178
　　　5.5.2　实现思路 178
　　　5.5.3　任务实施 178
　项目小结 200
　习题 200

项目6　多彩水果店 201

　6.1　最常用和最难用的控件——ListView 201
　　　6.1.1　ListView 的简单用法 201
　　　6.1.2　定制 ListView 的界面 204
　　　6.1.3　提升 ListView 的运行效率 208
　　　6.1.4　ListView 的单击事件 211
　6.2　更强大的滚动控件——RecyclerView 213
　　　6.2.1　RecyclerView 的简单用法 213
　　　6.2.2　实现横向滚动和瀑布流布局 218
　　　6.2.3　RecyclerView 的单击事件 223
　6.3　实战案例——多彩水果店 App 225
　　　6.3.1　界面分析 225
　　　6.3.2　实现思路 226

6.3.3　任务实施 …… 227

项目小结 …… 242

习题 …… 242

项目7　唱歌跳舞小管家——媒体播放器 …… 243

7.1　实战案例——简单音乐播放器 …… 243
　　7.1.1　界面分析 …… 243
　　7.1.2　实现思路 …… 245
　　7.1.3　任务实施 …… 245

7.2　实战案例——简单视频播放器 …… 264
　　7.2.1　界面分析 …… 264
　　7.2.2　实现思路 …… 264
　　7.2.3　任务实施 …… 264

项目小结 …… 276

习题 …… 276

项目8　我的第一桶金——理财通 App 的设计与实现 …… 278

8.1　使用数据库存储数据 …… 278
　　8.1.1　Android 系统的数据存储方式 …… 278
　　8.1.2　使用 SQLite 数据库存储数据 …… 279
　　8.1.3　创建和升级数据库 …… 279
　　8.1.4　添加数据 …… 284
　　8.1.5　查询数据 …… 290
　　8.1.6　修改数据 …… 294
　　8.1.7　删除数据 …… 300

8.2　实战案例——理财通 App 设计与实现 …… 302
　　8.2.1　作品分析 …… 303
　　8.2.2　制作 App 的欢迎界面 …… 307
　　8.2.3　注册界面的设计与功能 …… 312
　　8.2.4　登录界面的设计与功能 …… 318
　　8.2.5　主界面的设计与功能 …… 324
　　8.2.6　新增收入界面的设计与功能 …… 330
　　8.2.7　新增支出界面的设计与功能 …… 337
　　8.2.8　收入明细界面的设计与功能 …… 340
　　8.2.9　支出明细界面的设计与功能 …… 348
　　8.2.10　数据分析界面的设计与功能 …… 353
　　8.2.11　系统设置界面的设计与功能 …… 360
　　8.2.12　收入管理界面的设计与功能 …… 366
　　8.2.13　支出管理界面的设计与功能 …… 370

项目小结 ·· 375
习题 ·· 375

项目9 我的第一道菜——菜谱App的设计与实现 ·· 376

9.1 ViewPager控件 ·· 376
9.1.1 使用场景 ·· 377
9.1.2 ViewPager的基本用法 ·· 377
9.1.3 ViewPager结合Fragment的使用 ·· 380

9.2 使用HTTP访问网络 ·· 384
9.2.1 使用HttpURLConnection访问网络 ··· 384
9.2.2 使用OkHttp访问网络 ··· 389
9.2.3 网络访问框架的封装 ·· 394

9.3 解析JSON数据格式 ··· 398
9.3.1 使用JSONObject ··· 400
9.3.2 使用GSON ··· 406

9.4 实战案例——菜谱App的设计与实现 ·· 413
9.4.1 菜谱App引导界面设计 ··· 413
9.4.2 菜谱App主界面设计 ·· 419
9.4.3 菜谱App详情界面设计 ··· 430
9.4.4 菜谱收藏功能的设计与实现 ·· 437
9.4.5 菜谱搜索功能的设计与实现 ·· 445
9.4.6 个人中心的设计与实现 ··· 451

项目小结 ·· 454
习题 ·· 455

项目10 实战案例——移动互联网软件开发竞赛1 ·· 456

10.1 赛项简介 ··· 456
10.1.1 竞赛目的 ·· 456
10.1.2 竞赛内容 ·· 457
10.1.3 竞赛方式 ·· 457

10.2 比赛器材及技术平台 ··· 457
10.2.1 计算机配置 ··· 457
10.2.2 比赛平台 ·· 457
10.2.3 软件版本 ·· 458

10.3 考察知识点 ·· 458

10.4 地图导航案例 ··· 459
10.4.1 编码实现离线地图1 ··· 459

10.4.2　编码实现离线地图2 ……………………………………………………… 461
　项目小结 …………………………………………………………………………………… 462

项目11　实战案例——移动互联网软件开发竞赛2 ………………………………… 463
　11.1　编程案例——生活助手 …………………………………………………………… 463
　　　11.1.1　功能说明 …………………………………………………………………… 463
　　　11.1.2　题目要求 …………………………………………………………………… 464
　　　11.1.3　操作视频 …………………………………………………………………… 465
　11.2　编程案例——旅行助手 …………………………………………………………… 467
　　　11.2.1　功能说明 …………………………………………………………………… 467
　　　11.2.2　题目要求 …………………………………………………………………… 467
　　　11.2.3　操作视频 …………………………………………………………………… 468
　11.3　编程案例——数据分析 …………………………………………………………… 469
　　　11.3.1　功能说明 …………………………………………………………………… 469
　　　11.3.2　题目要求 …………………………………………………………………… 469
　　　11.3.3　操作视频 …………………………………………………………………… 472
　项目小结 …………………………………………………………………………………… 472

习题参考答案 ……………………………………………………………………………… 473

项目 1

第一个Android程序

【教学导航】

学习目标	(1) 了解手机操作系统。 (2) 了解 Android 的发展历史和版本。 (3) 了解 Android 的特征。 (4) 熟悉 Android 平台构架。 (5) 掌握 Android 开发环境。 (6) 掌握创建 Android 工程的方法。 (7) 了解 Android 工程的结构。 (8) 掌握创建 Android 模拟器的方法。 (9) 掌握运行与调试 Android 程序的方法
教学方法	任务驱动法、理论实践一体化、探究学习法、分组讨论法
课时建议	4 课时

1.1 Android 简介

Android 是 Google 公司开发的一个开源的、基于 Linux 的操作系统,主要应用在移动设备,如智能手机和平板电脑上。第一台 Android 智能手机发布于 2008 年 10 月,目前它已成为全球应用最广泛的手机操作系统。本节介绍 Android 的由来与发展、Android 的特征以及 Android 的平台架构。

1.1.1 手机操作系统

微课视频

在学习 Android 之前,先认识一下手机操作系统。操作系统是计算机系统的核心控制软件,其功能是控制和管理计算机硬件与软件资源。手机操作系统即移动终端操作系统,是在嵌入式操作系统的基础上发展而来的专门为手机设计的操作系统,它为用户使用手机提供统一的接口和友好的交互界面,也为手机功能的扩展、第三方软件的安装与运行提供平台。按照源代码、内核和应用环境等的开放程度,手机操作系统可分为开放型平台(基于 Linux 内核)和封闭型平台(基于 UNIX 和 Windows 内核)两大类。历史上流行的手机操作系统有 Symbian、Android、Windows Phone、iOS、Blackberry 等,手机操作系统发展历程如下。

1996 年 11 月,全球最大的计算机软件提供商微软公司发布了 Windows CE 1.0 操作系统,它是基于掌上计算机类的电子设备操作系统,为其后出现的手机操作系统做了铺垫。

1999 年 1 月,RIM 公司推出了首款搭载 Blackberry 系统的手机 Blackberry 850,这是智能手机革命的开始。Blackberry 操作系统主打安全,因此在推出早期很受商务人士的欢迎。

1999 年 3 月,塞班公司推出第一个 Symbian 操作系统 Symbian OS v5.x。2000 年 12 月,全球第一款 Symbian 系统手机——爱立信 R380 正式出售。Symbian 系统以其庞大的客户群和终端占有率称霸世界智能手机中低端市场。

2007 年 6 月,苹果公司的 iOS 系统登上了历史的舞台,iOS 将创新的移动电话、可触摸宽屏、网页浏览、手机游戏、手机地图等功能融合为一体,智能手机全新的概念开始进入人们的生活。

2008 年 12 月,诺基亚公司收购 Symbian 公司,Symbian 成为诺基亚独占系统。

2008 年 9 月,当苹果和诺基亚还沉溺于彼此的争斗之时,Android,这个由 Google 研发团队设计的"小机器人"悄然出现在世人面前,良好的用户体验和开放性的设计,让 Android 很快地进入了智能手机市场。

2010 年 10 月,面对 Android 和 iOS 带来的压力,微软发布第一个 Windows Phone 操作系统,展现了全新的 Windows 桌面设计风格。

2011 年 12 月,诺基亚官方宣布放弃 Symbian 品牌。

2013 年 9 月,微软收购诺基亚旗下手机业务,想要做大 Windows Phone,与此同时 Symbian 也彻底退出操作系统。

2019 年 12 月,由于 Android 和 iOS 的市场占有率越来越高,Blackberry 和 Windows Phone 停止更新。

由此可见,手机操作系统的发展在二十多年间发生了翻天覆地的变化,而在这二十多年存活下来的却为数不多,目前基本只剩下 Android 和 iOS 了。

微课视频

1.1.2　Android 发展史

Android 这个单词最早出现于法国作家利尔亚当在 1886 年发表的科幻小说《未来夏娃》中,他将外表像人的机器起名为 Android。Android 的 Logo 是由伊琳娜·布洛克设计的,设计灵感来源于男女厕所门上的图形符号,外加头上两根天线就构成了 Android 的 Logo,如图 1.1 所示。

Android 操作系统最初是由安迪·鲁宾(Andy Rubin)开发出的,2005 年被 Google 收购,并于 2007 年 11 月 5 日正式向外界展示了这款系统。在 Android 的发展过程当中,已经经历了十多个主要版本的变化。2009 年,Google 公司推出了 Android 1.5

图 1.1　Android Logo

(Donut),从这个版本开始,Android 的后续版本均用一个甜品来命名。随着后续的发展,越来越多的"甜品"(Android 版本)被 Google 公司陆续推出,如表 1.1 所示。

表 1.1　Android 系统版本及代号

发布时间	版本号	系统代号
2009.04	Android 1.5	Cupcake
2009.09	Android 1.6	Donut(甜甜圈)
2009.10	Android 2.0	Eclair(巧克力泡芙)
2010.05	Android 2.2	Froyo(冻酸奶)
2010.12	Android 2.3	Gingerbread(姜饼)
2011.02	Android 3.0	Honeycomb(蜂巢)
2011.10	Android 4.0	Ice Cream Sandwich(冰激凌三明治)
2012.10	Android 4.1	Jelly Bean(果冻豆)
2013.11	Android 4.4	KitKat(奇巧巧克力)
2014.10	Android 5.0	Lollipop(棒棒糖)
2015.09	Android 6.0	Marshmallow(棉花糖)
2016.07	Android 7.0	Nougat(牛轧糖)
2017.08	Android 8.0	Oreo(奥利奥)
2018.07	Android 9.0	Pie(派)
2019.08	Android 10.0	Q

从表 1.1 可以看出,从 Android 10 开始,以甜点命名的"传统"结束了。据 Google 描述,之所以改变这项传统,是这些代号的含义并不被全球所有人理解。在该版本中,除采取新的命名方式之外,为了适应对绿色文字很难辨识的视力障

图 1.2　Android 最新 Logo

碍类人群,Google 还对 Android 系统的 Logo 进行了更新,更新后的 Logo 由黑色文字和绿色机器人组成,整个 Logo 的对比度有所提高,如图 1.2 所示。

1.1.3　Android 特征

随着 Android 系统在全球范围内的流行,它的市场占有率已经超过了 iOS 系统。Android 系统以其易用性和开放性广受好评,但它除了这两个优势之外,还有一些其他的重要特征,Android 系统的主要特征如下。

1. 开放性

Android 的开放性主要表现在它的核心代码是对外开放的,这对开发者有强大的吸引力,用户和应用也因此迅速增长。另外,Android 的开放性允许任何移动终端厂商的加入,这也是它迅速强大的重要原因之一。

2. 挣脱束缚

在过去很长的一段时间,手机应用往往受到运营商制约,使用什么应用接入什么网络,几乎都受到运营商的控制。Android 打破了这种束缚,用户可以根据自己的需求来下载应用。

3. 丰富的硬件平台

由于 Android 的开放性,使得 Android 系统不仅可以应用在智能手机上,还能应用在平

板电脑、车载设备以及穿戴设备上,并且在不同的硬件平台上,Android 应用可能存在功能上的差异,却不会影响到数据同步。

4. 不受限制

与 iOS 相比,Android 给开发者提供了非常自由的环境,没有单一平台的限制,在遵循国家及各应用平台的审核制度情况下,开发的应用可以发布到国内国外的各种应用市场上。

5. 使用 Google 资源

由于 Android 由 Google 开发,因此使用 Android 平台可以便利地使用 Google 资源,如 Google 搜索、Google 地图等。

1.1.4　Android 平台架构

Android 是一种基于 Linux 的自由及开放源代码的操作系统。Android 的系统架构和其操作系统一样,采用分层架构,主要分为五层,从高到低分别是:系统应用层、Java API 框架层、系统运行库层、硬件抽象层、Linux 内核层,如图 1.3 所示。

1. 系统应用层(System APPs)

应用层放置了 Android 提供的核心应用程序,如电子邮件、短信、日历、浏览器、联系人等,用户自行下载的应用也被放置该层,所以它是用户实际接触 Android 的一层。

2. 接口框架层(Java API Framework)

接口框架层提供开发应用程序时需要使用到的接口,无论是 Android 自身提供的应用程序还是第三方的应用程序,都会使用到它。主要包括:视图系统(View System)、内容提供者(Content Provider)、通知管理器(Notification Manager)、活动管理器(Activity Manager)、资源管理器(Resource Manager)。通过使用它们,可以让程序更简洁,并且可以提高程序的复用性。

3. 系统运行库层(Native C/C++ Libraries & Android Runtime)

系统运行库层包含两部分:系统库及 Android 运行时库。系统库主要包括一组 C/C++ 库,用于 Android 系统中不同组件和服务的开发,如 ART 和 HAL。如果开发程序需要使用到 C/C++,则可以使用 Android NDK 直接地访问本机代码 C/C++库。

Android 运行时库包括核心库和 Dalvik 虚拟机两个部分。核心库中提供了 Java 语言核心库中包含的大部分功能。Dalvik 虚拟机负责运行程序,它使每一个 Android 应用都能运行在独立的进程中,同时,它只应用移动设备,不仅效率更高,而且占用更少的内存。

4. 硬件抽象层(Hardware Abstraction Layer)

Android 的硬件抽象层能以封闭源码形式提供硬件驱动模块。HAL 的目的是为了把接口框架层与 Linux 内核层隔开,让 Android 不过度依赖 Linux 内核层,以达成"内核独立"的概念,也让接口框架层的开发能在不考量驱动程序实现的前提下进行发展。

5. Linux 内核层(Linux Kernel)

Linux 内核层为搭载 Android 系统设备的硬件提供各种驱动,包括显示驱动、音频驱动、相机驱动、蓝牙驱动、电源管理驱动等。

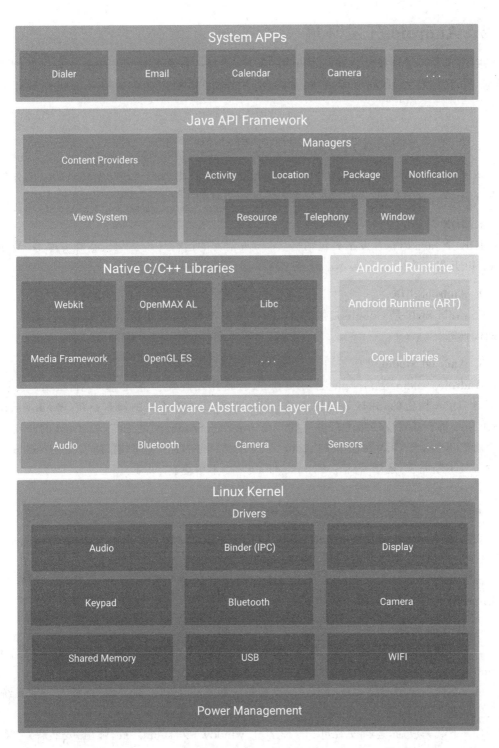

图 1.3 Android 平台架构

1.2 Android 开发环境

搭建开发环境是开发 Android 程序的前提,选择一款合适的模拟器能让程序的测试更顺利。本节介绍 Android 开发环境的搭建、Android 模拟器的创建以及真机测试环境的搭建。

微课视频

1.2.1 Android 开发所需的工具介绍

本书的 Android 程序都是由 Java 语言开发,因此搭建 Android 开发环境的一部分包括 JDK 的安装。总的来说,Android 开发所需要的工具如下。

1. JDK

JDK 是 Java 语言开发工具包,它包含 Java 语言的运行环境、工具、基础类库等,最新的 Android 开发要求的 JDK 版本必须是 JDK 8 及以上。

2. Android SDK

Android SDK 是 Google 提供的 Android 开发工具包,其中包括调试器、库、API 文档、示例代码等。

3. Android Studio

Android Studio 是 Google 官方提供的集成开发环境。它是基于 JetBrains 公司的 IntelliJ IDEA 构建的,并且专门用于 Android 开发。对于不同的操作系统 Windows、Mac OS、Linux 都有其对应的版本可供下载。在 Android Studio 问世之前,Android 的开发大多是使用 Eclipse,与 Google 提供的 Android 开发插件 ADT 一起使用,随着 Android Studio 的发布及不断完善,逐渐替代了 Eclipse,Google 也不再更新 ADT。

微课视频

1.2.2 搭建 Android 开发环境

在搭建 Android 开发环境之前,应该先了解计算机所使用的操作系统类型和版本,在 "此电脑"→"属性"中就可以找到。本教程的系统类型和版本是 64 位的 Windows 10 操作系统。安装步骤总结为:JDK 的安装→JDK 环境变量的配置→Android Studio 的安装→Android SDK 的下载,具体步骤如下。

下载链接

1. JDK 的安装

(1) 下载 JDK,打开浏览器,输入 JDK 的下载官网:https://www.oracle.com/java/technologies/javase-downloads.html,如图 1.4 所示。

(2) 本教程使用的版本是 JDK 13,找到对应版本进入到下载页面,选择与当前系统匹配的安装包下载,如图 1.5 所示。

(3) 下载完成后,运行安装包 jdk-13.0.1_windows-x64_bin.exe,弹出"安装程序"对话框,如图 1.6 所示。

(4) 单击"下一步"按钮,弹出"目标文件夹"对话框,如图 1.7 所示。单击"更改"按钮可改变安装路径,一般情况下默认即可。

图 1.4　JDK 下载官网

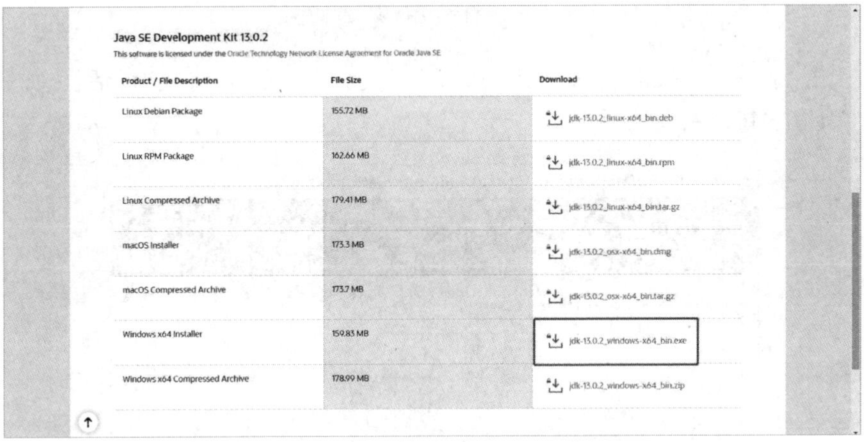

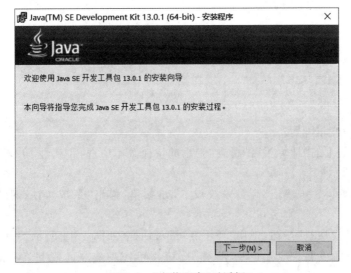

图 1.5　JDK 下载页面

图 1.6　"安装程序"对话框

图 1.7 "目标文件夹"对话框

（5）单击"下一步"按钮，JDK 进入安装，安装完成后，弹出"完成"对话框，如图 1.8 所示，单击"关闭"按钮结束安装。

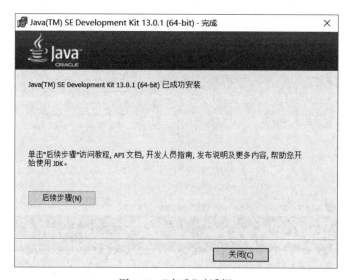

图 1.8 "完成"对话框

2. JDK 环境变量配置

（1）右击桌面上的"计算机"图标，在弹出的快捷菜单中选择"属性"子菜单，打开"系统"窗口，如图 1.9 所示。

（2）在左侧菜单栏中单击"高级系统设置"子菜单，弹出"系统属性"对话框，如图 1.10 所示。

（3）单击"环境变量"按钮，弹出"环境变量"对话框，如图 1.11 所示。

（4）单击"系统变量"标题下的"新建"按钮，弹出"新建系统变量"对话框，如图 1.12 所示。

图 1.9 "系统"窗口

图 1.10 "系统属性"对话框

图 1.11 "环境变量"对话框

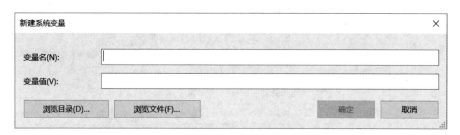

图 1.12 "新建系统变量"对话框

(5) 在"新建系统变量"对话框中填写变量名和变量值。变量名：JAVA_HOME；变量值：C:\Program Files\java\jdk-13.0.1(JDK 的安装路径)，如图 1.13 所示。

(6) 按照同样的方法在"新建系统变量"对话框中新建另一个变量，变量名：CLASSPATH；变量值：.;%JAVA_HOME%\lib;%JAVA_HOME%\lib\tools.jar;%JAVA_HOME%\lib\dt.jar(注意不要落下变量值前的标点符号)，如图 1.14 所示。

(7) 在如图 1.11 所示的"环境变量"对话框中，双击"系统变量"标题下 Path 变量，弹出"编辑环境变量"对话框，如图 1.15 所示。单击右方的"新建"按钮，在左侧编辑区输入两个变量：%JAVA_HOME%\bin 和%JAVA_HOME%\jre\bin，单击"确定"按钮保存。

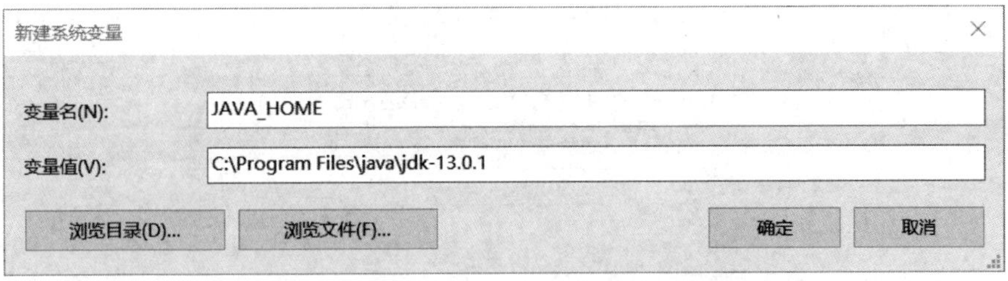

图 1.13 JAVA_HOME 环境变量配置

图 1.14 CLASSPATH 环境变量配置

图 1.15 "编辑环境变量"对话框

（8）在配置完环境变量之后，需要测试 JDK 是否安装成功，在桌面左下角任务栏右击 Windows 图标，在弹出的菜单中选择"运行"子菜单，弹出"运行"对话框，如图 1.16 所示。

（9）在"运行"对话框中输入命令"cmd"，单击"确定"按钮，弹出"系统命令提示符"窗口，如图 1.17 所示。

图 1.16 "运行"对话框

图 1.17 "系统命令提示符"窗口

（10）在"系统命令提示符"窗口中输入命令"java -version,"按 Enter 键,查看结果。如果 JDK 安装正确,会显示 JDK 版本信息,如图 1.18 所示。如果没有显示版本信息,则 JDK 没有安装正确,需要检查环境变量是否配置错误。

3. Android Studio 的安装

（1）下载 Android Studio,打开浏览器地址栏,输入 Android Studio 的下载链接：https://developer.android.google.cn/studio/archive#android-studio-3-0?utm_source=androiddevtools&utm_medium=website,如图 1.19 所示。

（2）本教程使用的是 Android Studio 3.5.3,找到对应版本打开,选择与当前系统匹配的安装包下载,如图 1.20 所示。

（3）下载完成后,运行安装包 android-studio-ide-191.6010548-windows.exe,弹出 Android Studio Setup 窗口,如图 1.21 所示。

Android Studio 3.5.3 下载

项目1 第一个Android程序 13

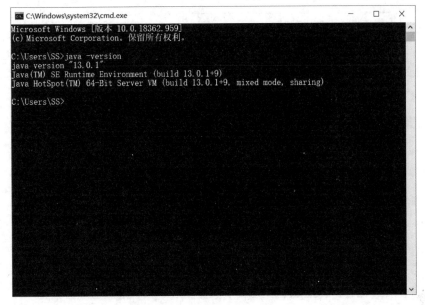

图 1.18 查看 JDK 版本

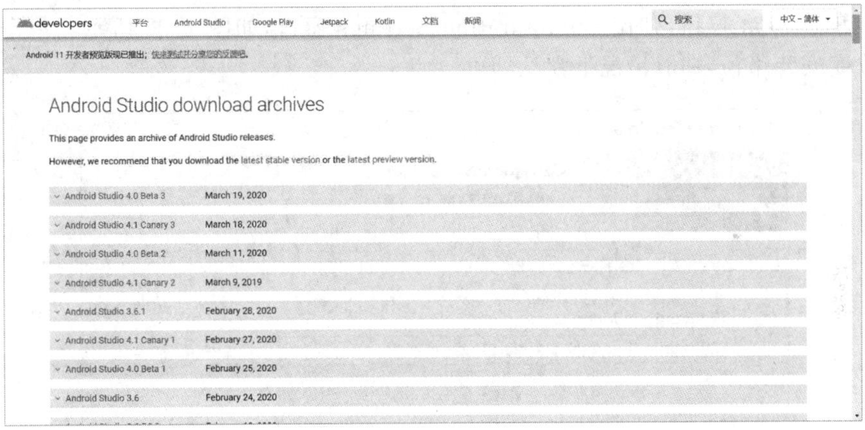

图 1.19 Android Studio 下载官网

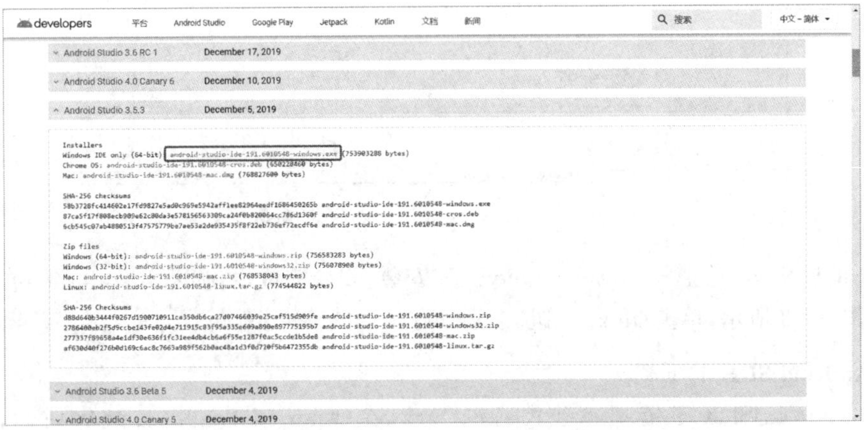

图 1.20 Android Studio 下载网页

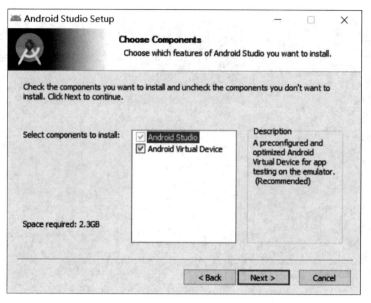

图 1.21　Android Studio Setup 窗口

（4）单击 Next 按钮，弹出 Configuration Settings 窗口，如图 1.22 所示。单击 Browse 按钮可改变安装路径，一般情况下默认即可。

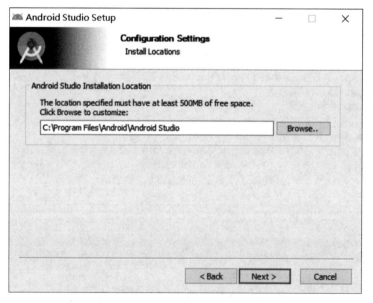

图 1.22　Configuration Settings 窗口

（5）单击 Next 按钮，Android Studio 进入安装，安装完成后，显示 Installation Complete 窗口，如图 1.23 所示，单击 Next 按钮。

4. Android SDK 的下载

（1）Android SDK 不需要单独去官网下载，在 Android Studio 安装完成后，会打开 Missing SDK 窗口，如图 1.24 所示。

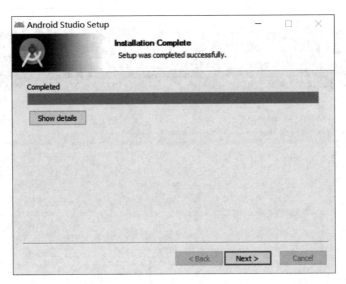

图 1.23 Installation Complete 窗口

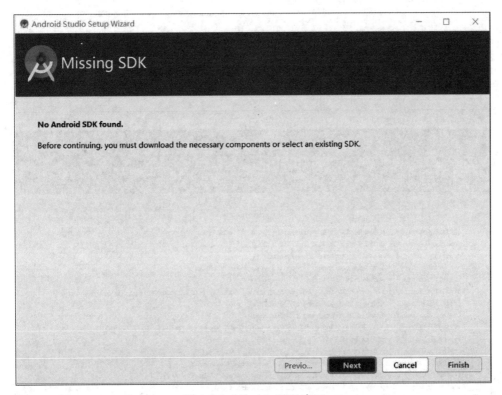

图 1.24 Missing SDK 窗口

(2) 单击 Next 按钮,打开 Downloading Components 窗口,如图 1.25 所示。

(3) 下载完成后,弹出如图 1.26 所示窗口,单击 Finish 按钮。

(4) 此时,Android Studio 开发环境全部搭建完成,打开 Welcome to Android Studio 窗口,如图 1.27 所示。

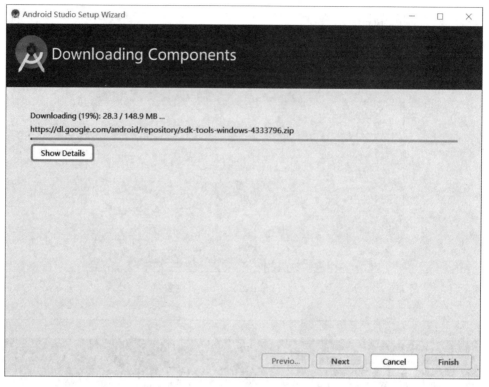

图1.25 Downloading Components 窗口

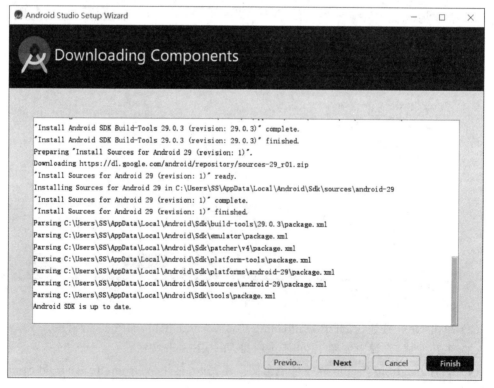

图1.26 下载完成

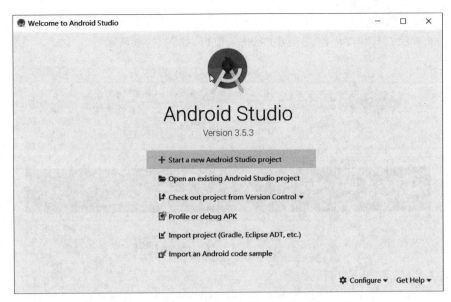

图 1.27　Welcome to Android Studio 窗口

1.2.3　创建 Android 原生模拟器

要想查看 Android 应用的运行效果，必须先创建一个可以运行 Android 应用的环境，这个环境可以是虚拟的也可以是真实的 Android 手机，前者称为模拟器，后者简称为真机。模拟器分为 Android 原生模拟器以及第三方模拟器。首先，介绍 Android 原生模拟器的创建过程。

（1）打开 Android Studio，如果是第一次打开，单击右下角 Configure 下拉列表，选择 AVD Manager 选项，如图 1.28 所示。

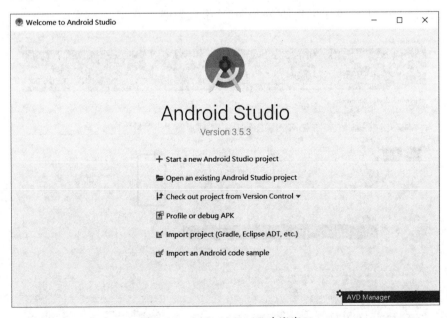

图 1.28　Android Studio 欢迎窗口

（2）如果不是第一次打开,则单击 Android Studio 主界面右上角 AVD Manager 图标（如图 1.29 所示),打开 Your Virtual Devices 窗口,如图 1.30 所示。

图 1.29　Android Studio AVD Manager

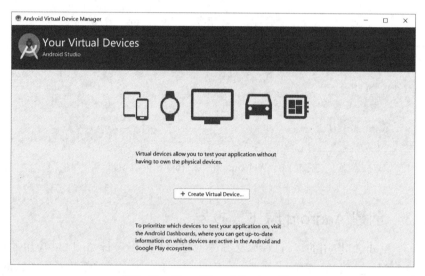

图 1.30　Your Virtual Devices 窗口

（3）单击 Create Virtual Device 按钮,弹出 Select Hardware 对话框,如图 1.31 所示,左侧 Category 标题下内容为模拟器的类型,选择 Phone 选项,表示创建手机类型的模拟器。中间区域选择手机型号,这里以选择 Nexus 5X 手机为例。

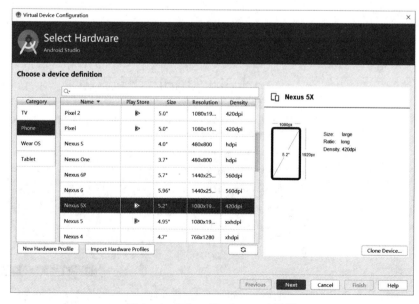

图 1.31　Select Hardware 对话框

（4）单击 Next 按钮，弹出 System Image 对话框，如图 1.32 所示。这里选择最新的系统 API 30，单击 API 30 后面的 Download 文字，下载并安装操作系统。

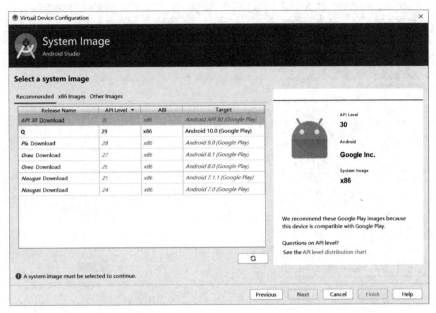

图 1.32　System Image 对话框

（5）在弹出的 License Agreement 对话框中，选择左侧的许可项，右侧下方选择 Accept 单选框，再单击 Next 按钮，如图 1.33 所示。

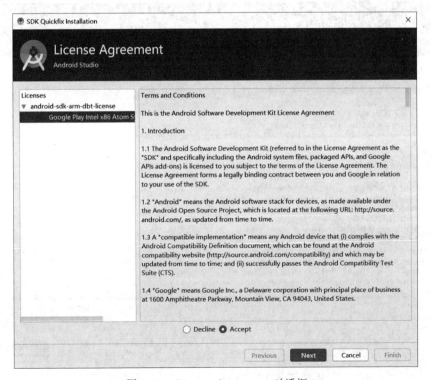

图 1.33　License Agreement 对话框

（6）在安装完成后，需要设置 RAM 所占的最大空间值，如图 1.34 所示，默认即可。

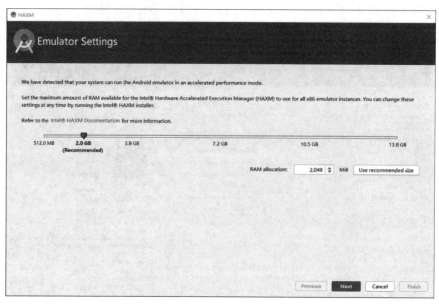

图 1.34　Emulator Settings 对话框

（7）单击 Next 按钮，弹出 SDK Quickfix Installation 对话框，如图 1.35 所示，单击 Finish 按钮，模拟器创建完成。

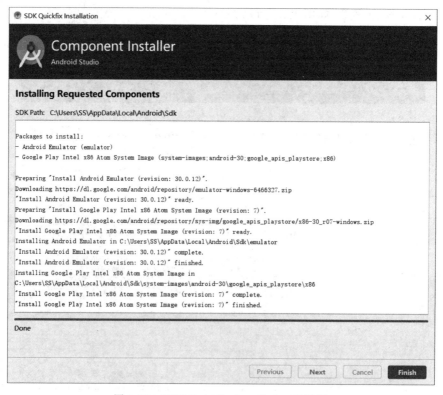

图 1.35　SDK Quickfix Installation 对话框

(8)模拟器创建完成后,进入 Virtual Device Configuration 对话框,如图 1.36 所示。在此可以配置模拟器的名称、默认启动方向(横屏/竖屏)等,一般为默认,确认信息后,单击 Finish 按钮。

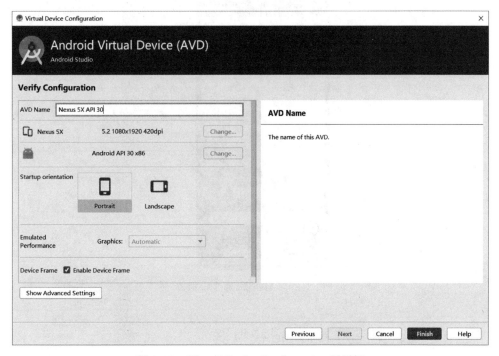

图 1.36　Virtual Device Configuration 对话框

(9)打开 Your Virtual Devices 窗口,如图 1.37 所示,前面添加的模拟器已经在此显示,单击右侧 Actions 列表的右三角形图标,模拟器就启动了。

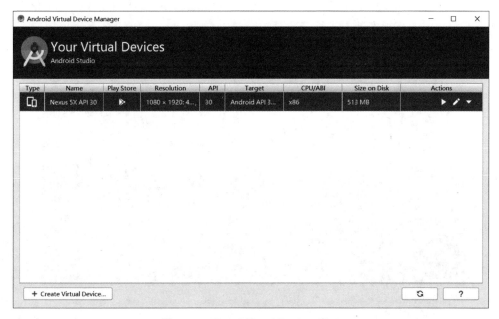

图 1.37　Your Virtual Devices 窗口

(10)启动后的模拟器如图 1.38 所示。在手机右侧有一些手机常用的功能键,如电源键、音量键等。现在,Android 程序就可以运行在该模拟器上了。

图 1.38　Android 原生模拟器

1.2.4　安装第三方 Android 模拟器

除了创建自带模拟器外,还可以安装第三方模拟器。市面上的第三方 Android 模拟器有很多,包括夜神、网易 MuMu、雷电等。本教程以夜神模拟器为例,它的安装及连接过程如下。

(1)使用浏览器打开夜神官网 https://www.yeshen.com/,如图 1.39 所示,根据计算机型号下载相应的夜神安装包进行安装。

图 1.39　夜神官网

（2）安装完成后，打开夜神模拟器，如图1.40所示。夜神模拟器默认为平板模式，单击右上角的"设置"图标。

图1.40　夜神模拟器启动界面

（3）弹出"系统设置"对话框，如图1.41所示，选择"性能设置"菜单，在"分辨率设置"标题下将"显示模式"下拉列表选择为"手机版"选项，并且选择需要的分辨率，单击"保存设置"按钮，此时，夜神模拟器切换为手机版。

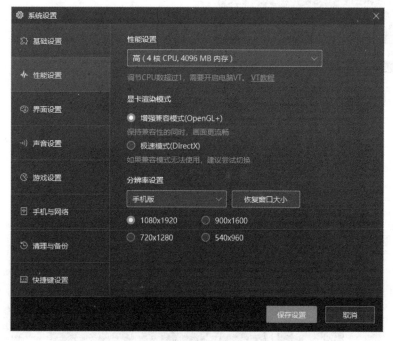

图1.41　夜神模拟器"系统设置"对话框

（4）下面需要将夜神模拟器与 Android Studio 连接。打开模拟器的"设置"应用，如图 1.42 所示。

（5）选择"开发者选项"菜单，进入到"开发者选项"设置界面，如图 1.43 所示，将"开发者选项"和"USB 调试"模式的开关图标都打开。

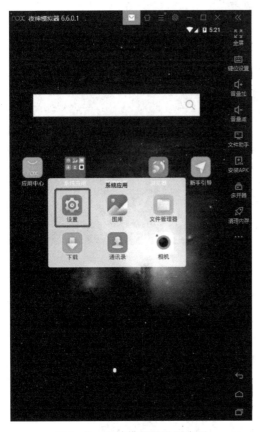

图 1.42　夜神模拟器设置应用　　　　　　图 1.43　夜神模拟器开发者选项

（6）在确认 Android Studio 已经启动的前提下，在桌面同时按 Win＋R 键，调出"运行"对话框，输入命令"cmd"，打开"系统命令提示符"窗口，通过 cd 命令进入到夜神模拟器的安装目录，如图 1.44 所示，本教程中夜神的安装目录为 D:\Program Files\Nox\bin。

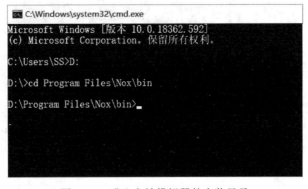

图 1.44　进入夜神模拟器的安装目录

（7）输入连接命令"nox_adb.exe connect 127.0.0.1:62001"，按 Enter 键，此时显示连接成功的提示信息，如图 1.45 所示，表示夜神模拟器与 Android Studio 已连接成功。

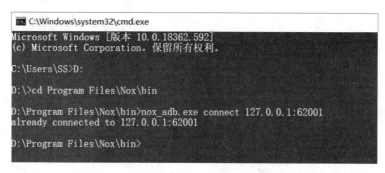

图 1.45　夜神模拟器与 Android Studio 连接

（8）打开 Android Studio，在右上方的模拟器下拉列表中显示了夜神模拟器的设备名称，如图 1.46 所示。此时，Android 程序就可以运行在该模拟器上了。

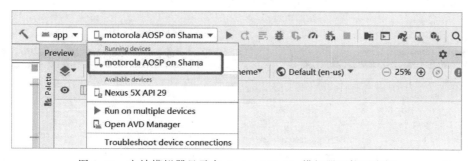

图 1.46　夜神模拟器显示在 Android Studio 模拟器下拉列表框

1.2.5　搭建真机测试环境

在初学 Android 时，通过模拟器就能满足大部分的测试需求。但模拟器和真实的手机存在着差别，例如，真实手机硬件的 API 无法通过模拟器模拟出来，在学习过程中需要调用这些 API 时，应该选择真机调试。本教程以华为手机为例，介绍真机测试环境的搭建过程。

（1）使用 USB 数据线将手机与计算机连接。

（2）打开手机的"设置"应用，选择"开发人员选项"菜单，如图 1.47 所示。

（3）开启"开发人员选项"和"USB 调试"，如图 1.48 和图 1.49 所示。

（4）打开 Android Studio，在右上方的模拟器下拉列表中显示出了手机的设备名称，如图 1.50 所示。此时，真机测试环境搭建完毕，Android 程序可以运行在手机上了。

图 1.47　开发人员选项

图 1.48　开启"开发人员选项"　　　　　图 1.49　开启"USB 调试"

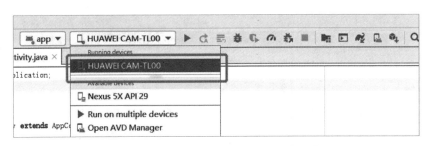

图 1.50　真机与 Android Studio 连接成功

1.3　实战案例——第一个 Android 程序

学习编程语言的第一个程序通常都是 HelloWorld,这已经是编程界的传统。本节利用项目 1 搭建好的 Android 开发环境以及 Android 模拟器,然后创建第一个 Android 程序:HelloWorld。通过第一个 Android 程序的创建,将学习 App 的工程结构、App 的运行测试以及 App 的发布。

微课视频

1.3.1　界面分析

本案例在界面左上角展示文字"HelloWorld",界面效果如图 1.51 所示。

图 1.51　第一个 Android 程序界面效果图

微课视频

1.3.2　实现思路

打开 Android Studio 新建一个工程,找到相应的文件添加界面显示的文字,最后通过模拟器运行。

1.3.3　任务实施

【任务 1-1】　创建第一个 App 项目

(1) 打开 Android,如果是第一次打开,单击 Start a new Android Studio project 按钮创建一个新工程,如图 1.52 所示。如果不是第一次打开,则在 Android 主界面左上角执行 File→New→New Project 命令创建一个新工程,如图 1.53 所示。

(2) 弹出 Choose your project 对话框,如图 1.54 所示。标签栏可以选择工程运行的硬件环境:手机和平板、穿戴设备、电视、汽车驾驶、物联网。本书讲述的是手机 App 开发,因此单击默认 Phone and Tablet 标签。接下来选择活动模板,Android 提供了很多内置的活动模板,初学时,页面都是自行编写,因此,选择 Empty Activity 选项,单击 Next 按钮。

(3) 弹出 Configure your project 对话框,如图 1.55 所示。Name 文本框中填写工程名称,默认为"My Application",为了让工程便于寻找,通常依据工程内容设置工程名称,本案例设置名称为"HelloWorld";Package name 文本框中填写包名,系统将根据包名来区分不同的应用,通常是公司域名的反写,Android Studio 会根据工程名称自动生成一个 Package name,本案例采用自动生成的 Package name;Save location 用来选择工程存储的位置,可自行选择;Language 下拉列表框用来选择编程语言,目前有 Java 和 Kotlin,本书的

图 1.52　创建 Android 工程(首次打开 Android Studio)

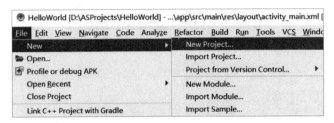

图 1.53　创建 Android 工程(非首次打开 Android Studio)

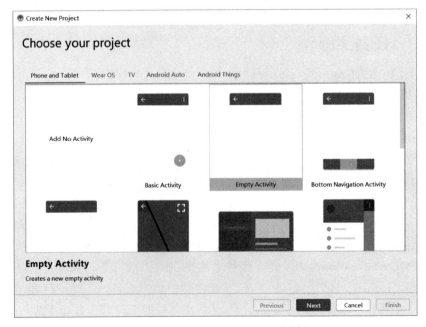

图 1.54　Choose your project 对话框

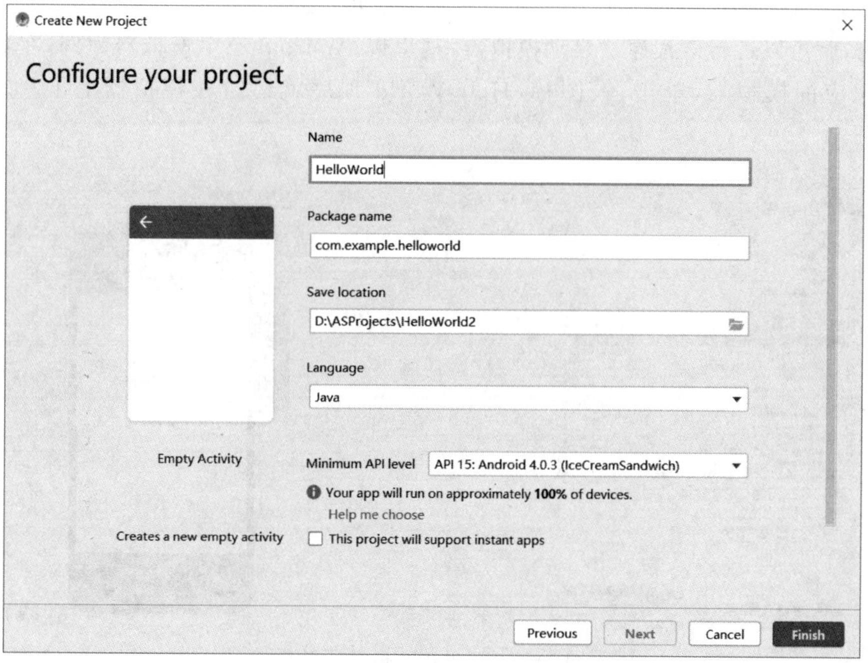

图 1.55 Configure your project 对话框

Android 应用都是基于 Java 开发,因此选择 Java;Minimum API level 下拉列表框用来选择最低兼容的 API 版本,该选择是根据每个 Android 版本的市场活动份额来决定的,本书默认选择 API 15,这样就可以运行在几乎 100% 的设备上了。以上几项内容填写完成后,单击 Finish 按钮。

(4) 正式跳转到 Android 开发界面,如图 1.56 所示,此时,项目正在同步,第一次新建项目的同步过程会比较长,需要耐心等待,当页面下方 Sync 窗口的内容都加载成功时,同步完成。

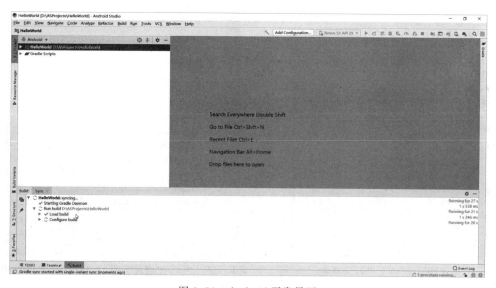

图 1.56 Android 开发界面

（5）Android Studio 首次创建工程时就是默认的 HelloWorld 程序，因此，不需要编写任何代码，就可以运行该程序。启动模拟器，这里以 Android 原生模拟器为例，在页面右上方选择相应的模拟器，单击 ▶ 图标，HelloWorld 程序就运行在模拟器上了，如图 1.57 所示。

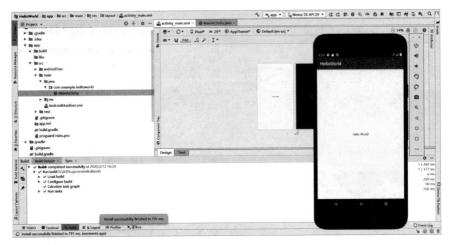

图 1.57　HelloWorld 运行结果

【任务 1-2】　认识 App 工程结构

工程创建成功后，在主界面左上角区域可以看到工程的目录结构。Android Studio 默认的目录结构为 Android 模式的结构，在此模式下无法查看到工程所有的文件，可以通过如图 1.58 所示的下拉列表框切换到 Project 模式。

展开 Project，如图 1.59 所示，这便是工程真实的目录结构。

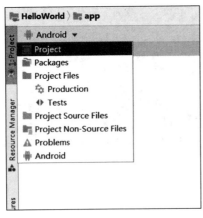

图 1.58　切换工程目录模式

图 1.59　Project 模式下的 HelloWorld 目录结构

以 HelloWorld 项目为例，各目录和文件的介绍如下。

（1）.gradle 目录。

Android Studio 默认是采用 Gradle 作为构建工具的，.gradle 目录下存放 Gradle 的版本等文件，是系统自动生成的，很少会使用。

（2）.idea 目录。

存放开发所需的相关配置文件，由系统自动生成。

（3）app 目录。

存放项目的源代码、资源及其他相关文件，后面的开发工作基本都是在这个目录下进行的，展开 app 目录，如图 1.60 所示。app 各目录和文件的作用如下。

- build：存放项目编译后的文件，包括最终生成的 APK 文件。
- libs：存放项目中所使用的第三方 jar 包。
- src/androidTest：存放测试用例。
- src/main：存放项目的源代码等相关文件。
- src/main/java：存放 Java 代码。其中，com.example.helloword 是项目的包名，MainActivity 是源代码文件。
- src/main/res：存放页面相关的资源文件。其中，drawable 目录通常用来存放图片资源；layout 目录存放布局文件，activity_main.xml 文件就是 HelloWorld 项目的页面布局文件；mipmap 目录用来存放各种分辨率的图标资源文件；values 目录存放颜色、字符串和样式等资源文件；AndroidManifest.xml 文件是提供项目配置的清单文件。

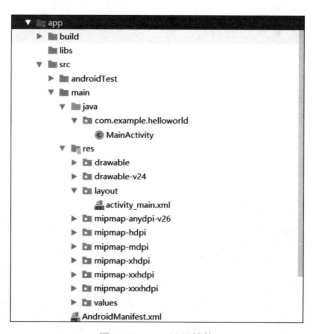

图 1.60　app 目录结构

（4）gradle 目录。

存放 gradle 的配置文件。

(5) gitignore 文件。

gitignore 文件用来将指定的目录或文件排除在版本控制之外。

(6) build.gradle 文件。

build.gradle 文件项目的 Gradle 编译文件。

(7) gradle.properties 文件。

gradle.properties 文件用来设置 Gradle 相关的全局属性。

(8) gradlew 文件。

Linux、Mac 系统命令行执行的 Gradle 编译脚本。

(9) gradlew.bat 文件。

和 gradlew 文件类似，gradlew.bat 文件是 Windows 系统命令行执行的 Gradle 编译脚本。

(10) HelloWorld.iml 文件。

所有 IDEA 项目都会自动生成的一个文件，用于标识这是一个 IDEA 项目。

(11) local.properties 文件。

local.properties 文件用来配置 Android SDK/NDK 所在的路径。

(12) settings.gradle 文件。

settings.gradle 文件用于指定项目中所有引入的模块，通常情况下模块的引入都是自动完成的。

认识完项目的目录结构后，继续完成任务。目前 HelloWorld 运行的结果（图 1.57）和任务的效果图（图 1.51）存在差别，任务中的 HelloWorld 文字位于左上角，而目前是位于中间，修改的过程如下。

(1) 在 Project 模式下，打开 HelloWorld/app/src/main/res/layout 目录下的布局文件 activity_main.xml，可以看到如下代码：

```xml
<?xml version="1.0" encoding="utf-8"?>
<android.support.constraint.ConstraintLayout xmlns:android="http://schemas.android.com/apk/res/android"
    xmlns:app="http://schemas.android.com/apk/res-auto"
    xmlns:tools="http://schemas.android.com/tools"
    android:layout_width="match_parent"
    android:layout_height="match_parent"
    tools:context=".MainActivity">

    <TextView
        android:layout_width="wrap_content"
        android:layout_height="wrap_content"
        android:text="Hello World!"
        app:layout_constraintBottom_toBottomOf="parent"
        app:layout_constraintLeft_toLeftOf="parent"
        app:layout_constraintRight_toRightOf="parent"
        app:layout_constraintTop_toTopOf="parent" />

</android.support.constraint.ConstraintLayout>
```

(2) 将第二行的 android.support.constraint.ConstraintLayout 修改为 LinearLayout，意为将系统默认的布局修改为线性布局，不同的布局方式会影响控件的排列方式。关于布局的相关知识,会在项目 3 中详细讲到。修改后的代码如下。

```xml
<?xml version = "1.0" encoding = "utf-8"?>
<LinearLayout xmlns:android = "http://schemas.android.com/apk/res/android"
    xmlns:app = "http://schemas.android.com/apk/res-auto"
    xmlns:tools = "http://schemas.android.com/tools"
    android:layout_width = "match_parent"
    android:layout_height = "match_parent"
    tools:context = ".MainActivity">

    <TextView
        android:layout_width = "wrap_content"
        android:layout_height = "wrap_content"
        android:text = "Hello World!"
        app:layout_constraintBottom_toBottomOf = "parent"
        app:layout_constraintLeft_toLeftOf = "parent"
        app:layout_constraintRight_toRightOf = "parent"
        app:layout_constraintTop_toTopOf = "parent" />

</LinearLayout>
```

(3) 在页面右上方选择相应的模拟器,单击 ▶ 图标,会发现 HelloWorld 位置移动到左上角了,也就达到了任务的效果。

【任务 1-3】 App 运行测试

在程序开发过程中难免会遇到一些问题,此时就需要对程序进行测试与调试,下面是 Android 调试工具 Logcat 的介绍。

Logcat 是 Android 中的命令行工具,通过它可以查看程序从启动到关闭的日志信息。要在程序中输出日志,需要调用 Android 提供的 Log 类相关的静态方法,Log 类将日志信息按照重要程度划分为六个级别,由低到高分别是 Verbose、Debug、Info、Warning、Error、Assert。除了 Assert 外,其余五个都有其对应的静态方法,分别如下。

(1) Log.v():输出 Verbose 级别的日志信息。
(2) Log.d():输出 Debug 级别的日志信息。
(3) Log.i():输出 Info 级别的日志信息。
(4) Log.w():输出 Warning 级别的日志信息。
(5) Log.e():输出 Error 级别的日志信息。

这些方法中包含两个参数,第一个参数为日志的标签,第二个参数为日志的内容。以 HelloWorld 项目为例,要实现在程序创建时输出不同类型的日志信息,并且在 Logcat 中查看,操作过程如下。

(1) 打开 HelloWorld/app/src/main/java/com.example.helloworld 目录下的类文件 MainActivity.java,在 onCreate()方法中添加 Log 不同级别的日志输出代码,如下。

```java
public class MainActivity extends AppCompatActivity {
    @Override
    protected void onCreate(Bundle savedInstanceState) {
```

```
        super.onCreate(savedInstanceState);
        setContentView(R.layout.activity_main);
        //在程序创建时输出不同类型的日志信息
        Log.v("MainActivity", "这是一条 Verbose 信息");
        Log.d("MainActivity", "这是一条 Debug 信息");
        Log.i("MainActivity", "这是一条 Info 信息 ");
        Log.w("MainActivity", "这是一条 Warning 信息" );
        Log.e("MainActivity", "这是一条 Error 信息" );
    }
}
```

（2）运行项目，在主界面左下角单击 Logcat 菜单打开 Logcat 窗口，如图 1.61 所示。可以发现输出了多条日志信息，其中，Warning 级别的日志信息显示为蓝色，Error 级别的日志信息显示为红色。

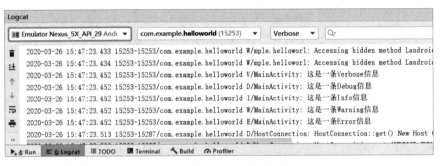

图 1.61　Logcat 日志窗口

（3）在 Logcat 窗口下，可以对日志进行筛选和搜索。如图 1.62 所示，单击日志上方的 `Verbose` 下拉列表框，可以筛选不同级别的日志。选择 Verbose 选项，显示所有日志；选择 Debug 选项，显示除 Verbose 之外的所有日志；选择 Info 选项，显示除 Verbose、Debug 之外的所有日志；选择 Warn 选项，显示 Warn、Error、Assert 类型的日志；选择 Error 选项，显示 Error、Assert 的日志；选择 Assert 选项，只显示 Assert 日志。

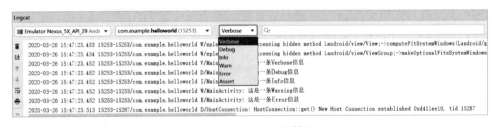

图 1.62　Logcat 日志筛选

（4）在下拉列表框右方是搜索框，在搜索框中可以通过输入关键字进行搜索，例如，输入"MainActivity"，结果如图 1.63 所示。

（5）除了上述两种方式外，还可以使用 Logcat 提供的日志过滤器筛选日志信息，单击 Logcat 窗口右上角的"过滤器"下拉列表框，如图 1.64 所示。选择 Edit Filter Configuration 选项编辑过滤器配置。

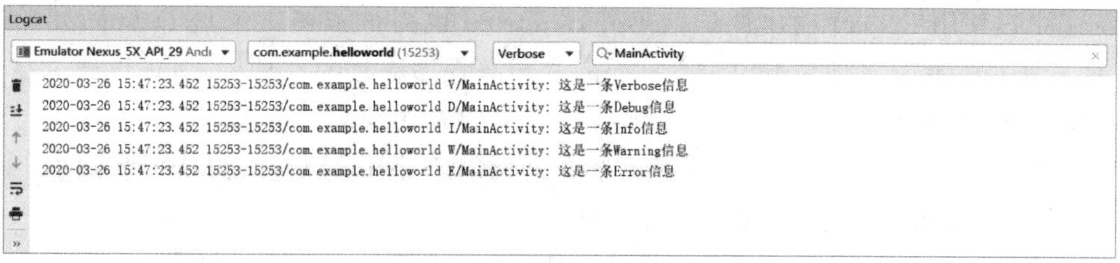

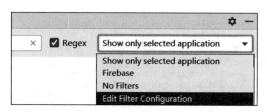

图 1.63　Logcat 日志关键字搜索

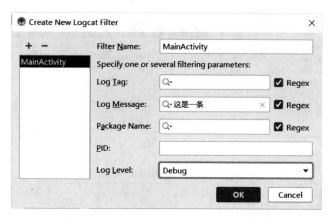

图 1.64　Logcat 日志过滤器

（6）此时弹出 Create New Logcat Filter 对话框，在此可以配置自己的过滤器，如图 1.65 所示。Filter Name 文本框中输入过滤器名称，如输入"MainActivity"；Log Tag 文本框中输入标签关键字；Log Message 文本框中输入信息关键字，如输入"这是一条"；Package Name 文本框中输入筛选的包名；PID 文本框中输入筛选的进程 ID；Log Level 下拉列表框是日志等级选择，如可以选择 Debug 选项；对话框左上角的＋和－图标可以添加或删除过滤器。

图 1.65　Create New Logcat Filter 对话框

（7）单击 OK 按钮，通过此过滤器筛选的结果如图 1.66 所示。

图 1.66　Logcat 日志过滤器筛选结果

除了可以使用Logcat输出调试信息外,Android也提供了调试工具Debug,使用过其他IDE工具的读者应该对它不陌生。Debug是跟踪程序流程的一种模式,可以通过在代码处设置断点,再利用Debug窗口查看。断点是在代码行中设置停止标记,当程序执行到标记行时会暂停,开发者可以从Debug窗口查看到此时程序中各变量的数值、状态等。下面以HelloWorld为例,介绍Debug的使用方法。

(1)打开MainActivity.java,假设要在第15行设置断点,则用鼠标在第15行左侧灰色处单击设置断点,此时会出现一个红色圆圈,为断点标记,如图1.67所示。如果想设置多个断点,则以同样的方式操作。

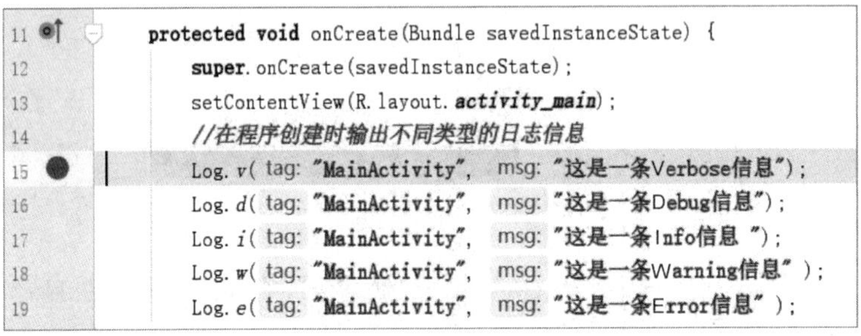

图1.67 设置断点

(2)在保证模拟器运行成功的情况下,单击Android Studio主界面右上方的 图标,进入Debug模式,会发现断点标记上方多了一个√,表明已进入Debug模式。此时,打开主界面下方的Debug窗口,可以查看程序的变量等信息,如图1.68所示。

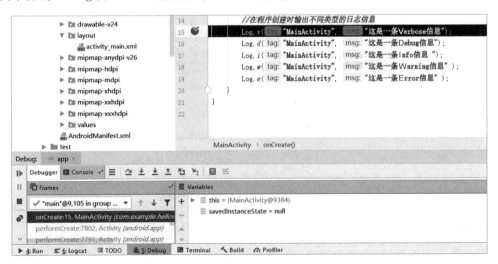

图1.68 Debug窗口

在Debug窗口中有很多操作按钮,常用按钮的功能如下。

　　(Step Over):执行下一行代码(遇到方法时,默认不进入方法),快捷键为F8。

　　(Step Into):进入某一个方法,快捷键为F7。

　　(Step Out):跳出某一个方法,快捷键为Shift+F8。

　　(Resume Program):执行下一个断点,快捷键为F9。

■（Stop App）：停止调试，快捷键为 Ctrl+F2。

【任务 1-4】 App 项目发布

在程序开发完成后，如果要发布到互联网上供别人使用，就需要将自己的程序打包成正式的 Android 安装包文件，即 APK 文件。以 HelloWorld 为例，打包的过程如下。

（1）执行菜单栏的 Bulid→Generate Signed Bundle/APK 命令，弹出如图 1.69 所示对话框，选择 APK 单选框，单击 Next 按钮。

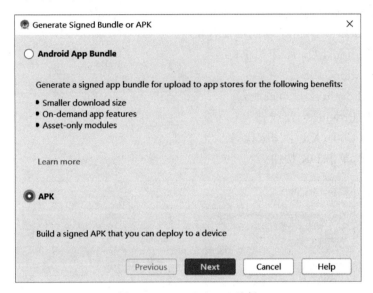

图 1.69 Bundle/APK 选择

（2）弹出 Generate Signed Bundle or APK 对话框，如图 1.70 所示，Key store path 文本框是密钥库文件地址，由于是第一次发布，需要新建一个密钥库文件。

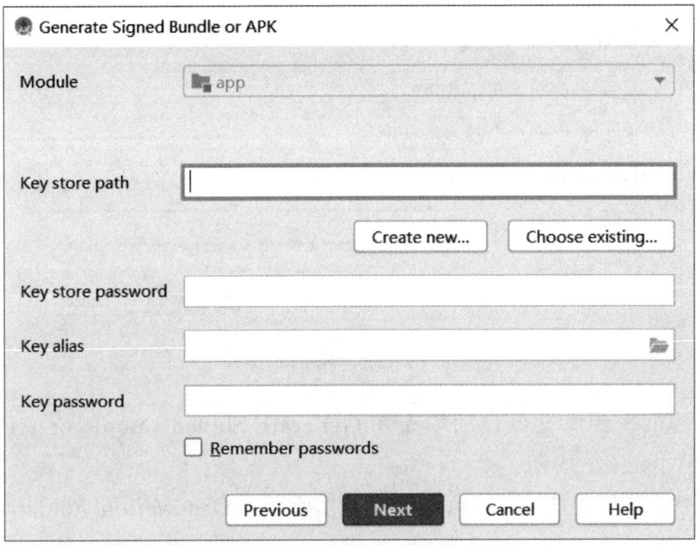

图 1.70 生成签名 APK

（3）单击 Create new 按钮，弹出 New Key Store 对话框，如图 1.71 所示。填写各个文本框内容，如下。

- Key store path：密钥库文件路径。
- Password：密码。
- Confirm：确认密码。
- Alias：别名。
- Validity(years)：有限期(年)，默认值为 25。
- First and Last Name：密钥颁发者全名。
- Organizational Unit：组织单位。
- Organization：组织。
- City or Locality：城市或地方。
- State or Province：州或省。
- Country Code(XX)：国家代码。

填写完成后，单击 OK 按钮。

图 1.71 新建密钥库文件

（4）刚刚新建的密钥库文件自动填充在 Generate Signed Bundle or APK 对话框中，如图 1.72 所示。单击 Next 按钮。

（5）进行 APK 路径及版本选择，如图 1.73 所示。Destination Folder 表示 APK 文件保存路径；Build Variants 表示版本类型，选择 release 选项表示生成的 APK 是正式版本，选择 debug 选项表示测试版本；Signature Versions 表示签名版本，V1 和 V2 复选框都勾选。单击 Finish 按钮，等待打包。

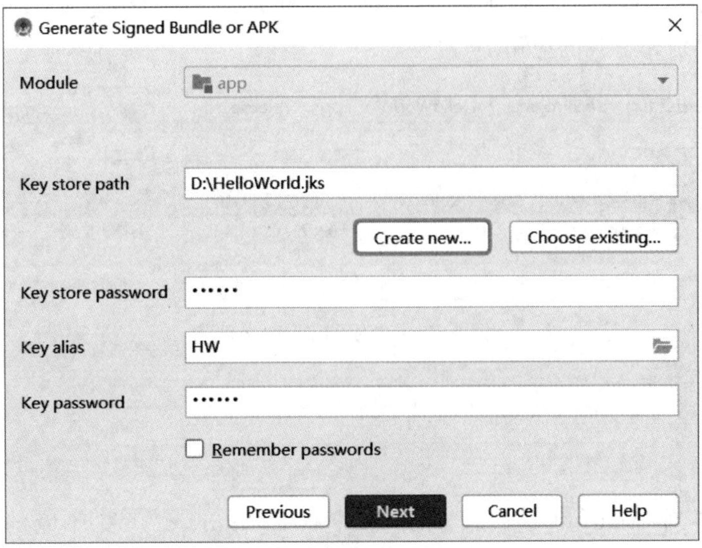

图1.72　新建的密钥库自动填充

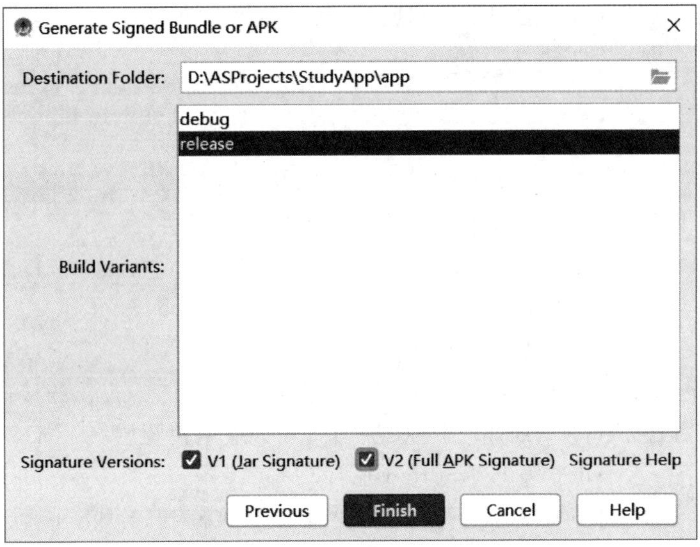

图1.73　APK路径及版本选择

（6）打包完成后，在右下角会弹出打包成功的信息提示框，如图1.74所示，单击提示信息框中的locate文字。

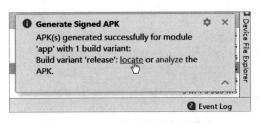

图1.74　APK打包成功提示信息

（7）进入 APK 所在的目录,如图 1.75 所示。此时,APK 打包完成,这个程序就是可以发布在市场上供别人下载的应用了。

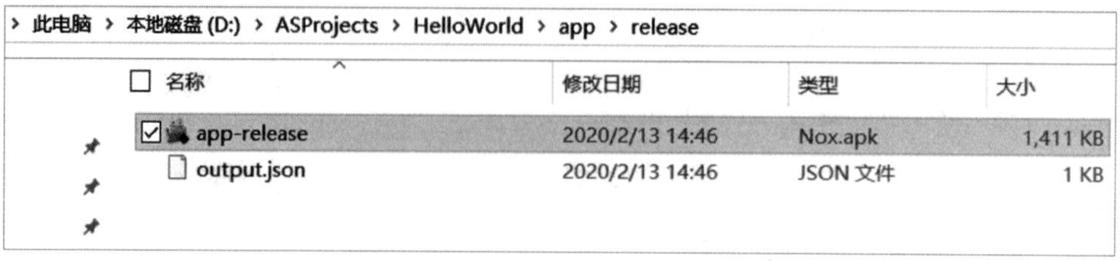

图 1.75 生成的 APK 文件

项目小结

本项目主要对 Android 进行了简介并讲解了 Android 开发环境的搭建。通过对 Android 基础知识的介绍,初学者了解了 Android 开发的入门知识。最后,在搭建好 Android 开发环境以及 Android 模拟器的基础上,实现第一个 Android 案例程序。通过第一个 Android 程序的创建,帮助初学者掌握运行、调试及发布 Android 程序的方法,为后续学习做铺垫。

习题

1. Android 四层架构中,应用框架层使用的是(　　)语法。
   A. C　　　　　　B. C++　　　　　　C. Java　　　　　　D. Android
2. Android 四层架构中,系统库层使用的是(　　)语法。
   A. VB　　　　　　B. C /C++　　　　　　C. Java　　　　　　D. Android
3. 应用程序员编写的 Android 应用程序,主要是调用(　　)提供的接口进行实现。
   A. 应用程序层　　　B. 应用框架层　　　C. 应用视图层　　　D. 系统库层
4. Google 于(　　)正式发布 Android 平台。
   A. 2007 年 11 月 5 日　　　　　　B. 2008 年 11 月 5 日
   C. 2007 年 1 月 10 日　　　　　　D. 2009 年 4 月 30 日
5. Android 应用程序需要打包成(　　)文件格式在手机上安装运行。
   A. .class　　　　　B. .xml　　　　　C. .apk　　　　　D. .dex
6. Log 类中表示调试信息的方法是(　　)。
   A. Log.v()　　　　B. Log.d()　　　　C. Log.w()　　　　D. Log.e()
7. Android Studio 开发工具对安装环境没有任何要求。(　　)
8. Android SDK 不需要提前下载,在安装完 Android Studio 后会提示下载。(　　)
9. Android Studio 开发的程序只能运行在模拟器上,不可以运行在真机上。(　　)
10. Windows Mobile 不是手机操作系统。(　　)

项目 2

猜猜我的生肖

【教学导航】

学习目标	(1) 了解 TextView 控件的属性和方法。 (2) 熟悉 EditText 控件的属性和方法。 (2) 掌握 Button 控件的监听器。 (4) 熟悉 ImageView 控件的属性和方法。 (5) 理解 DatePicker 控件的属性和方法。 (6) 掌握 App 之间的通信方式。 (7) 培养简单案例的设计开发能力。
教学方法	任务驱动法、理论实践一体化、探究学习法、分组讨论法
课时建议	8 课时

2.1 Android 常用 UI 控件

　　控件是界面组成的主要元素,界面中的控件有序排放和完美组合,便可在用户眼前呈现出丰富多彩的页面。Android 系统提供了种类繁多的控件,例如 TextView(文本框)、EditText(编辑框)、Button(按钮)等,用户通过这些控件与界面进行交互,因此掌握这些控件的使用对日后开发工作至关重要。本节将针对 Android 中的常用控件进行讲解。

2.1.1 TextView 控件

　　在使用手机时,经常会看见一些文本信息,这些文本信息通常是由 TextView 控件显示的。TextView 控件是 Android 中很常用的控件,该控件常被用来显示一段文字、电话号码、URL 链接、E-mail 地址等,所以称为文本控件。

微课视频

1. TextView 控件常用属性

TextView 控件属性较多,为了让初学者更好地掌握,下面列举一些常用属性,如表 2.1 所示。

表 2.1 TextView 常用属性

属性	功能描述	实 例
android:id	控件的唯一标识 ID	android:id="@+id/textView1"
android:layout_width	控件的宽度	"wrap_content":根据需要显示的内容进行调整 "match_parent":扩充至其父控件的宽度 android:layout_width="200dp":宽度为固定值
android:layout_height	控件的高度	与 layout_width 使用方式类似,推荐单位为 dp
android:text	显示的内容	android:text="Hello Android"
android:textColor	文本颜色	android:textColor="#FF0B07"
android:textSize	字体大小	android:textSize="20sp",推荐单位为 sp
android:singleLine	是否单行显示	android:singleLine="true"为单行显示,值为"false"表示不是单行显示
android:gravity	文本的对齐方式	android:gravity="center",本文在控件内居中显示
android:layout_gravity	控件的对齐方式	android:layout_gravity="center",控件在父容器中居中显示

微课视频

TextView 控件还有很多属性,这里就不再一一列举了。

2. TextView 控件常用方法

控件添加了属性之后,在布局界面中可以直接预览效果,有时用户需要在程序运行后动态改变文本内容,这时就可以借助控件的方法。控件的方法可以动态、实时地改变文本控件的内容。下面列举 TextView 控件常用的几种方法,如表 2.2 所示。

表 2.2 TextView 常用方法

方 法	功能描述	实 例
setText(CharSequence text)	设定控件显示内容	textview.setText("Hello Android")
getText()	获得控件的显示文本	String str = textview.getText().toString()
setTextSize()	设置显示字体的大小	textview.setTextSize(40)
setTextColor()	设置显示字体的颜色	textview.setTextColor(Color.rgb(0,0,255))

3. TextView 控件案例

1) 案例分析

(1) 界面分析。本案例共有一个布局界面,在该布局界面中添加两个文本控件,第一个控件显示文本内容,第二个控件在运行时会动态更改显示的内容。

(2) 设计思路。布局界面中的一个控件通过添加属性的方式达到用户需求,另一个控件在类文件中通过控件的方法动态更改显示内容,达到用户的需求。

2) 实现步骤

(1) 新建工程。启动 Android Studio,默认会打开一个 My Application 文件,单击界面左上角的 File→New→New Project 子菜单,打开 Choose your project 对话框。在对话框中

默认选中第一个选项卡 Phone and Table，同时默认选中 Empty Activity 页面，单击 Next 按钮，打开 Configure your project 对话框。输入工程名称为"ZSTextView"，工程的包名为"com.example.zstextview"，选择保存位置，单击 Finish 按钮，如图 2.1 所示。

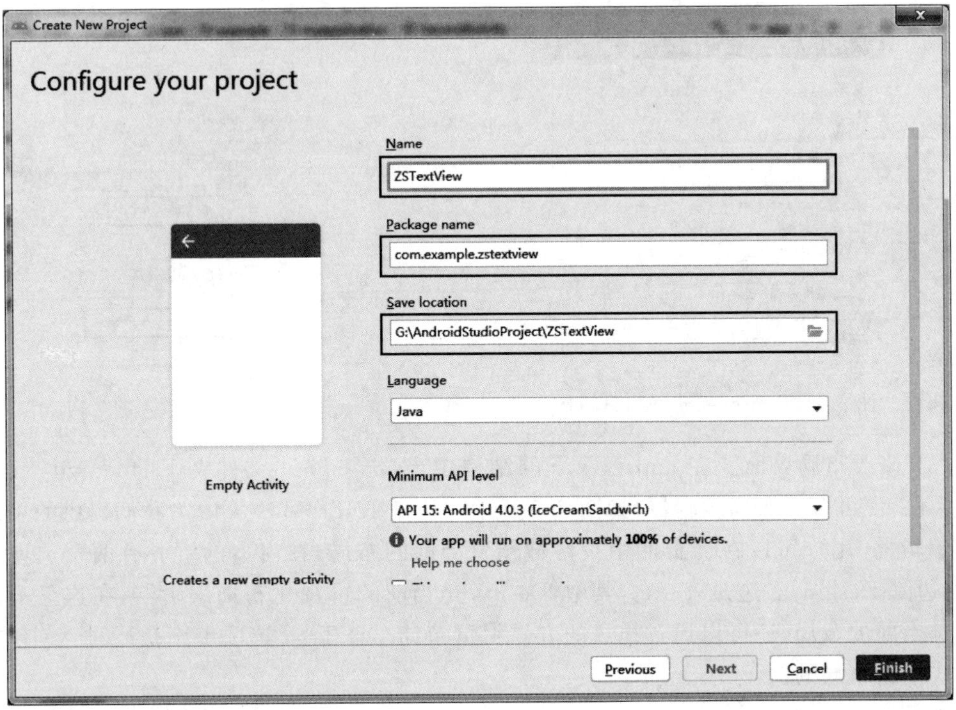

图 2.1　新建工程

（2）切换工程的项目结构。任何一个新建的项目，都会默认使用 Android 模式的项目结构，但这并不是项目真实的目录结构，而是被 Android Studio 转换过的，这种项目结构简洁明了，适合进行快速开发，但是对于新手来说可能并不易于理解。单击图 2.2 中的 Android 可以切换至 Project 项目结构模式，如图 2.3 所示。

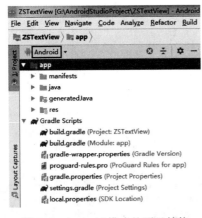

图 2.2　Android 模式的项目结构

图 2.3　切换项目结构模式

这里将项目结构模式切换为 Project，这就是项目真实的目录结构了，如图 2.4 所示。打开该模式下方的 app 文件夹，依次展开，便看到工程的布局界面和工程的类文件，其中，activity_main.xml 是布局界面，MainActivity.java 为类文件，如图 2.5 所示。

图 2.4　Project 模式的项目结构　　　　图 2.5　app 目录下的结构

（3）修改布局界面。在 app 目录下的结构中，双击 layout 文件夹下的 activity_main.xml 文件，便可打开布局编辑器，如图 2.6 所示。这是 Android Studio 为我们提供的可视化布局编辑器，用户可以在屏幕的中央区域预览当前的布局，在窗口的下方有两个选项卡标签，左边是 Design，右边是 Text。其中，Design 是当前的可视化布局编辑器，在这里不仅可以预览当前的布局，还可以通过拖放的方式编辑布局；而 Text 选项卡则是通过 XML 文件的方式编辑布局的。

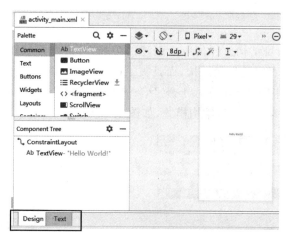

图 2.6　布局编辑器

单击 Text 标签，切换到 Text 选项卡，可以看到如下代码。

```
< android.support.constraint.ConstraintLayout xmlns:android = "http://schemas.android.com/apk/res/android"
    xmlns:app = "http://schemas.android.com/apk/res-auto"
    xmlns:tools = "http://schemas.android.com/tools"
    android:layout_width = "match_parent"
```

```
        android:layout_height = "match_parent"
        tools:context = ".MainActivity">
        <TextView
            android:layout_width = "wrap_content"
            android:layout_height = "wrap_content"
            android:text = "Hello World!"
            app:layout_constraintBottom_toBottomOf = "parent"
            app:layout_constraintLeft_toLeftOf = "parent"
            app:layout_constraintRight_toRightOf = "parent"
            app:layout_constraintTop_toTopOf = "parent" />
</android.support.constraint.ConstraintLayout>
```

修改布局类型为LinearLayout,添加方向属性为垂直。删除现有的TextView控件,如图2.7所示。

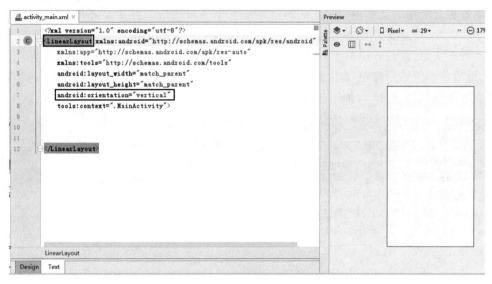

图2.7 线性布局界面

在图2.7中,如果在界面右侧看不到界面的预览视图,可单击Android工程页面左下角的 按钮,选中列表中的Preview,在布局界面中便可看到预览视图。

(4)添加TextView控件。在布局界面中添加一个TextView控件,并在TextView控件的内部添加几个属性。

```
<?xml version = "1.0" encoding = "utf-8"?>
<LinearLayout xmlns:android = "http://schemas.android.com/apk/res/android"
    xmlns:app = "http://schemas.android.com/apk/res-auto"
    xmlns:tools = "http://schemas.android.com/tools"
    android:layout_width = "match_parent"
    android:layout_height = "match_parent"
    android:orientation = "vertical"
    tools:context = ".MainActivity">
    <TextView
```

```
        android:id = "@ + id/textView_1"
        android:layout_width = "match_parent"
        android:layout_height = "wrap_content"
        android:text = "中国梦" />
</LinearLayout>
```

其中,android:id 是给控件定义了一个唯一的标识符 ID,本行代码 android:id="@+id/textView_1"中,ID 为 textView_1,之后可以在类文件代码中通过 ID 对该元素进行操作。android:layout_width 指定了当前控件的宽度,这里使用 match_parent 表示让当前控件的宽度和父容器一样宽。android:layout_height 指定了当前控件的高度。这里使用 wrap_content 表示当前控件的高度只要能刚好包含里面的内容就行。android:text 指定了控件中显示的文字内容。

继续在界面中添加第二个 TextView 控件,这时宽度值使用 wrap_content 表示控件的宽度刚好包含里面的文字内容。一般称这种宽度值为自适应。使用 android:textSize 属性设置字体大小,字体大小的单位是 sp。使用 android:textColor 属性设置文本的颜色,颜色值以#号开头,后面是 6 位的十六进制数(0~9,A~F)。前两位表示红色,中间的两位表示绿色,最后的两位表示蓝色,6 位任意组合就会出来一种颜色。使用 android:textStyle 属性设置文本字体风格,bold 为加粗。代码如下。

```
<?xml version = "1.0" encoding = "utf-8"?>
<LinearLayout xmlns:android = "http://schemas.android.com/apk/res/android"
    xmlns:app = "http://schemas.android.com/apk/res-auto"
    xmlns:tools = "http://schemas.android.com/tools"
    android:layout_width = "match_parent"
    android:layout_height = "match_parent"
    android:orientation = "vertical"
    tools:context = ".MainActivity">
    <TextView
        android:id = "@ + id/textView_1"
        android:layout_width = "match_parent"
        android:layout_height = "wrap_content"
        android:text = "中国梦" />
    <TextView
        android:id = "@ + id/textView_2"
        android:layout_width = "wrap_content"
        android:layout_height = "wrap_content"
        android:text = "梦之蓝"
        android:textSize = "40sp"
        android:textColor = "#0000ff"
        android:textStyle = "bold" />
</LinearLayout>
```

现在两个 TextView 控件已经添加完成,可以预览一下当前布局,如图 2.8 所示。

(5) 类文件中动态修改 TextView 控件的内容。还可以在程序中通过代码动态地更改 TextView 中的文字大小、文字颜色、文字内容等。打开 MainActivity.java 类文件,修改 MainActivity 的代码如下。

图2.8 预览当前布局

```
public class MainActivity extends AppCompatActivity {
//第一步:定义对象
private TextView textView_11;
private TextView textView_12;
    @Override
    protected void onCreate(Bundle savedInstanceState) {
        super.onCreate(savedInstanceState);
        setContentView(R.layout.activity_main);
        //第二步:绑定控件
        //findViewById方法:作用是把类文件中定义的对象与界面中的控件关联起来
        //关联的作用:对类文件中对象的操作就等于对布局界面中控件的操作,反之一样
        textView_11 = findViewById(R.id.textView_1);
        textView_12 = findViewById(R.id.textView_2);
        //第三步:控件方法的使用
        textView_11.setTextSize(40);  //方法 1:setTextSize()方法,设置文字的大小
        textView_11.setTextColor(Color.rgb(255,0,0)); //方法 2:setTextColor()方法,设置文
                                                     //字的颜色
        String str1 = textView_12.getText().toString();//方法 3:getText()方法,用来获取控
                                                     //件的内容
        textView_11.setText(str1 + "中国你好");        //方法 4:setText()方法,设定控件显
                                                     //示的文本内容
    }
}
```

在类文件中,输入以上代码。第一步:定义对象,以 TextView textView_11 为例,其中的 TextView 是对象类型,textView_11 是对象名。如果对象类型为红色时,说明需要导入该类,此时鼠标定位在 TextView 处,按 Alt+Enter 组合键,便可导入 TextView 类。第二步:绑定控件,可以通过 findViewById()方法获取布局文件中的控件。第三步:控件方法的调用,这些方法是 TextView 控件最常用的几种方法,需要注意的是在 setText()方法中,

括号里的内容一定是字符串类型,否则会出错。

(6) 运行程序,效果如图 2.9 所示。

图 2.9　TextView 运行效果

微课视频

2.1.2　EditText 控件

EditText 是程序用于和用户进行交互的另一个重要控件,它允许用户在控件里输入和编辑内容,并可以在程序中对这些内容进行处理。EditText 的应用场景非常广泛,在进行发短信、发微博、聊 QQ 等操作时,就会用到文本编辑控件 EditText。

1. EditText 控件常用属性

EditText 除了具有 TextView 的一些属性外,还有自己的特有属性,如表 2.3 所示。

表 2.3　EditText 常用属性

属　　性	功能描述	实　　例
android:hint	提示编辑框中要输入的内容	android:hint="请输入用户名"
android:lines	输入内容的行数	android:lines="4",设置固定行数来决定控件的高度
android:maxLines	最大行数	android:maxLines="10"
android:minLines	最小行数	android:minLines="2"
android:inputType	指定当前文本框显示内容的文本类型	android:inputType="textPassword":输入文本格式密码 android:inputType="numberPassword":输入数字格式密码 android:inputType="number":数字格式 android:inputType="phone":拨号键盘 android:inputType="datetime":日期时间键盘
android:textSize	文字大小	android:textSize="20sp",推荐单位为 sp

2. EditText 控件常用方法

EditText 控件方法如表 2.4 所示。

表 2.4 EditText 常用方法

方法	功能描述	实例
getText()	获取 EditText 控件中输入的信息	String str = edittext.getText().toString()
setText(CharSequence text)	设定输入的内容	edittext.setText("Hello Android")
setTextSize()	设置输入文字的大小	edittext.setTextSize(40)
setTextColor()	设置输入文字的颜色	edittext.setTextColor(Color.rgb(0,0,255))

3. EditText 控件案例

1) 案例分析

(1) 界面分析。布局界面中有四个 EditText 控件,第一个提示输入用户名,第二个提示输入密码,第三个提示输入数字密码,第四个提示输入爱好。程序运行后第四个输入框内容已显示输入,用户获取第四个控件的内容并连接其他字符串显示到第一个控件上。

(2) 设计思路。布局界面中的控件通过添加属性的方式达到用户需求,第一个控件和第四个控件在类文件中通过调用控件的方法,达到用户的需求。

2) 实现步骤

(1) 创建一个新的工程,工程名为 ZSEditText,如图 2.10 所示。

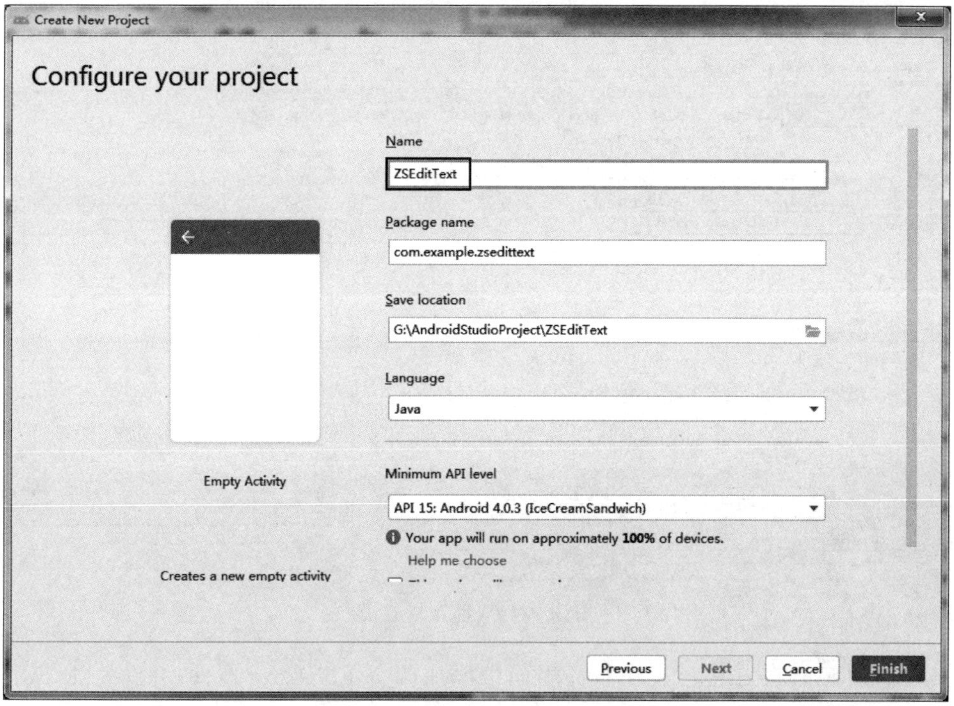

图 2.10 工程名称

(2) 切换工程的 Project 项目结构。选择该模式下方的 app，依次展开，便看到工程的布局界面和工程的类文件，其中，activity_main.xml 是布局界面，MainActivity.java 为类文件，如图 2.11 所示。

图 2.11　Project 项目结构

(3) 修改布局界面。双击 layout 文件夹下的 activity_main.xml 文件，便可打开布局编辑器，修改布局类型为 LinearLayout，添加方向属性为垂直。删除现有的 TextView 控件，如图 2.12 所示。

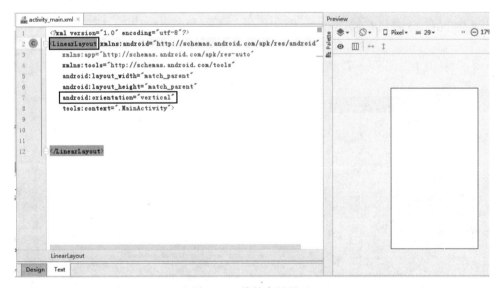

图 2.12　线性布局界面

(4) 添加 EditText 控件。在布局界面中添加四个 EditText 控件，并在每个 EditText 控件的内部添加属性，代码如下。

```xml
<?xml version = "1.0" encoding = "utf-8"?>
<LinearLayout xmlns:android = "http://schemas.android.com/apk/res/android"
    xmlns:app = "http://schemas.android.com/apk/res-auto"
    xmlns:tools = "http://schemas.android.com/tools"
    android:layout_width = "match_parent"
    android:layout_height = "match_parent"
    android:orientation = "vertical"
    tools:context = ".MainActivity">
    <EditText
        android:id = "@+id/editText_1"
        android:layout_width = "match_parent"
        android:layout_height = "wrap_content"
        android:hint = "请输入用户名"
        android:maxLines = "2"/>
    <EditText
        android:id = "@+id/editText_2"
        android:layout_width = "match_parent"
        android:layout_height = "wrap_content"
        android:inputType = "textPassword"
        android:hint = "请输入密码"/>
    <EditText
        android:id = "@+id/editText_3"
        android:layout_width = "match_parent"
        android:layout_height = "wrap_content"
        android:inputType = "numberPassword"
        android:hint = "请输入数字密码"/>
    <EditText
        android:id = "@+id/editText_4"
        android:layout_width = "match_parent"
        android:layout_height = "wrap_content"
        android:hint = "请输入你的爱好"
        android:textSize = "40sp"
        android:textColor = "#0000ff"
        android:textStyle = "bold"
        android:singleLine = "true"
        android:background = "#ffff00"/>
</LinearLayout>
```

这里使用 android:hint 属性指定了一段提示性的文本。一旦用户输入了任何内容,这些提示性的文字就会消失。这种提示功能在 Android 里是非常容易实现的,甚至不需要做任何的逻辑控制,因为系统已经帮我们都处理好了。android:inputType 属性用来设置文本的类型,用于帮助输入法显示合适的键盘类型。如果设置 android:inputType = "number",则默认弹出的输入键盘为数字键盘,且输入的内容只能为数字。当属性值取 android:inputType = "textPassword"时,输入的内容以密码形式显示。

(5) EditText 方法的调用。可以在程序中通过代码动态地更改 EditText 中的文字大小、文字颜色、文字内容等。打开 MainActivity.java 类文件,修改 MainActivity 的代码如下。

```
public class MainActivity extends AppCompatActivity {
//第一步:定义对象
    private EditText edit_1,edit_4;
    @Override
    protected void onCreate(Bundle savedInstanceState) {
        super.onCreate(savedInstanceState);
        setContentView(R.layout.activity_main);
        //第二步:把定义的对象 edit_1 与界面中的控件 ID 为 editText_1 的控件关联起来
        edit_1 = findViewById(R.id.editText_1);
        edit_4 = findViewById(R.id.editText_4);
        //第三步:EditText 控件的方法
        edit_1.setTextSize(35);                              //方法 1:设置输入文字的大小
        edit_1.setTextColor(Color.rgb(0,130,255));//方法 2:设置输入文字的颜色
        edit_4.setText("梅花");                              //方法 3:设定输入的文本内容
        String str1 = edit_4.getText().toString();   //方法 4:获取输入的文本内容
        edit_1.setText(str1 + "你好吗?");            //将获取的内容显示到 edit_1 中
    }
}
```

（6）运行程序，效果如图 2.13 所示。

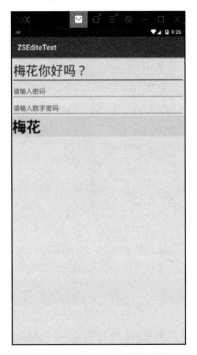

图 2.13　EditText 运行效果

微课视频

2.1.3　Button 控件

Button 是程序开发中必不可少的一个控件，其作用是用于响应用户的一系列单击事件，使程序更加流畅和完整，常被称为"按钮控件"。它可配置的属性和 TextView 是差不多

的,它可调用的方法和 TextView 也是类似的。Button 控件常用的单击方式有多种,最常用的是使用匿名内部类和在当前 Activity 中实现 OnClickListener 接口这两种方式,接下来针对这两种单击事件进行详细讲解。

1. Button 控件常用的两种单击事件

1) 使用匿名内部类方式

首先创建一个布局界面,添加按钮控件,属性如下。

```xml
<LinearLayout xmlns:android="http://schemas.android.com/apk/res/android"
    xmlns:app="http://schemas.android.com/apk/res-auto"
    xmlns:tools="http://schemas.android.com/tools"
    android:layout_width="match_parent"
    android:layout_height="match_parent"
    android:orientation="vertical"
    tools:context=".MainActivity">
    <Button
        android:id="@+id/button_one"
        android:layout_width="match_parent"
        android:layout_height="wrap_content"
        android:text="按钮1"        />
</LinearLayout>
```

上述代码中,给按钮设置了 ID 属性,方便对按钮进行查找以及设置相关事件。类文件代码如下。

```java
public class MainActivity extends AppCompatActivity {
    private Button btn_one;
    @Override
    protected void onCreate(Bundle savedInstanceState) {
        super.onCreate(savedInstanceState);
        setContentView(R.layout.activity_main);
        btn_one = findViewById(R.id.button_one);
        btn_one.setOnClickListener(new View.OnClickListener() {
            @Override
            public void onClick(View view) {
                btn_one.setText("按钮1已被单击");
            }
        });
    }
}
```

上述代码中,通过 findViewById()方法初始化控件,然后给按钮1添加单击事件。用 setOnClickListener()方法对按钮1进行绑定,然后采用匿名内部类作为监听器对单击事件进行监听,最后实现 onClick()方法,在此方法中编写按钮被单击后的代码逻辑。界面中按钮比较少时,使用该方法比较便捷。

2) 采用实现 OnClickListener 接口方式

首先创建一个布局界面,添加按钮控件,属性如下。

```xml
<LinearLayout xmlns:android = "http://schemas.android.com/apk/res/android"
    xmlns:app = "http://schemas.android.com/apk/res-auto"
    xmlns:tools = "http://schemas.android.com/tools"
    android:layout_width = "match_parent"
    android:layout_height = "match_parent"
    android:orientation = "vertical"
    tools:context = ".MainActivity">
    <Button
        android:id = "@+id/button_two"
        android:layout_width = "match_parent"
        android:layout_height = "wrap_content"
        android:text = "按钮2"         />
</LinearLayout>
```

上述代码中,给按钮设置了 ID 属性,方便对按钮进行查找以及设置相关事件。类文件代码如下。

```java
public class MainActivity extends AppCompatActivity implements View.OnClickListener{
    private Button btn_two;
    @Override
    protected void onCreate(Bundle savedInstanceState) {
        super.onCreate(savedInstanceState);
        setContentView(R.layout.activity_main);
        btn_two = findViewById(R.id.button_two);
        btn_two.setOnClickListener(this);

    }
    @Override
    public void onClick(View view) {
        switch (view.getId()){
            case R.id.button_two:
                btn_two.setText("按钮2已被单击");
                break;
                default:
                    break;
        }
    }
}
```

上述代码中,Activity 实现了 OnClickListener 接口并重写 onClick()方法,然后通过 switch 语句判断哪个按钮被单击,显然这种形式在按钮较多的情况下可以降低代码的重复率,界面中按钮较多时使用该方法比较便捷。需要注意的是,btn_two.setOnClickListener(this)语句中有一个 this 参数,这个 this 代表的是该 Activity 的引用,由于 Activity 实现了 OnClickListener 接口,所以在这里 this 代表了 OnClickListener 的引用。

2. Button 控件案例

1) 案例分析

(1) 界面分析。布局界面中有 7 个控件,其中,两个 TextView 控件内容为请输入用户名和密码,两个 EditText 控件用于输入用户名和密码,"登录"按钮用于获取用户名和密码,"取消"按钮用于清除用户名和密码。最下方的 TextView 控件用于显示获取的用户名的密码。

(2) 设计思路。布局界面中的控件通过添加属性的方式达到用户需求,"登录"和"取消"按钮的功能通过按钮单击事件达到用户的需求。

2) 实现步骤

(1) 创建一个新的工程,工程名为 ZSButton。

(2) 切换工程的 Project 项目结构。选择该模式下方的 app,依次展开,便看到工程的布局界面和工程的类文件,其中,activity_main.xml 是布局界面,MainActivity.java 为类文件。

(3) 设计布局界面。双击 layout 文件夹下的 activity_main.xml 文件,便可打开布局编辑器,修改布局类型为 LinearLayout,添加方向属性为垂直。依次添加 7 个控件,代码如下。

```xml
<?xml version = "1.0" encoding = "utf - 8"?>
<LinearLayout xmlns:android = "http://schemas.android.com/apk/res/android"
    xmlns:app = "http://schemas.android.com/apk/res - auto"
    xmlns:tools = "http://schemas.android.com/tools"
    android:layout_width = "match_parent"
    android:layout_height = "match_parent"
    android:orientation = "vertical"
    tools:context = ".MainActivity">
    <TextView
        android:layout_width = "wrap_content"
        android:layout_height = "wrap_content"
        android:text = "请输入用户名:"
        android:textSize = "30sp"
        android:textColor = "#030303"
        android:textStyle = "bold"
        android:layout_marginTop = "10dp"/>
    <EditText
        android:id = "@ + id/editText_inputname"
        android:layout_width = "match_parent"
        android:layout_height = "wrap_content"
        android:hint = "用户名"
        android:textStyle = "bold"/>
    <TextView
        android:layout_width = "wrap_content"
        android:layout_height = "wrap_content"
        android:text = "请输入密码:"
        android:textSize = "30sp"
        android:textColor = "#030303"
        android:textStyle = "bold"
        android:layout_marginTop = "10dp"/>
    <EditText
        android:id = "@ + id/editText_inputpwd"
        android:layout_width = "match_parent"
        android:layout_height = "wrap_content"
```

```xml
            android:hint = "密    码"
            android:inputType = "textPassword"
            android:textStyle = "bold"/>
    <Button
            android:id = "@+id/button_login"
            android:layout_width = "match_parent"
            android:layout_height = "wrap_content"
            android:text = "登录"
            android:textSize = "30sp"
            android:textColor = "#000000"
            android:textStyle = "bold"
            android:layout_marginTop = "50dp" />
    <Button
            android:id = "@+id/button_cancle"
            android:layout_width = "match_parent"
            android:layout_height = "wrap_content"
            android:text = "取消"
            android:textSize = "30sp"
            android:textColor = "#000000"
            android:textStyle = "bold"
            android:layout_marginTop = "10dp" />
    <TextView
            android:id = "@+id/textView_display"
            android:layout_width = "wrap_content"
            android:layout_height = "wrap_content"
            android:text = "显示获取的用户名和密码"
            android:textSize = "20sp"
            android:textColor = "#0E0D0E"
            android:textStyle = "bold"
            android:layout_marginTop = "10dp"
            android:visibility = "gone"/>
</LinearLayout>
```

上述代码看似很多,其实并不复杂,只是将之前学到的知识结合在一起而已,再利用各控件的属性调整它们的位置和样式。最下面的 TextView 文本控件中,使用 android：visibility 属性设置控件的显示和隐藏,其属性有 3 个,分别为"visible""invisible""gone"。属性值为 invisible 表示不显示控件,但是控件所占用的空间不会消失,所以会显示一块和控件一样大小的背景色；visible 表示显示控件；gone 表示不显示控件,控件所占用的空间也消失。在 Android 开发中,大部分控件都有 visibility 这个属性。

(4) 按钮单击事件功能。可以在程序中通过代码动态地更改 Button 控件上的文字大小、文字颜色、文字内容等。打开 MainActivity.java 类文件,修改 MainActivity 的代码如下。

```java
public class MainActivity extends AppCompatActivity {
    //第一步:定义对象
    private EditText edit_name,edit_pwd;
```

```java
        private Button btn_login,btn_cancle;
        private TextView txt_display;
        @Override
        protected void onCreate(Bundle savedInstanceState) {
            super.onCreate(savedInstanceState);
            setContentView(R.layout.activity_main);
            //第二步:绑定控件
            edit_name = findViewById(R.id.editText_inputname);
            edit_pwd = findViewById(R.id.editText_inputpwd);
            btn_login = findViewById(R.id.button_login);
            btn_cancle = findViewById(R.id.button_cancle);
            txt_display = findViewById(R.id.textView_display);
            //第三步:给"登录"按钮添加单击事件
            //1:谁发生了事件(按钮),2:发生了什么事情(被单击),3:谁来管理这个事情(监听器)
            btn_login.setOnClickListener(new View.OnClickListener() {
                @Override
                public void onClick(View view) {
                    //写按钮单击事件的代码
                    String strname = edit_name.getText().toString();   //获取输入的用户名
                    String strpwd = edit_pwd.getText().toString();     //获取输入的密码
                    txt_display.setVisibility(View.VISIBLE);           //让文本控件变得可见
                    //显示到文本控件里面,其中:+连接字符串,用于把字符串连接在一起,\n 表示换行
                    txt_display.setText("哈哈,我偷偷地盗取了你的秘密信息\n用户名为:" +
strname + "\n密码为:" + strpwd);
                }
            });
            //第四步:"取消"按钮单击事件
            btn_cancle.setOnClickListener(new View.OnClickListener() {
                @Override
                public void onClick(View view) {
                    txt_display.setVisibility(View.GONE);   //让文本控件变得不可见
                    edit_name.setText("");//设定用户名输入框内容为空,即清空用户名输入框的内容
                    edit_pwd.setText("");   //设定密码输入框内容为空,即清空密码输入框的内容
                }
            });
        }
    }
```

上述代码中,第一步:定义了 EditText、Button、TextView 三种类型,共计 5 个对象。第二步:绑定控件,通过 findViewById()方法分别获取到布局文件中的 5 个控件。第三步:"登录"按钮单击事件,采用匿名内部类作为监听器对"登录"按钮单击事件进行监听,最后实现 onClick()方法。"登录"按钮内部代码逻辑为:获取输入的用户名,获取输入的密码,设置 TextView 文本控件为可见,最后将获取的内容显示到文本控件里面。同理,"取消"按钮内部代码逻辑为:设置文本控件不可见,同时清空输入的用户名和密码。

(5) 运行程序,效果如图 2.14 所示。

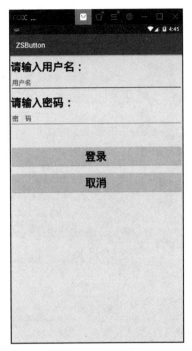 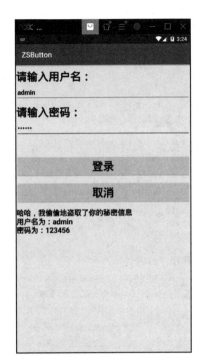

图 2.14　程序运行前后效果对比

微课视频

2.1.4　ImageView 控件

ImageView 是用于在界面上展示图片的一个控件,它可以让程序界面变得更加丰富多彩。学习这个控件需要提前准备好一些图片,图片通常都是放在以 drawable 开头的目录下。

1. ImageView 控件常用属性

ImageView 除了具有 TextView 的一些属性外,还有自己的特有属性,如表 2.5 所示。

表 2.5　ImageView 常用属性

属　性	功能描述	实　例
android:src	给 ImageView 控件指定一张图片	android:src="@drawable/rose"

2. ImageView 控件常用方法

ImageView 控件方法,如表 2.6 所示。

表 2.6　ImageView 常用方法

方　法	功能描述	实　例
setImageResource()	给控件设置图片但不改变背景色	imageView1.setImageResource(R.drawable.img)
setBackgroundResource()	改变控件的背景色	imageView2.setBackgroundResource(R.drawable.img)

3. ImageView 控件案例

1) 案例分析

(1) 界面分析。布局界面中有3个控件：上方的 ImageView 控件显示一朵玫瑰花；中间的 TextView 控件显示"玫瑰"文字；下方为 Button 按钮控件，程序运行后，单击按钮，ImageView 控件中的图片变成了一张动物图片，TextView 控件显示的文字变成了"动物"文字。

(2) 设计思路。布局界面中的控件通过添加属性的方式达到用户需求，可以通过调用控件的方法更改 ImageView 控件和 TextView 控件的文字，达到用户的需求。

2) 实现步骤

(1) 创建一个新的工程，工程名为 ZSImageView。

(2) 切换工程的 Project 项目结构。选择该模式下方的 app，依次展开，便看到工程的布局界面和工程的类文件，其中，activity_main.xml 是布局界面，MainActivity.java 为类文件。

(3) 准备一张玫瑰图片 rose.jpg 和一张动物图片 dongwu.jpg，将其粘贴到 app 目录结构中 res 下方的 drawable 文件夹下，如图 2.15 所示。

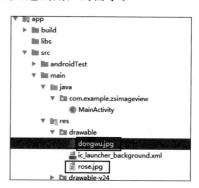

图 2.15 图片在工程中的保存位置

(4) 设计布局界面。双击 layout 文件夹下的 activity_main.xml 文件，便可打开布局编辑器，修改布局类型为 LinearLayout，添加方向属性为垂直。依次添加3个控件，代码如下。

```xml
<?xml version = "1.0" encoding = "utf-8"?>
<LinearLayout xmlns:android = "http://schemas.android.com/apk/res/android"
    xmlns:app = "http://schemas.android.com/apk/res-auto"
    xmlns:tools = "http://schemas.android.com/tools"
    android:layout_width = "match_parent"
    android:layout_height = "match_parent"
    android:orientation = "vertical"
    tools:context = ".MainActivity">
    <ImageView
        android:id = "@+id/imageView_1"
        android:layout_width = "wrap_content"
        android:layout_height = "300dp"
        android:src = "@drawable/rose"/>
    <TextView
        android:id = "@+id/textView_1"
        android:layout_width = "wrap_content"
        android:layout_height = "wrap_content"
        android:text = "玫瑰"
        android:textSize = "30sp"
        android:textColor = "#9C27B0"
        android:textStyle = "bold"
        android:layout_gravity = "center"/>
```

```xml
<Button
    android:id = "@+id/button_1"
    android:layout_width = "match_parent"
    android:layout_height = "wrap_content"
    android:text = "单击可以给你惊喜"
    android:textSize = "30sp"
    android:textColor = "#FFEB3B"
    android:textStyle = "bold"
    android:layout_gravity = "center"
    android:background = "#000000"    />
</LinearLayout>
```

其中，android:src 属性给 ImageView 控件指定了一张图片，由于有些图片的宽度是未知的，所以将图片的宽度设置为 wrap_content，为了防止某些图片过高而导致下方的文字和按钮看不到，所以将图片的高度值设置为 300dp。下方的两个控件都是一些简单的属性调整了位置和样式。

（5）实现案例功能。可以在程序中通过代码动态地更改 ImageView 控件中的图片和 TextView 控件显示的文字内容。打开 MainActivity.java 类文件，修改 MainActivity 的代码如下。

```java
public class MainActivity extends AppCompatActivity {
//第一步:定义对象
    private ImageView img_1;
    private TextView txt_1;
    private Button btn_1;
    @Override
    protected void onCreate(Bundle savedInstanceState) {
        super.onCreate(savedInstanceState);
        setContentView(R.layout.activity_main);
        //第二步:绑定控件
        img_1 = findViewById(R.id.imageView_1);
        txt_1 = findViewById(R.id.textView_1);
        btn_1 = findViewById(R.id.button_1);
        //第三步:按钮单击事件
        btn_1.setOnClickListener(new View.OnClickListener() {
            @Override
            public void onClick(View v) {
                img_1.setImageResource(R.drawable.dongwu);   //更改 ImageView 控件中的图片
                txt_1.setText("动物");                        //更改 TextView 控件的内容
            }
        });
    }
}
```

（6）运行程序，效果如图 2.16 所示。

图 2.16 程序运行前后效果对比

2.1.5 DatePicker 控件

DatePicker 是 Android 中的日期选择控件,DatePicker 可以通过设置属性来确定日期选择范围,也可以通过定义好的方法获取到当前选中的时间,并且在修改日期的时候,有相应的事件对其进行响应。

1. DatePicker 控件常用属性

DatePicker 除了具有 ID、宽度、高度等一些属性外,还有自己的特有属性,如表 2.7 所示。

表 2.7 DatePicker 常用属性

属 性	功能描述	实 例
android:calendarViewShown	是否显示日历	android:calendarViewShown="false"
android:startYear	设置可选开始年份	android:startYear="1970"　　默认:1900
android:endYear	设置可选结束年份	android:endYear="2019"　　默认:2100
android:maxDate	设置可选最大日期	android:maxDate="01/01/2030"
android:minDate	设置可选最小日期	android:minDate="12/30/1980"

2. DatePicker 控件常用方法

DatePicker 控件方法,如表 2.8 所示。

表 2.8 DatePicker 常用方法

方 法	功能描述	实 例
intgetYear()	获得当前控件选择的年份	datepicker_1.getYear()
intgetMonth()	获得当前控件选择的月份(0~11)	date_picker.getMonth()+1(所获月份比实际月份小 1)
intgetDayOfMonth()	获得当前控件选择的日	date_picker.getDayOfMonth()

除了常用的获取属性的 setter、getter 方法之外,还需要特别注意一个初始化的方法 init(),用于做 DatePicker 控件的初始化,并且设置日期被修改后回调的响应事件。此方法的签名如下。

```
date_picker.init(int year, int monthOfYear, int dayOfMonth, new DatePicker.OnDateChangedListener() {
    @Override
    public void onDateChanged(DatePicker datePicker, int i, int i1, int i2) {

    }
});
```

从上面的 init()方法可以看到,括号内的前 3 个参数 year、monthOfYear、dayOfMonth 是日期控件所显示的系统日期,DatePicker 被修改时响应的事件是 DatePicker.OnDateChangedListener 事件,如果要响应此事件,需要实现其中的 onDateChanged()方法。其内部的参数 int i、int i1、int i2 则表示日期控件被修改后的年月日。

3. DatePicker 控件案例

1) 案例分析

(1) 界面分析。布局界面中有四个控件,上方的 TextView 控件显示 DatePicker 系统日期,中间的 DatePicker 控件可以随意修改日期,下方的 TextView 控件显示修改后的日期,特殊的日期如 3 月 12 日、3 月 15 日等,除了显示日期和节日外,同时显示一张节日图片。

(2) 设计思路。布局界面中的控件通过添加属性的方式达到用户需求,上方的 TextView 控件通过调用 DatePicker 控件的 getYear()、getMonth()、getDayOfMonth()三种方法得到内容,上方的两个控件通过调用 DatePicker 控件的 init()方法,并实现其中的 onDateChanged()方法,将修改后的日期和对应的图片显示出来,达到用户的需求。

2) 实现步骤

(1) 创建一个新的工程,工程名为 ZSDatePicker。

(2) 切换工程的 Project 项目结构。选择该模式下方的 app,依次展开,便看到工程的布局界面和工程的类文件,其中,activity_main.xml 是布局界面,MainActivity.java 为类文件。

(3) 准备一张植树节图片 threeot.jpg 和一张消费者权益图片 threeof.jpg,将其粘贴到 app 目录结构中 res 下方的 drawable 文件夹下。

(4) 设计布局界面。双击 layout 文件夹下的 activity_main.xml 文件,便可打开布局编辑器,修改布局类型为 LinearLayout,添加方向属性为垂直。依次添加 4 个控件,代码如下。

```xml
<?xml version = "1.0" encoding = "utf - 8"?>
<LinearLayout xmlns:android = "http://schemas.android.com/apk/res/android"
    xmlns:app = "http://schemas.android.com/apk/res - auto"
    xmlns:tools = "http://schemas.android.com/tools"
    android:layout_width = "match_parent"
    android:layout_height = "match_parent"
    android:orientation = "vertical"
    tools:context = ".DatePickerActivity">
    <TextView
        android:id = "@ + id/textView_11"
        android:layout_width = "match_parent"
        android:layout_height = "wrap_content"
        android:text = "显示原始的日期"
        android:textSize = "30sp"
        android:textColor = "#000000"
        android:textStyle = "bold"
        android:gravity = "center"
        android:layout_margin = "10dp"/>
    <DatePicker
        android:id = "@ + id/datePicker_11"
        android:layout_width = "wrap_content"
        android:layout_height = "wrap_content"
        android:layout_gravity = "center"/>
    <TextView
        android:id = "@ + id/textView_12"
        android:layout_width = "match_parent"
        android:layout_height = "wrap_content"
        android:text = "显示更改之后的日期"
        android:textSize = "30sp"
        android:textColor = "#000000"
        android:textStyle = "bold"
        android:gravity = "center"
        android:layout_margin = "10dp"/>
    <ImageView
        android:id = "@ + id/img_1"
        android:layout_width = "wrap_content"
        android:layout_height = "wrap_content"
        android:layout_gravity = "center"/>
</LinearLayout>
```

上述代码中，其中，android:layout_gravity 属性指定了日期控件的对齐方式，属性取值为 center 表示控件居中对齐。ImageView 没有添加 android:src 属性，说明程序运行后，图片控件中不显示图片。

（5）实现案例功能。我们可以在程序中通过代码动态地更改 TextView 控件显示的文字内容和 ImageView 控件中显示的图片，打开 MainActivity.java 类文件，修改 MainActivity 的代码如下。

```java
public class DatePickerActivity extends AppCompatActivity {
//定义对象
    private DatePicker date_picker;
    private TextView txt_ysdate;
    private TextView txt_gghdate;
    private ImageView img_display;
    @Override
    protected void onCreate(Bundle savedInstanceState) {
        super.onCreate(savedInstanceState);
        setContentView(R.layout.activity_date_picker);

        //绑定控件
        date_picker = findViewById(R.id.datePicker_11);
        txt_ysdate = findViewById(R.id.textView_11);
        txt_gghdate = findViewById(R.id.textView_12);
        img_display = findViewById(R.id.img_1);

        //获取时间 获取日期控件默认的日期
        int ysyear = date_picker.getYear();
        int ysmonth = date_picker.getMonth() + 1;    //月份 0→1月,1→2月,2→3月(月份是从
                                                     //0开始的,0~11,代表生活中1~12月)

        int ysday = date_picker.getDayOfMonth();

        //把获取的时间显示出来——文本控件
        txt_ysdate.setText(ysyear + "年" + ysmonth + "月" + ysday + "日");
    //获取当前系统时间
        Time t = new Time();
        t.setToNow();
//能够获取修改后的日期——给日期添加了一个监听器,new DatePicker.OnDateChangedListener
date_picker.init(t.year, t.month , t.monthDay,new DatePicker.OnDateChangedListener() {
    @Override
  public void onDateChanged(DatePicker datePicker, int i, int i1, int i2) {
            // int i, int i1, int i2,就是更改之后的年月日
txt_gghdate.setText(i + "年" + (i1 + 1) + "月" + i2 + "日");
img_display.setVisibility(View.GONE);     //让图片控件不可见
if((i1 + 1) == 3&&i2 == 12){
    txt_gghdate.setText(i + "年" + (i1 + 1) + "月" + i2 + "日" + ",植树节");
                                                    //设置TextView控件的文本内容
   img_display.setVisibility(View.VISIBLE);         //设置ImageView控件可见
   img_display.setImageResource(R.drawable.threeot); //设置ImageView控件显示的图片
   }else if((i1 + 1) == 3&&i2 == 15){
txt_gghdate.setText(i + "年" + (i1 + 1) + "月" + i2 + "日" + ",国际消费者权益日");
     img_display.setVisibility(View.VISIBLE);
   img_display.setImageResource(R.drawable.threeof);
      }
     }
        });
    }
}
```

上述代码中，获取日期控件的年月日，显示到上方 TextView 控件中。通过 date_picker.getMonth()获取当前月,返回数值为 0～11,代表生活中的 1～12 月,需要+1 来显示。在修改日期选择控件的事件响应中,采用 if 语句,根据修改后的月和日进行判断,如果月等于 3 并且日等于 12,将下方的 TextView 控件显示当天为植树节,设置 ImageView 控件为可见,并将 ImageView 控件中显示一张植树节图片。同理,如果是 3 月 15 日,也会显示对应的节日和节日的图片,选择其他日期时,TextView 控件中只显示修改后的日期,调用图片控件的 setVisibility(View.GONE)方法,设置图片控件不可见。

(6) 运行程序,效果如图 2.17 所示。

图 2.17　程序运行前后效果对比

2.2　App 之间的通信

微课视频

在 Android 系统中,每个应用程序通常都由多个界面组成,每个界面就是一个Activity,在这些界面进行跳转时,实际上也就是 Activity 之间的跳转。Activity 之间的跳转需要用到 Intent(意图)组件,通过 Intent 可以开启新的 Activity 实现界面跳转功能。本节将针对 Activity 之间的跳转进行详细讲解。

Intent 被称为意图,是程序中各组件进行交互的一种重要方式,它不仅可以指定当前组件要执行的动作,还可以在不同组件之间进行数据传递。一般用于启动 Activity、Service 以及发送广播等。根据开启目标组件的方式不同,Intent 被分为显式意图和隐式意图两种类型,接下来分别针对这两种意图进行详细讲解。

2.2.1 显式 Intent

1. 基本语法

显式 Intent 可以直接通过名称开启指定的目标组件,其方法如表 2.9 所示。

表 2.9 显式 Intent 基本语法

方　法	功能描述	实　例
Intent(Context packageContext, Class<?> cls)	创建一个 Intent 对象,明确 Intent 跳转时的源 Activity 和目标 Activity	Intent intent = new Intent(MainActivity.this,SecondActivity.class); 参数 1 表示当前的 Activity 对象,参数 2 表示要启动的目标 Activity
startActivity(intent)	启动目标组件,根据 Intent 启动某个 Activity	startActivity(intent);开启 SecondActivity 页面

2. 向下一个页面传递数据

在通信技术不发达时,人与人之间的消息传递往往是通过信函的方式。在 Android 系统中,组件之间也可以进行消息传递或者数据传递,此时使用的是 Intent。在 Activity 启动活动时传递数据的思路很简单,因为 Intent 提供了一系列的重载的 PutExtra()方法,可以把我们想要传递的数据暂存在 Intent 中,启动了一个活动后,只需要把这些数据再从 Intent 中取出来就可以了。传递数据的方法如表 2.10 所示。

表 2.10 显式 Intent 数据传递方法

方　法	功能描述	实　例
putExtra(String name,String value)	传递字符串类型的数据	intent.putExtra("name","张三"); 第一个参数是键,第二个参数是值
putExtra(String name,int value)	传递整型的数据	intent.putExtra("phone",15006258956)
putExtra(String name,boolean value)	传递布尔型的数据	intent.putExtra("friend",true)
putExtra(String name,Serializable value)	传递序列化的对象	intent.putExtra("data",courseInfo)

3. 下一个页面接收数据

首先通过 getIntent()方法获取 Intent 对象,然后调用 getXXXExtra(String name)方法,根据传入的键名,取出相应的数据。接收数据的方法如表 2.11 所示。

表 2.11 接收 Intent 传递过来的数据方法

方　法	功能描述	实　例
getStringExtra(String name)	接收字符串类型的数据	String myname=getIntent().getStringExtra("name")
getIntExtra(String name, int defaultValue)	接收整型的数据	int myphone=getIntent().getIntExtra("phone",0)
getBooleanExtra(String name, boolean defaultValue)	接收布尔型的数据	Boolean myfriend = getIntent().getBooleanExtra("friend",false)
getSerializableExtra(String name)	接收序列化的对象	CourseInfo onecourseinfo = (CourseInfo) getIntent().getSerializableExtra("data")

4. 显式 Intent 案例

1) 案例分析

(1) 界面分析。Intent 是用来实现页面跳转的,因此,工程中只要有两个页面。第一个页面中有三个 EditText 控件,用来输入用户的信息,此外还有一个 Button 控件,单击该控件可以跳转到第二页。第二个页面中有 4 个 TextView 控件,用来获取第一个页面传递的数据并显示出来。

(2) 设计思路。布局界面中的控件通过添加属性的方式达到用户需求,页面跳转时通过 putExtra() 方法传递用户输入的数据,第二个页面通过 getXXXExtra() 方法获取数据并显示,达到用户的需求。

2) 实现步骤

(1) 创建一个新的工程,工程名为 ZSIntent。

(2) 切换工程的 Project 项目结构。选择该模式下方的 app,依次展开,便看到工程的布局界面和工程的类文件,其中,activity_main.xml 是布局界面,MainActivity.java 为类文件。

(3) 在工程中添加一个新的页面。展开 app 目录结构:app→src→main→java→com.example.zsintent,右击 com.example.zsintent 包→New→Activity→Empty Activity,会弹出一个创建活动的对话框,将活动命名为 SecondActivity,默认勾选 Generate Layout File 关联布局界面,布局界面名称为 activity_second,但不要勾选 Launcher Activity。单击 Finish 按钮,便可在工程中完成第二个页面的添加。

(4) 准备两张图片,图片名为 eight.jpg 和 five.jpg,将其粘贴到 app 目录结构中 res 下方的 drawable 文件夹下,两张图片作为两个页面的背景图片。

(5) 设计第一个 Activity 的布局界面。双击 layout 文件夹下的 activity_main.xml 文件,便可打开布局编辑器,修改布局类型为 LinearLayout,添加方向属性为垂直。依次添加 3 个 EditText 控件和 1 个 Button 控件,代码如下。

```xml
<?xml version = "1.0" encoding = "utf-8"?>
<LinearLayout xmlns:android = "http://schemas.android.com/apk/res/android"
    xmlns:app = "http://schemas.android.com/apk/res-auto"
    xmlns:tools = "http://schemas.android.com/tools"
    android:layout_width = "match_parent"
    android:layout_height = "match_parent"
    android:orientation = "vertical"
    android:background = "@drawable/eight"
    tools:context = ".MainActivity">
    <EditText
        android:id = "@+id/editText_name"
        android:layout_width = "match_parent"
        android:layout_height = "wrap_content"
        android:hint = "请输入用户名"
        android:layout_marginTop = "200dp"
        android:textSize = "30sp"/>
    <EditText
```

```xml
        android:id = "@ + id/editText_age"
        android:layout_width = "match_parent"
        android:layout_height = "wrap_content"
        android:hint = "年龄"
        android:textSize = "30sp"/>
    <EditText
        android:id = "@ + id/editText_sno"
        android:layout_width = "match_parent"
        android:layout_height = "wrap_content"
        android:hint = "学号"
        android:textSize = "30sp"/>
    <Button
        android:id = "@ + id/button_next"
        android:layout_width = "match_parent"
        android:layout_height = "wrap_content"
        android:text = "跳转到下一页"
        android:textSize = "30sp"
        android:textColor = "#000000"/>
</LinearLayout>
```

上述代码中，使用 android：background 给线性布局界面添加背景，其属性值为 @drawable/eight 表示引用了 res 下方的 drawable 文件夹下的 eight.jpg 图片作为界面的背景。

（6）实现第一个 Activity 页面的功能。可以在程序中通过 EditText 控件将内容传递到第二页，打开 MainActivity.java 类文件，修改 MainActivity 的代码，如下。

```java
public class MainActivity extends AppCompatActivity {
    //定义对象
    private Button btn_next;
    private EditText edit_name,edit_age,edit_sno;
    @Override
    protected void onCreate(Bundle savedInstanceState) {
        super.onCreate(savedInstanceState);
        setContentView(R.layout.activity_main);
        //绑定控件
        btn_next = findViewById(R.id.button_next);
        edit_name = findViewById(R.id.editText_name);
        edit_age = findViewById(R.id.editText_age);
        edit_sno = findViewById(R.id.editText_sno);
        //单击按钮，跳到下一页
        btn_next.setOnClickListener(new View.OnClickListener() {
            @Override
            public void onClick(View view) {
                //跳转到下一页：指定了跳转的源位置和目的位置
                Intent intent = new Intent(MainActivity.this,SecondActivity.class);
                //传值
                intent.putExtra("name",edit_name.getText().toString());//传递的值的类型：字符串
```

```
            intent.putExtra("age",Integer.valueOf(edit_age.getText().toString()) );
//传递的值的类型：整数
            intent.putExtra("sno",edit_sno.getText().toString());//传递的值的类型：字符串
            intent.putExtra("friend",true);//传递了一个布尔类型的值
                //启动意图,完成跳转
                startActivity(intent);
            }
        });
    }
}
```

上述代码中，intent 传递数据是使用 putExtra("name", edit_name.getText().toString())方法，方法中的第一个参数是"键"，第二个参数是"值"，这里获取编辑框输入的用户名就是需要传递的值。下方的一行采用 Integer.*valueOf*(edit_age.getText().toString())方法，将输入的内容转换为整型。intent.putExtra("friend",true)中传递的数据类型为布尔型。

（7）设计第二个 Activity 的布局界面。双击 layout 文件夹下的 activity_second.xml 文件，便可打开布局编辑器，修改布局类型为 LinearLayout，添加方向属性为垂直。依次添加4 个 TextView 控件代码如下。

```xml
<?xml version="1.0" encoding="utf-8"?>
<LinearLayout xmlns:android="http://schemas.android.com/apk/res/android"
    xmlns:app="http://schemas.android.com/apk/res-auto"
    xmlns:tools="http://schemas.android.com/tools"
    android:layout_width="match_parent"
    android:layout_height="match_parent"
    android:orientation="vertical"
    android:background="@drawable/five"
    tools:context=".SecondActivity">
    <TextView
        android:id="@+id/textView_name"
        android:layout_width="match_parent"
        android:layout_height="wrap_content"
        android:text="接收用户名"
        android:layout_marginTop="250dp"
        android:layout_marginLeft="100dp"
        android:textSize="30sp"/>
    <TextView
        android:id="@+id/textView_age"
        android:layout_width="match_parent"
        android:layout_height="wrap_content"
        android:text="接收年龄"
        android:textSize="30sp"
        android:layout_marginTop="20dp"
        android:layout_marginLeft="100dp"/>
    <TextView
        android:id="@+id/textView_sno"
```

```xml
            android:layout_width = "match_parent"
            android:layout_height = "wrap_content"
            android:text = "接收学号"
            android:layout_marginTop = "20dp"
            android:layout_marginLeft = "100dp"
            android:textSize = "30sp"/>
    <TextView
            android:id = "@+id/textView_friend"
            android:layout_width = "match_parent"
            android:layout_height = "wrap_content"
            android:text = "接收是否有女朋友"
            android:layout_marginTop = "20dp"
            android:layout_marginLeft = "100dp"
            android:textSize = "30sp"/>
</LinearLayout>
```

（8）实现第二个 Activity 页面的功能。获取传递过来的数据并显示出来，打开 SecondActivity.java 类文件，修改 SecondActivity 的代码如下。

```java
public class SecondActivity extends AppCompatActivity {
    //定义对象
    TextView txt_name,txt_age,txt_sno,txt_friend;
    @Override
    protected void onCreate(Bundle savedInstanceState) {
        super.onCreate(savedInstanceState);
        setContentView(R.layout.activity_second);
    //绑定控件
        txt_name = findViewById(R.id.textView_name);
        txt_age = findViewById(R.id.textView_age);
        txt_sno = findViewById(R.id.textView_sno);
        txt_friend = findViewById(R.id.textView_friend);
    //接收传递过来的值
        String myname = getIntent().getStringExtra("name");   //接收字符串类型的值
        int myage = getIntent().getIntExtra("age",0);  //接收数字类型的数据时,要加默认值:0
        String mysno = getIntent().getStringExtra("sno");   //接收字符串类型的值
        Boolean myfriend = getIntent().getBooleanExtra("friend",false);  //接收布尔型数据时,
                                                                          //要加默认值:false
    //把接收的数据显示出来
        txt_name.setText("姓名:" + myname);      //将接收的名字显示出来
        txt_age.setText("年龄:" + myage);         //将接收的年龄显示出来
        txt_sno.setText("学号:" + mysno);         //将接收的学号显示出来
        if (myfriend){                            //如果接收的myfriend值为真,显示有对象
            txt_friend.setText("是否有对象:有");
        }else{                                    //否则显示无对象
            txt_friend.setText("是否有对象:无");
        }
    }
}
```

（9）运行程序，效果如图 2.18 所示。

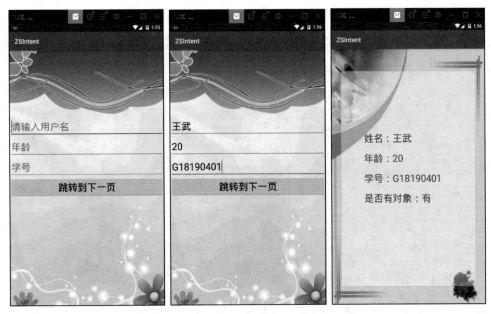

图 2.18　程序运行效果图

2.2.2　隐式 Intent

相比于显式 Intent，隐式 Intent 则含蓄了许多，它并不明确指出我们想要启动哪一个活动，而是指定了一系列更为抽象的 action 和 category 等信息，然后交由系统去分析这个 Intent，并帮我们找出合适的活动去启动。

1. 基本用法

使用隐式 Intent，不仅可以启动自己程序内的活动，还可以启动其他程序的活动，这使得 Android 多个应用程序之间的功能共享成为可能。例如，你的应用程序中需要展示一个网页，这时没有必要自己去实现一个浏览器，而是只需要调用系统的浏览器来打开这个网页就行了。用法如表 2.12 所示。

表 2.12　隐式 Intent 基本语法

方　　法	功能描述	实　　例
Intent.setAction(String action)	设置 Action 属性	Intent intent = new Intent(); Intent.setAction(android.intent.action.VIEW)
Intent.setData(String data)	设置 Data 属性	intent.setData(Uri.parse("www.baidu.com"))
Intent.setType(String type)	设置 Type 属性	intent.setType(vnd.android.cursor.dir/contact)

Intent.putExtras(Bundle bundle)	设置 Extra 属性	Bundle bundle=newBundle();
		bundle.putString("KEY_HEIGHT","180");
		bundle.putString("KEY_WEIGHT","80");
		intent.setExtra(bundle)

2. 隐式 Intent 跳转举例

建一个新的工程 ZSIntentTest，在布局界面上添加一个按钮，按钮代码如下。

```xml
<Button
    android:id = "@+id/button_view"
    android:layout_width = "match_parent"
    android:layout_height = "wrap_content"
    android:text = "打开百度"
    android:textColor = "#000000"
    android:textStyle = "bold"
    android:textSize = "30sp"        />
```

类文件 MainActivity 中按钮单击事件的代码如下。

```java
public class MainActivity extends AppCompatActivity {
//定义对象
    private Button btn_view;
    @Override
    protected void onCreate(Bundle savedInstanceState) {
        super.onCreate(savedInstanceState);
        setContentView(R.layout.activity_main);
        //绑定控件
        btn_view = findViewById(R.id.button_view);
        //按钮单击事件——打开百度
        btn_view.setOnClickListener(new View.OnClickListener() {
            @Override
            public void onClick(View view) {
                Intent intent1 = new Intent(Intent.ACTION_VIEW);
                intent1.setData(Uri.parse("http://www.baidu.com"));
                startActivity(intent1);
            }
        });
    }
}
```

这里首先指定了 Intent 的动作是 Intent.ACTION_VIEW，这是一个 Android 系统内置的动作，其常量值为 android.intent.action.VIEW。然后通过 Uri.parse() 方法，将一个网址字符串解析成一个 Uri 对象，再调用 Intent 的 setData() 方法将这个 Uri 对象传递进去。

运行程序，在界面上单击按钮，就可以看到打开了系统浏览器，打开了百度界面，如图 2.19 所示。

3. 返回数据给上一个活动

在 Activity 中，使用 Intent 既可以将数据传给下一个 Activity，还可以将数据回传给上一个 Activity，通过查阅 API 文档可以发现，Activity 中提供了一个 startActivityForResult (Intent intent, int requestCode) 方法，该方法也用于启动 Activity，并且这个方法可以在当前 Activity 销毁时返回一个结果给上一个 Activity。这种功能在实际开发中很常见，例如，

图 2.19 系统浏览器界面

打开图库,选择好照片后,会返回到页面并带回所选的图片信息。

在上个例子 ZSIntentTest 的界面中,再添加一个按钮,代码如下。

```
< Button
android:id = "@ + id/button_photo"
android:layout_width = "match_parent"
android:layout_height = "wrap_content"
android:text = "打开相册"
android:textColor = "#000000"
android:textStyle = "bold"
android:textSize = "30sp"          />
```

在类文件 MainActivity 中,完善代码如下。

```
public class MainActivity extends AppCompatActivity {
//定义对象
    private Button btn_view,btn_photo;
    @Override
    protected void onCreate(Bundle savedInstanceState) {
        super.onCreate(savedInstanceState);
        setContentView(R.layout.activity_main);
//绑定控件
        btn_view = findViewById(R.id.button_view);
        btn_photo = findViewById(R.id.button_photo);
```

```java
//按钮单击事件——打开百度
btn_view.setOnClickListener(new View.OnClickListener() {
    @Override
    public void onClick(View view) {
        Intent intent = new Intent(Intent.ACTION_VIEW);
        intent.setData(Uri.parse("http://www.baidu.com"));
        startActivity(intent);
    }
});
//按钮单击事件——打开相册
btn_photo.setOnClickListener(new View.OnClickListener() {
    @Override
    public void onClick(View view) {
        Intent intent = new Intent(Intent.ACTION_GET_CONTENT);
        intent.setType("image/*");
        startActivityForResult(intent,123);
    }
});
```

startActivityForResult(Intent intent,int requestCode)方法接收两个参数,第一个参数是 Intent 对象,第二个参数是请求码,请求码在之后的回调中判断数据的来源,输入一个唯一的值即可。单击按钮之后,会打开一个相册,选择完一张图片之后,系统会回调给上一个活动的 onActivityResult()方法,因此在本页面中会重写这个方法来得到返回的数据,代码如下。

```java
@Override
    protected void onActivityResult(int requestCode, int resultCode, @Nullable Intent data) {
    switch (requestCode){
    case 123:
        if(resultCode == RESULT_OK){
    Toast.makeText(MainActivity.this,data.getData().getPath(),Toast.LENGTH_SHORT).show();
        }
        break;
         default:
            break;
        }
    }
```

onActivityResult()方法带有三个参数,第一个参数 requestCode,即我们在启动活动时传入的请求码;第二个参数 resultCode,即我们在返回数据时传入的处理结果码;第三个参数 data,即携带着返回数据的 Intent。由于在一个活动中有可能调用 startActivityForResult()方法去启动很多不同的活动,每一个活动返回的数据都会回调到 onActivityResult()这个方法中,因此首先要做的就是通过检查 requestCode 的值来判断数据来源,确定数据是从哪个页面返回的之后,再通过 resultCode 的值来判断处理结果是否成功,最后从 data 中取值并打印出来,这样就完成了向上一个活动返回数据的工作。

运行程序,单击"打开相册"按钮,打开系统图库页面,从图库中选择一张照片,在系统回调方法中,便可收到图库页面返回的数据,这里采用 Toast 提示将图片缩略图的路径打印了出来。效果如图 2.20 所示。

图 2.20 系统图库界面

2.3 实战案例——猜猜我的生肖

微课视频

作为中国传统吉祥文化的十二生肖:子鼠丑牛,寅虎卯兔……一岁一个瑞兽。生肖文化,华夏一家。生肖是每一个人的吉祥物,十二生肖艺术形象具有独特的文化意义。本节将利用项目 2 所学过的知识,引导学生设计并实现自己喜欢的 App,希望通过本案例的学习,不仅能够提高学生简单案例的设计能力,也希望通过生肖 App,将人们心目中的神奇动物,悠久的民族文化符号,成为维系我们民族的情感和文化的纽带。

2.3.1 界面分析

本案例中包含两个界面:主界面和生肖查询界面,效果如图 2.21 所示。

2.3.2 实现思路

在第一个界面中,输入用户名,选择出生年月日,单击"生肖查询"按钮,把输入的用户名和出生日期这两个值传递到下一页。第二页将接收的值显示到对应的控件中,并根据接收到的年份进行判断,从 12 张生肖图片和 12 段生肖象征寓意中找出对应的图片和性格特点并显示出来。

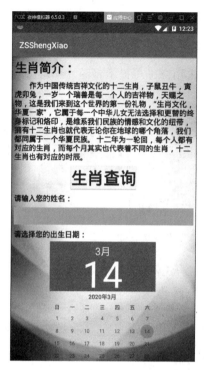 

图 2.21 猜猜我的生肖界面效果图

2.3.3 任务实施

【任务 2-1】 主界面的设计

(1) 创建一个新的工程,工程名为 ZSShengXiao。

(2) 切换工程的 Project 项目结构。选择该模式下方的 app,依次展开,便看到工程的布局界面和工程的类文件,其中,activity_main.xml 是布局界面,MainActivity.java 为类文件。

(3) 在工程中添加一个新的页面。右击 com.example.zsintent 包→New→Activity→Empty Activity,会弹出一个创建活动的对话框,将活动命名为 SecondActivity,默认勾选 Generate Layout File 关联布局界面,布局界面名称为 activity_second 但不要勾选 Launcher Activity。单击 Finish 按钮,便可在工程中完成第二个页面的添加。

(4) 准备两张图片,图片名为 xingkongone.jpg 和 xingkongtwo.jpg,将其粘贴到 app 目录结构中 res 下方的 drawable 文件夹下,两张图片作为两个页面的背景图片。

(5) 准备 12 张生肖图片,将其粘贴到 app 目录结构中 res 下方的 drawable 文件夹下。以子鼠的图片 zishu.png 图片为例,在类文件中,可以通过 "R.drawable.zishu" 来获取子鼠的图片。

(6) 准备 12 个生肖的象征寓意的文字描述。展开 app 目录结构中 res 下方的 value 文件夹,双击 strings.xml 文件将其打开,按照子鼠内容的方式,将另外 11 个生肖的寓意特点向下依次排列。

```xml
<resources>
    <string name="app_name">ZSShengXiao</string>
    <string name="子鼠">鼠:象征机警应变,善处逆境,子孙繁衍,家业兴旺.有生生不息,繁盛不衰之吉祥寓意.
        (1)对应的生肖月:阴历十一月下雪,屋内多鼠患.
        (2)对应的时辰:子时:是为半夜,又名子夜,(北京时间晚上23-01时)古人认为老鼠在这个时间段最为活跃,故而称为子鼠.
        (3)对应的节气时间段:子鼠:大雪-小寒,(阴历)十一月,(阳历)12月,(时刻)23时-1时,(五行)水阳(阴阳).
    </string>
    …
</resources>
```

上述代码中,定义一个字符串常量,name="子鼠"中的"name"是字符串常量名,"子鼠"是常量值。<string name="子鼠">是字符串的开始节点,</string>是字符串的结束节点,性格特点的文字内容要放置在开始节点与结束节点之间。在类文件中,通过"R.string.子鼠"方式便可获取子鼠生肖的象征寓意内容。

(7) 设计第一个Activity的布局界面。双击layout文件夹下的activity_main.xml文件,便可打开布局编辑器,修改布局类型为LinearLayout,添加方向属性为垂直。依次添加控件,代码如下。

```xml
<?xml version="1.0" encoding="utf-8"?>
<LinearLayout xmlns:android="http://schemas.android.com/apk/res/android"
    xmlns:app="http://schemas.android.com/apk/res-auto"
    xmlns:tools="http://schemas.android.com/tools"
    android:layout_width="match_parent"
    android:layout_height="match_parent"
    android:orientation="vertical"
    android:background="@drawable/xingkongone"
    tools:context=".MainActivity">
    <TextView
        android:layout_width="wrap_content"
        android:layout_height="wrap_content"
        android:text="生肖简介:"
        android:textStyle="bold"
        android:textColor="#000000"
        android:textSize="30sp"
        android:layout_margin="5dp"/>
    <TextView
        android:layout_width="wrap_content"
        android:layout_height="wrap_content"
        android:text="作为中国传统吉祥文化的十二生肖,子鼠丑牛,寅虎卯兔……一岁一个瑞兽是每一个人的吉祥物,天赐之物,这是我们来到这个世界的第一份礼物,"生肖文化,华夏一家",它是每一个中华儿女无法选择和更替的终身标记和烙印,是维系我们民族的情感和文化的纽带,拥有十二生肖也就代表无论在地球的哪个角落,我们都同属于一个华夏民族.十二年为一轮回,每个人都有对应的生肖,而每个月其实也代表着不同的生肖,十二生肖也有对应的时辰."
        android:textStyle="bold"
        android:textColor="#000000"
```

```xml
        android:textSize = "18sp"
        android:layout_margin = "5dp"/>
    <Button
        android:id = "@+id/button_search"
        android:layout_width = "wrap_content"
        android:layout_height = "wrap_content"
        android:text = "生肖查询"
        android:textStyle = "bold"
        android:textColor = "#000000"
        android:textSize = "40sp"
        android:background = "#ffff00"
        android:layout_gravity = "center"
        android:layout_margin = "5dp"          />
    <TextView
        android:layout_width = "wrap_content"
        android:layout_height = "wrap_content"
        android:text = "请输入您的姓名:"
        android:textStyle = "bold"
        android:textColor = "#000000"
        android:textSize = "18sp"
        android:layout_margin = "5dp"/>
    <EditText
        android:id = "@+id/edit_inputname"
        android:layout_width = "match_parent"
        android:layout_height = "wrap_content"
        android:textStyle = "bold"
        android:textSize = "30sp"
        android:layout_margin = "5dp"
        android:background = "#D7a3f5"          />
    <TextView
        android:layout_width = "wrap_content"
        android:layout_height = "wrap_content"
        android:text = "请选择您的出生日期:"
        android:textStyle = "bold"
        android:textColor = "#000000"
        android:textSize = "18sp"
        android:layout_margin = "5dp"/>
    <DatePicker
        android:id = "@+id/datePicker_birth"
        android:layout_width = "wrap_content"
        android:layout_height = "wrap_content"
        android:layout_gravity = "center">
    </DatePicker>
</LinearLayout>
```

【任务 2-2】 主界面到生肖界面的跳转

实现第一个 Activity 页面的功能。可以在程序中通过代码将 EditText 控件的内容和 DatePicker 控件选择的日期传递到第二页,打开 MainActivity.java 类文件,修改 MainActivity 的代码如下:

```java
public class MainActivity extends AppCompatActivity {
//第一步:定义对象
    Button btn_search;
    EditText edit_inputname;
    DatePicker date_birth;
    @Override
    protected void onCreate(Bundle savedInstanceState) {
        super.onCreate(savedInstanceState);
        setContentView(R.layout.activity_main);
        //第二步:绑定控件
        btn_search = findViewById(R.id.button_search);
        edit_inputname = findViewById(R.id.edit_inputname);
        date_birth = findViewById(R.id.datePicker_birth);
        //第三步:按钮单击事件
        btn_search.setOnClickListener(new View.OnClickListener() {
            @Override
            public void onClick(View v) {
                Intent intent = new Intent( MainActivity.this , SecondActivity.class );
                intent.putExtra("name",edit_inputname.getText().toString());
                intent.putExtra("year",date_birth.getYear());
                intent.putExtra("yue",date_birth.getMonth());
                intent.putExtra("ri",date_birth.getDayOfMonth());
                startActivity(intent);
            }
        });
    }
}
```

上述代码中,在按钮单击事件的内部,明确了跳转的页面,同时获取了输入的名字和选择的年月日,传递到了第二页。

【任务 2-3】 生肖查询界面的设计

设计第二个 Activity 的布局界面。双击 layout 文件夹下的 activity_second.xml 文件,便可打开布局编辑器,修改布局类型为 LinearLayout,添加方向属性为垂直。依次添加所需控件,代码如下。

微课视频

```xml
<?xml version = "1.0" encoding = "utf-8"?>
<LinearLayout xmlns:android = "http://schemas.android.com/apk/res/android"
    xmlns:app = "http://schemas.android.com/apk/res-auto"
    xmlns:tools = "http://schemas.android.com/tools"
    android:layout_width = "match_parent"
    android:layout_height = "match_parent"
    android:orientation = "vertical"
    android:background = "@drawable/xingkongtwo"
    tools:context = ".SecondActivity">
    <TextView
        android:id = "@+id/textView_getname"
```

```xml
        android:layout_width = "wrap_content"
        android:layout_height = "wrap_content"
        android:text = "显示获取到的用户名"
        android:textStyle = "bold"
        android:textColor = "#ffffff"
        android:textSize = "20sp"
        android:layout_margin = "5dp"/>
    <TextView
        android:id = "@+id/textView_getbirth"
        android:layout_width = "wrap_content"
        android:layout_height = "wrap_content"
        android:text = "显示获取到的出生日期"
        android:textStyle = "bold"
        android:textColor = "#ffffff"
        android:textSize = "20sp"
        android:layout_margin = "5dp"/>
    <ImageView
        android:id = "@+id/imageView_imgzodiac"
        android:layout_width = "wrap_content"
        android:layout_height = "wrap_content"
        android:src = "@drawable/ic_launcher_background"
        android:layout_gravity = "center"
        android:layout_margin = "20dp"/>
    <TextView
        android:id = "@+id/textView_contentzodiac"
        android:layout_width = "wrap_content"
        android:layout_height = "wrap_content"
        android:text = "显示生肖的象征寓意"
        android:textStyle = "bold"
        android:textColor = "#ffffff"
        android:textSize = "20sp"
        android:layout_margin = "5dp"/>
</LinearLayout>
```

【任务 2-4】 生肖查询功能的实现

实现第二个 Activity 页面的功能。获取传递过来的姓名和日期显示到对应的控件里面，同时根据日期选择对应的生肖图片和寓意文字并显示出来。打开 SecondActivity.java 类文件，修改 SecondActivity 的代码如下。

```java
public class SecondActivity extends AppCompatActivity {
    //第一步:定义对象
    private TextView txt_getname,txt_getbirth,txt_contentzodiac;
    private ImageView img_imgzodiac;
    @Override
    protected void onCreate(Bundle savedInstanceState) {
        super.onCreate(savedInstanceState);
        setContentView(R.layout.activity_second);
        //第二步:绑定控件
```

```java
        txt_getname = findViewById(R.id.textView_getname);
        txt_getbirth = findViewById(R.id.textView_getbirth);
        txt_contentzodiac = findViewById(R.id.textView_contentzodiac);
        img_imgzodiac = findViewById(R.id.imageView_imgzodiac);
        //第三步:接收第一页传递过来的值并且显示到对应的文本控件里面
        String str1 = getIntent().getStringExtra("name");
        int myyear = getIntent().getIntExtra("year",0);
        int mymonth1 = getIntent().getIntExtra("yue",0);
        int mymonth = mymonth1 + 1;
        int myday = getIntent().getIntExtra("ri",0);
        txt_getname.setText("你好:" + str1);
        txt_getbirth.setText("您的出生日期为:" + myyear + "年" + mymonth + "月" + myday + "日");
        //第四步:从12张生肖图片中选择与日期对应的一张图片 下标:0~11
        int[] imgarr = {R.drawable.zishu,R.drawable.chouniu,R.drawable.yinhu,
                        R.drawable.maotu,R.drawable.chenlong,R.drawable.sishe,
                        R.drawable.wuma,R.drawable.weiyang,R.drawable.shenhou,
                        R.drawable.youji,R.drawable.xugou,R.drawable.haizhu };
        int[] contentarr = { R.string.子鼠,R.string.丑牛,R.string.寅虎,
                        R.string.卯兔,R.string.辰龙,R.string.巳蛇,
                        R.string.午马,R.string.未羊,R.string.申猴,
                        R.string.酉鸡,R.string.戌狗,R.string.亥猪 };
        //如何根据选择的年从数组里面找对应的图片和文字呢?
        int i = find(mymonth,myday);
        img_imgzodiac.setImageResource(imgarr[i]);      //根据数组下标显示图片
        txt_contentzodiac.setText(contentarr[i]); //根据数组下标显示生肖寓意文字
    }
    private int find(int mymonth, int myday) {
        int i = 0;
        if(myear % 12 == 0){i = 8;}
        if(myear % 12 == 1){i = 9;}
        if (myear % 12 == 2) { i = 10; }
        if (myear % 12 == 3) { i = 11; }
        if (myear % 12 == 4) { i = 0; }
        if (myear % 12 == 5) { i = 1; }
        if (myear % 12 == 6) { i = 2; }
        if (myear % 12 == 7) { i = 3; }
        if (myear % 12 == 8) { i = 4; }
        if (myear % 12 == 9) { i = 5; }
        if (myear % 12 == 10) { i = 6; }
        if (myear % 12 == 11) { i = 7; }
        return i;
    }
}
```

上述代码中,获取了从第一页传递过来的姓名和出生日期,显示到了对应的控件上,然后根据出生日期从12生肖里面寻找对应的生肖图片和生肖象征寓意。当 Android 应用程序被编译,会自动生成一个 R 类,其中包含所有 res/目录下资源的 ID,如布局文件、资源文件、图片文件、values 下所有文件的 ID 等。在写 Java 代码需要用这些资源的时候,可以使

用 R 类,通过子类+资源名或者直接使用资源 ID 来访问资源。因此,用 R.drawable.zishu 等来引用生肖图片,同时 R.drawable.zishu 是特殊的 int 型,因此 12 张生肖图片放到了一个 int 类型的数组里面,同理,12 个生肖的象征寓意的描述也放到了 int 类型的数组里面,然后根据不同年份找出数组下标。通过数组下标从数组中取出对应的图片和文字并显示出来。

项目小结

本项目内容紧紧围绕 Android 当中的 UI 控件和 App 之间的通信内容来展开,通过 5 个常用的 UI 控件,同学们对常用控件的属性和方法有了简单的了解,同时也学习了多页面之间的跳转及不同页面之间的数据传递。最后通过猜猜我的生肖案例,培养学生简单 App 的设计与开发能力。

习题

1. 以下( )控件用来显示文本。
   A. ImageView   B. TextView   C. EditText   D. Button
2. 以下( )属性用来设置控件的宽度。
   A. width   B. height   C. text   D. id
3. 以下( )属性用来设置控件字体的颜色。
   A. textStyle   B. textSize   C. textColor   D. Color
4. 在 Android 中,导入类文件的快捷键是( )。
   A. Alt+Enter   B. Ctrl+Alt+Delete
5. 下列( )可作 EditText 编辑框的提示信息。
   A. android:inputType   B. android:text
   C. android:digits   D. android:hint
6. 按钮属性 android:layout_gravity="center"的含义是( )。
   A. 控件左对齐   B. 控件右对齐
   C. 控件居中对齐   D. 控件上的文字居中对齐
7. 对于创建 Intent 对象,下列( )写法是正确的。
   A. Intent myintent=new Intent(MainActivity.this,SecondActivity.class);
   B. intent myintent=new intent(MainActivity.this,SecondActivity.class);
   C. Intent myintent=new Intent(MainActivity.class,SecondActivity.this);
   D. intent myintent=new intent(MainActivity.class,SecondActivity.this);
8. Android 项目工程下面的 drawable 目录的作用是( )。
   A. 放置应用到的图片资源
   B. 主要放置一些文件资源

C. 放置字符串、颜色、数组等常量数据

D. 放置一些与 UI 相应的布局文件

9. android:src 是(　　)控件的专有属性。

　　A. ImageView　　　　　　　　B. TextView

　　C. EditText　　　　　　　　　D. Button

10. @drawable/abc 和 R.drawable.abc 分别用在(　　)。

　　A. 类文件和布局界面　　　　　B. 布局界面和类文件

　　C. 类文件　　　　　　　　　　D. 布局界面

项目 3

智能计算器

【教学导航】

学习目标	（1）理解 Android 的 UI 的含义。 （2）掌握 Android 的几种常见布局。 （3）熟悉 Android 系统中的样式和主题。 （4）培养简单案例的设计开发能力。
教学方法	任务驱动法、理论实践一体化、探究学习法、分组讨论法
课时建议	6 课时

3.1 Android UI 常用布局

一个丰富的界面总是要由很多控件组成的，如何才能让各个控件都有条不紊地摆放在界面上呢？这就需要借助布局来实现了。布局是一种可用于放置很多控件的容器，它可以按照一定的规律调整内部控件的位置，从而编写出精美的界面。当然，布局的内部除了放置控件外，也可以放置布局，通过多层布局的嵌套，就能够完成一些比较复杂的界面。本节详细讲解 Android 中最基本的 3 种布局：LinearLayout（线性布局）、RelativeLayout（相对布局）和 FrameLayout（帧布局）。

3.1.1 UI 简介

UI 即 User Interface（用户界面）的简称，是人和设备之间交互的工具，小到手机端的应用，大到计算机上的桌面程序，用户所触摸到的屏幕就是用户界面。

1. UI 界面

用户界面并不是一个新鲜的概念，早在中国古代使用的算盘，它是由珠子组成的最早的人机交互界面。后来，MS DOS 操作系统走进了大众视野，但它呈现出来的是黑色屏幕和白色代码的命令窗口,界面既不美观也不友好。后来图形化用户界面兴起,这种界面上有很多图标,计算机端用鼠标就可以进行简单的操作,手机端应用手指触摸便可响应。

UI 设计就是设计界面美观、舒适的人机交互界面,让软件开发变得有品位、有个性,让用户操作变得简单、舒适。

2. UI 设计相关的几个概念

在 Android 中,进行界面设计时,经常会用到 View、ViewGroup、Padding、Margins 等概念,对于初学 Android 的人来说,一般不好理解。下面对这几个概念进行详细的介绍。

1) View

View 在 Android 中可以理解为视图,它占据屏幕上的一块矩形区域,负责提供组件绘制和事件处理的方法,如图 3.1 所示。View 相当于窗户上的玻璃,设计窗户时,可以设计自己所需要的玻璃数,可以是 4 块玻璃,也可以是 6 块玻璃。也就是说,在手机屏幕中可以有很多个 View,在 Android 手机中,使用 View 是通过 View 类来实现的,View 类就是所有控件的基类,某一个控件如 TextView 是 View 的子类。

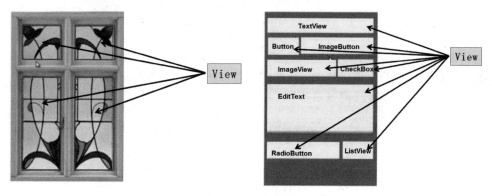

图 3.1　UI 界面上的 View

2) ViewGroup

ViewGroup 在 Android 中可以理解为容器。ViewGroup 类继承自 View 类,它是 View 类的扩展,是用来容纳其他组件的容器,但是由于 ViewGroup 是一个抽象类,所以在实际应用中通常总是使用 ViewGroup 的子类来作为容器的,如图 3.2 所示。

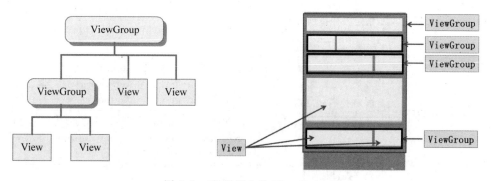

图 3.2　UI 界面上的 ViewGroup

3) Padding 和 Margins

Padding 表示在控件的顶部、底部、左侧和右侧的填充像素,也称为内边距。它设置的是控件内容到控件边缘的距离。Padding 将占据控件的宽度和高度。设置指定的内边距后,内容将偏离控件边缘指定的距离。

Margins 表示控件的顶部、底部、左侧和右侧的空白区域,称为外边距。它设置的是控件与其父容器的距离。Margins 不占据控件的宽度和高度。为控件设置外边距后,该控件将远离父容器指定的距离,如果还有相邻组件,那么也将远离其相邻组件指定距离。

内边距和外边距示意图如图 3.3 所示。

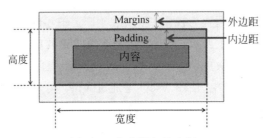

图 3.3 内边距和外边距

4) dp 和 sp

dp 是宽度、高度的单位。例如:按钮的宽度为 100dp,按钮的高度为 60dp。

sp 是文字大小的单位,例如:文字大小为 32sp 等。

3.1.2 LinearLayout 布局

LinearLayout 又称作线性布局,是一种最常用的布局,正如它的名字一样,这个布局会将它所包含的控件在线性方向上依次排列。

1. LinearLayout 常用属性

LinearLayout 属性较多,下面列举一些常用属性,如表 3.1 所示。

表 3.1 LinearLayout 常用属性

属　　性	功能描述	实　　例
android:orientation	指定控件排列方向	android:orientation="vertical" 内部控件在垂直方向上排列
		android:orientation="horizontal" 内部控件在水平方向上排列
android:layout_gravity	指定控件在布局中的对齐方式	android:layout_gravity="top" 顶端对齐(方向:水平) android:layout_gravity="center_vertical" 垂直居中 android:layout_gravity="bottom" 底端对齐
		android:layout_gravity="left" 左对齐(方向:垂直) android:layout_gravity="center_horizontal" 水平居中 android:layout_gravity="right" 右对齐
android:layout_weight	允许控件使用比例方式指定控件的大小	android:layout_weight="1"

1) orientation 属性

orientation 属性可以用来控制线性布局中控件的排列方向。新建一个工程 LinearLayoutTest,打开 activity_main.xml 布局界面,输入代码如下。

```xml
<?xml version = "1.0" encoding = "utf-8"?>
<LinearLayout xmlns:android = "http://schemas.android.com/apk/res/android"
    xmlns:app = "http://schemas.android.com/apk/res-auto"
    xmlns:tools = "http://schemas.android.com/tools"
    android:layout_width = "match_parent"
    android:layout_height = "match_parent"
    android:orientation = "vertical"
    tools:context = ".MainActivity">
    <Button
        android:layout_width = "wrap_content"
        android:layout_height = "wrap_content"
        android:text = "按钮 1"            />
    <Button
        android:layout_width = "wrap_content"
        android:layout_height = "wrap_content"
        android:text = "按钮 2"            />
    <Button
        android:layout_width = "wrap_content"
        android:layout_height = "wrap_content"
        android:text = "按钮 3"            />
</LinearLayout>
```

上述代码中,界面布局采用 LinearLayout。在 LinearLayout 中添加了 3 个 Button,每个 Button 的高度和宽度都是 wrap_content,并通过 android:orientation 属性指定了排列方向是 vertical,现在运行程序,效果如图 3.4 所示。

然后修改 LinearLayout 的排列方向,将 android:orientation 属性的值修改为 horizontal,这就意味着要让 LinearLayout 中的控件在水平方向上排列。重新运行程序,效果如图 3.5 所示。

2) layout_gravity 属性

该属性与 android:gravity 属性看起来有些相似,它们的区别在于 android:gravity 用于指定文字在控件中的对齐方式,而 android:layout_gravity 用于指定控件在布局中的对齐方式,当 LinearLayout 的排列方向为 horizontal 时,只有垂直方向上的对齐方式才会生效,修改 activity_main.xml 中的代码如下。

```xml
<?xml version = "1.0" encoding = "utf-8"?>
<LinearLayout xmlns:android = "http://schemas.android.com/apk/res/android"
    xmlns:app = "http://schemas.android.com/apk/res-auto"
    xmlns:tools = "http://schemas.android.com/tools"
    android:layout_width = "match_parent"
    android:layout_height = "match_parent"
    android:orientation = "horizontal"
    tools:context = ".MainActivity">
```

```xml
<Button
    android:layout_width = "wrap_content"
    android:layout_height = "wrap_content"
    android:text = "按钮 1"
    android:layout_gravity = "top"/>
<Button
    android:layout_width = "wrap_content"
    android:layout_height = "wrap_content"
    android:text = "按钮 2"
    android:layout_gravity = "center_vertical"/>
<Button
    android:layout_width = "wrap_content"
    android:layout_height = "wrap_content"
    android:text = "按钮 3"
    android:layout_gravity = "bottom"/>
</LinearLayout>
```

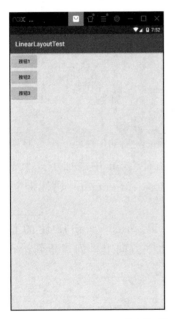

图 3.4　LinearLayout 垂直排列　　　　　　图 3.5　LinearLayout 水平排列

由于目前 LinearLayout 的排列方向是 horizontal,每添加一个控件,水平方向上的长度都会改变,因而无法指定控件左右对齐,只能指定垂直方向上的排列方向。将第一个按钮对齐方式指定为 top,第二个按钮对齐方式指定为 center_vertical,第三个按钮对齐方式指定为 bottom。重新运行程序,效果如图 3.6 所示。

这里需要注意,如果 LinearLayout 的排列方向是 horizontal,内部的控件就绝对不能将宽度指定为 match_parent,因为这样的话,单独一个控件就会将整个水平方向占满,其他的控件就没有可放置的位置了。同样的道理,如果 LinearLayout 的排列方向为 vertical,内部的控件就不能将高度指定为 match_parent。而且排列方向为 vertical 时,只有水平方向上的对齐方式才会生效,修改 activity_main.xml 中的代码如下:

```xml
<?xml version = "1.0" encoding = "utf-8"?>
<LinearLayout xmlns:android = "http://schemas.android.com/apk/res/android"
    xmlns:app = "http://schemas.android.com/apk/res-auto"
    xmlns:tools = "http://schemas.android.com/tools"
    android:layout_width = "match_parent"
    android:layout_height = "match_parent"
    android:orientation = "vertical"
    tools:context = ".MainActivity">
    <Button
        android:layout_width = "wrap_content"
        android:layout_height = "wrap_content"
        android:text = "按钮 1"
        android:layout_gravity = "left"/>
    <Button
        android:layout_width = "wrap_content"
        android:layout_height = "wrap_content"
        android:text = "按钮 2"
        android:layout_gravity = "center_horizontal"/>
    <Button
        android:layout_width = "wrap_content"
        android:layout_height = "wrap_content"
        android:text = "按钮 3"
        android:layout_gravity = "right"/>
</LinearLayout>
```

再次运行程序,效果如图 3.7 所示。

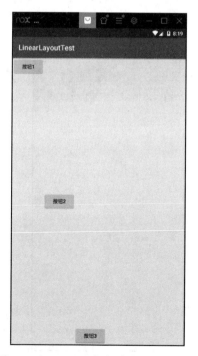
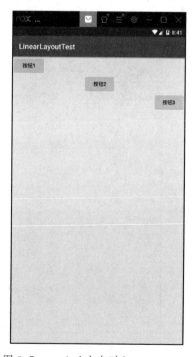

图 3.6　horizontal 方向时 layout_gravity　　　图 3.7　vertical 方向时 layout_gravity

3) layout_weight 属性

这个属性允许使用比例的方式指定控件的大小,它在手机适配性方面可以起到非常重要的作用。例如,编写一个消息发送界面,需要一个文本编辑框和一个"发送"按钮,修改 activity_main.xml 中的代码如下。

```xml
<?xml version = "1.0" encoding = "utf-8"?>
<LinearLayout xmlns:android = "http://schemas.android.com/apk/res/android"
    xmlns:app = "http://schemas.android.com/apk/res-auto"
    xmlns:tools = "http://schemas.android.com/tools"
    android:layout_width = "match_parent"
    android:layout_height = "match_parent"
    android:orientation = "horizontal"
    tools:context = ".MainActivity">
    <EditText
        android:layout_width = "wrap_content"
        android:layout_height = "wrap_content"
        android:hint = "请输入内容"
        android:layout_weight = "1"/>
    <Button
        android:layout_width = "wrap_content"
        android:layout_height = "wrap_content"
        android:text = "发送"
        android:layout_weight = "1"/>
</LinearLayout>
```

上述代码中,在 EditText 和 Button 里都将 android:layout_weight 属性指定为 1,表示 EditText 和 Button 将在水平方向上平分宽度,运行效果如图 3.8 所示。如果删除 Button 控件里的 android:layout_weight="1"这行代码,表示 EditText 会占满屏幕的所有剩余空间,效果如图 3.9 所示。

图 3.8　指定 layout_weight 平分宽度　　　图 3.9　指定 layout_weight 效果

2. LinearLayout 布局案例

1) 案例分析

（1）界面分析。本案例共有一个布局界面，自上而下的垂直方向，用户名、密码、按钮这三行是嵌套的水平线性布局，自左向右的水平方向，两个按钮控件权重为1。

（2）设计思路。布局界面中通过 android:orientation 属性设定线性布局的方向达到用户需求，通过线性布局的嵌套实现用户名、密码和按钮这三行控件的添加，通过 layout_weight 属性设定控件所占的宽度比例，达到用户的需求。

2) 实现步骤

（1）创建一个新的工程，工程名为 ZSLinearLayout。

（2）切换工程的 Project 项目结构。选择该模式下方的 app，依次展开，便看到工程的布局界面和工程的类文件，其中，activity_main.xml 是布局界面，MainActivity.java 为类文件。

（3）准备一张图片，图片名为 background.png，将其粘贴到 app 目录结构中 res 下方的 drawable 文件夹下，这张图片作为页面的背景图片。

（4）设计布局界面。双击 layout 文件夹下的 activity_main.xml 文件，便可打开布局编辑器，修改布局类型为 LinearLayout，添加方向属性为 vertical。代码如下。

```xml
<?xml version = "1.0" encoding = "utf-8"?>
<LinearLayout xmlns:android = "http://schemas.android.com/apk/res/android"
    xmlns:app = "http://schemas.android.com/apk/res-auto"
    xmlns:tools = "http://schemas.android.com/tools"
    android:layout_width = "match_parent"
    android:layout_height = "match_parent"
    android:orientation = "vertical"
    android:background = "@drawable/background"
    tools:context = ".MainActivity">
    <TextView
        android:layout_width = "wrap_content"
        android:layout_height = "wrap_content"
        android:text = "用户登录"
        android:textSize = "40sp"
        android:textColor = "#000000"
        android:textStyle = "bold"
        android:layout_marginTop = "100dp"
        android:layout_marginBottom = "100dp"
        android:layout_gravity = "center"/>
    <LinearLayout
        android:layout_width = "match_parent"
        android:layout_height = "wrap_content"
        android:orientation = "horizontal"
        android:layout_marginTop = "20dp"
        android:layout_marginLeft = "10dp"
        android:layout_marginRight = "10dp"
```

```xml
        android:gravity = "center">
        <TextView
            android:layout_width = "wrap_content"
            android:layout_height = "wrap_content"
            android:text = "用户名:"
            android:textSize = "30sp"
            android:textColor = "#000000" />
        <EditText
            android:id = "@+id/edit_inputname"
            android:layout_width = "match_parent"
            android:layout_height = "wrap_content"
            android:hint = "请输入用户名"
            android:textSize = "25sp" />
    </LinearLayout>
    <LinearLayout
        android:layout_width = "match_parent"
        android:layout_height = "wrap_content"
        android:orientation = "horizontal"
        android:layout_marginTop = "20dp"
        android:layout_marginLeft = "10dp"
        android:layout_marginRight = "10dp"
        android:gravity = "center">
        <TextView
            android:layout_width = "wrap_content"
            android:layout_height = "wrap_content"
            android:text = "密    码:"
            android:textSize = "30sp"
            android:textColor = "#000000"       />
        <EditText
            android:id = "@+id/edit_inputpwd"
            android:layout_width = "match_parent"
            android:layout_height = "wrap_content"
            android:hint = "请输入密码"
            android:inputType = "textPassword"
            android:textSize = "25sp" />
    </LinearLayout>
    <CheckBox
        android:id = "@+id/check_remeber"
        android:layout_width = "wrap_content"
        android:layout_height = "wrap_content"
        android:text = "记住密码"
        android:textStyle = "bold"
        android:textSize = "20sp"
        android:layout_gravity = "right"
        android:layout_margin = "10dp"/>

    <LinearLayout
        android:layout_width = "match_parent"
        android:layout_height = "wrap_content"
```

```xml
            android:orientation = "horizontal"
            android:layout_marginTop = "20dp"
            android:layout_marginLeft = "10dp"
            android:layout_marginRight = "10dp"
            android:gravity = "center">
            <Button
                android:id = "@+id/button_no"
                android:layout_width = "wrap_content"
                android:layout_height = "wrap_content"
                android:text = "取消"
                android:textSize = "30sp"
                android:textColor = "#050505"
                android:layout_weight = "1"/>
            <Button
                android:id = "@+id/button_yes"
                android:layout_width = "wrap_content"
                android:layout_height = "wrap_content"
                android:text = "登录"
                android:textSize = "30sp"
                android:textColor = "#050505"
                android:layout_weight = "1"         />
    </LinearLayout>
</LinearLayout>
```

从上述代码中可以看出，布局的内部除了放置控件外，也可以放置布局，通过多层布局的嵌套，完成了比较复杂的界面。最外层布局界面是 LinearLayout 线性布局，使用 android:orientation 属性指定布局的方向，属性值为 vertical 表示布局内的控件自上而下垂直排列。用户名所在行的两个控件是水平排列的，为了实现效果，采用了嵌套线性布局，将 android:orientation 属性取值 horizontal 表示自左向右的水平方向。将显示用户名的 TextView 控件和输入用户名的 EditText 控件放置于嵌套的线性布局的内部。下方又是一个水平方向的线性布局，里面放置了密码行的两个控件。密码行下方又是一个水平方向的线性布局，两个按钮控件放到布局的内部，为了让两个按钮控件宽度相等，添加了 layout_weight 属性，属性值为 1 表示等分宽度。运行程序，效果如图 3.10 所示。

3.1.3　RelativeLayout 布局

RelativeLayout 又称作相对布局，也是一种常用的布局，该布局和 LinearLayout 的排列规则不同，RelativeLayout 显得更加随意一些，它可以通过相对定位的方式让控件出现在布局中的任何位置。也正因为如此，RelativeLayout 中的属性非常多。

1. RelativeLayout 常用属性

RelativeLayout 的属性非常多，不过这些属性都是有规律可循的，并不难理解和记忆。控件相对于父容器进行定位时。常用属性如表 3.2 所示。

图 3.10　程序运行效果

表 3.2　RelativeLayout 常用属性——控件相对于父容器进行定位

属　　性	功能描述	实　　例
android:layout_alignParentTop	其属性值为 boolean 值，用于指定控件是否与布局管理器顶端对齐	android:layout_alignParentTop = "true"
android:layout_alignParentBottom	其属性值为 boolean 值，用于指定控件是否与布局管理器底端对齐	android:layout_alignParentBottom = "true"
android:layout_alignParentLeft	其属性值为 boolean 值，用于指定控件是否与布局管理器左边对齐	android:layout_alignParentLeft = "true"
android:layout_alignParentRight	其属性值为 boolean 值，用于指定控件是否与布局管理器右边对齐	android:layout_alignParentRight = "true"
android:layout_centerHorizontal	其属性值为 boolean 值，用于指定控件是否位于布局管理器水平居中的位置	android:layout_centerVertical = "true"
android:layout_centerVertical	其属性值为 boolean 值，用于指定控件是否位于布局管理器垂直居中的位置	android:layout_centerHorizontal = "true"
android:layout_centerInParent	其属性值为 boolean 值，用于指定控件是否位于布局管理器的中央位置	android:layout_centerInParent = "true"

相对布局定位有两种方式,第一种是控件相对于父容器进行定位,第二种是控件相对于控件进行定位。表 3.2 中的属性值是第一种定位方式常用的几个属性,当定位方式为控件相对于控件进行定位时,常用属性如表 3.3 所示。

表 3.3 RelativeLayout 常用属性——控件相对于控件进行定位

属　　性	功能描述	实　　例
android:layout_above	其属性值为其他 UI 组件的 ID 属性,用于指定该组件位于哪个组件的上方	android:layout_above="@id/button_5"
android:layout_below	其属性值为其他 UI 组件的 ID 属性,用于指定该组件位于哪个组件的下方	android:layout_below="@id/button_5"
android:layout_toLeftOf	其属性值为其他 UI 组件的 ID 属性,用于指定该组件位于哪个组件的左侧	android:layout_toLeftOf="@id/button_5"
android:layout_toRightOf	其属性值为其他 UI 组件的 ID 属性,用于指定该组件位于哪个组件的右侧	android:layout_toRightOf="@id/button_5"
android:layout_alignTop	其属性值为其他 UI 组件的 ID 属性,用于指定该组件与哪个组件的上边界对齐	android:layout_alignTop="@id/button_5"
android:layout_alignBottom	其属性值为其他 UI 组件的 ID 属性,用于指定该组件与哪个组件的下边界对齐	android:layout_alignBottom="@id/button_5"
android:layout_alignLeft	其属性值为其他 UI 组件的 ID 属性,用于指定该组件与哪个组件的左边界对齐	android:layout_alignLeft="@id/button_5"
android:layout_alignRight	其属性值为其他 UI 组件的 ID 属性,用于指定该组件与哪个组件的右边界对齐	android:layout_alignRight="@id/button_5"

2. RelativeLayout 布局案例

1) 案例分析

(1) 界面分析。本工程中一共有两个布局界面,第一个布局界面中有 5 个 Button 控件,其中 4 个按钮位于布局的 4 个角,第 5 个按钮位于布局的正中心。第二个布局界面中有 1 个图片和 7 个 Button 控件,其中一个按钮位于布局的正中心,4 个按钮位于中心按钮的上下左右,另外两个按钮位于中心按钮的左下角和右上角处。

(2) 设计思路。布局界面一通过控件相对于父容器进行定位时的 RelativeLayout 常用属性达到用户需求,布局界面二通过控件相对于控件进行定位时的属性达到用户的需求。

2) 实现步骤

(1) 创建一个新的工程,工程名为 ZSRelativeLayout。

(2) 切换工程的 Project 项目结构。选择该模式下方的 app,依次展开,便看到工程的布局界面和工程的类文件,其中,activity_main.xml 是布局界面,MainActivity.java 为类文件。

(3) 在工程中添加一个新的页面。右击 com. example. zsintent 包→New→Activity→Empty Activity,会弹出一个创建活动的对话框,将活动命名为 SecondActivity,默认勾选 Generate Layout File 关联布局界面,布局界面名称为 activity_second 但不要勾选 Launcher Activity。单击 Finish 按钮,便可在工程中完成第二个页面的添加。

(4) 准备 6 张图片,将其粘贴到 app 目录结构中 res 下方的 drawable 文件夹下。第二个布局界面中,playbg.png 是页面上方的一张游戏图片,start.jpg 图片位于布局界面的中心位置,up.jpg、down.jpg、left.jpg、rigth.jpg 是周围的四张图片。

(5) 设计第一个 Activity 的布局界面。双击 layout 文件夹下的 activity_main.xml 文件,便可打开布局编辑器,修改布局类型为 RelativeLayout,代码如下。

```xml
<?xml version="1.0" encoding="utf-8"?>
<RelativeLayout xmlns:android="http://schemas.android.com/apk/res/android"
    xmlns:app="http://schemas.android.com/apk/res-auto"
    xmlns:tools="http://schemas.android.com/tools"
    android:layout_width="match_parent"
    android:layout_height="match_parent"
    tools:context=".MainActivity">
    <Button
        android:layout_width="wrap_content"
        android:layout_height="wrap_content"
        android:text="按钮 1"
        android:textSize="40sp"
        android:layout_alignParentTop="true"
        android:layout_alignParentLeft="true" />
    <Button
        android:layout_width="wrap_content"
        android:layout_height="wrap_content"
        android:text="按钮 2"
        android:textSize="40sp"
        android:layout_alignParentBottom="true"
        android:layout_alignParentLeft="true" />
    <Button
        android:layout_width="wrap_content"
        android:layout_height="wrap_content"
        android:text="按钮 3"
        android:textSize="40sp"
        android:layout_alignParentTop="true"
        android:layout_alignParentRight="true" />
    <Button
        android:layout_width="wrap_content"
        android:layout_height="wrap_content"
        android:text="按钮 4"
        android:textSize="40sp"
        android:layout_alignParentBottom="true"
        android:layout_alignParentRight="true" />
    <Button
```

```
        android:layout_width = "wrap_content"
        android:layout_height = "wrap_content"
        android:text = "按钮 5"
        android:textSize = "40sp"
        android:layout_centerInParent = "true"/>
</RelativeLayout>
```

上述代码中，Button1 和父布局的左上角对齐，Button2 和父布局的左下角对齐，Button3 和父布局的右上角对齐，Button4 和父布局的右下角对齐，Button5 位于父布局中心位置。运行程序，效果如图 3.11 所示。其中每个控件都是相对于父布局进行定位的，控件还可以相对于控件进行定位。

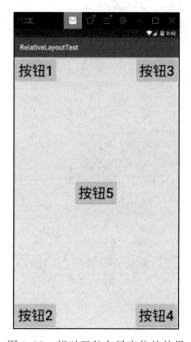

图 3.11　相对于父布局定位的效果

（6）设计第二个 Activity 的布局界面。双击 layout 文件夹下的 activity_second.xml 文件，便可打开布局编辑器，修改布局类型为 RelativeLayout，依次添加所需控件，代码如下。

```
<?xml version = "1.0" encoding = "utf - 8"?>
<RelativeLayout xmlns:android = "http://schemas.android.com/apk/res/android"
    xmlns:app = "http://schemas.android.com/apk/res - auto"
    xmlns:tools = "http://schemas.android.com/tools"
    android:layout_width = "match_parent"
    android:layout_height = "match_parent"
    tools:context = ".SecondActivity">
    <ImageView
        android:layout_width = "wrap_content"
        android:layout_height = "210dp"
        android:src = "@drawable/background"            />
```

```xml
<Button
    android:id = "@+id/button_5"
    android:layout_width = "100dp"
    android:layout_height = "100dp"
    android:layout_centerInParent = "true"
    android:background = "@drawable/start" />
<Button
    android:layout_width = "100dp"
    android:layout_height = "100dp"
    android:background = "@drawable/up"
    android:layout_above = "@+id/button_5"
    android:layout_alignLeft = "@+id/button_5" />
<Button
    android:layout_width = "100dp"
    android:layout_height = "100dp"
    android:background = "@drawable/down"
    android:layout_below = "@id/button_5"
    android:layout_alignLeft = "@id/button_5" />
<Button
    android:layout_width = "100dp"
    android:layout_height = "100dp"
    android:background = "@drawable/left"
    android:layout_toLeftOf = "@id/button_5"
    android:layout_alignTop = "@id/button_5" />
<Button
    android:layout_width = "100dp"
    android:layout_height = "100dp"
    android:background = "@drawable/right"
    android:layout_toRightOf = "@id/button_5"
    android:layout_alignTop = "@id/button_5" />
<Button
    android:layout_width = "50dp"
    android:layout_height = "50dp"
    android:background = "#ff0000"
    android:layout_toRightOf = "@id/button_5"
    android:layout_above = "@id/button_5" />
<Button
    android:layout_width = "50dp"
    android:layout_height = "50dp"
    android:background = "#0000ff"
    android:layout_toLeftOf = "@id/button_5"
    android:layout_below = "@id/button_5" />
</RelativeLayout>
```

这个代码稍微复杂一点儿，不过还是有规律可循的。android:layout_above 属性可以让一个控件位于另一个控件的上方，需要为这个属性指定相对控件 ID 的引用，这里填入了 @id/button_5，表示让该控件位于 ID 号为 5 的按钮的上方。其他属性也是类似的。android:layout_below 表示让一个控件位于另一个控件的下方，android:layout_toLeftOf 表示让一个控件位于另一个控件的左侧，android:layout_toRightOf 表示让一个控件位于另一个控件的右侧。android:layout_alignTop 表示让一个控件与另一个控件顶部对齐。注意，当一个控件去引用另一个控件的 ID 时，该控件一定要定义在引用控件的后面，否则会出现找不到 ID 的情况。

（7）修改页面的启动顺序。依次展开 app 项目结构，双击下方的 Android 配置文件 AndroidManifest.xml，配置文件代码中，<activity android:name=".MainActivity">是第一个页面的开始节点，下方的</activity>是第一个页面的结束节点，中间的四行代码就是决定页面启动顺序的关键代码，将这四行代码剪切、粘贴到第二个页面开始节点<activity android:name=".SecondActivity">与结束节点</activity>的中间位置。修改后代码如下。

```
<application
    android:allowBackup = "true"
    android:icon = "@mipmap/ic_launcher"
    android:label = "@string/app_name"
    android:roundIcon = "@mipmap/ic_launcher_round"
    android:supportsRtl = "true"
    android:theme = "@style/AppTheme">
    <activity android:name = ".SecondActivity">
        <intent-filter>
            <action android:name = "android.intent.action.MAIN" />
            <category android:name = "android.intent.category.LAUNCHER" />
        </intent-filter>
    </activity>
    <activity android:name = ".MainActivity">
    </activity>
</application>
```

（8）重新运行程序，效果如图 3.12 所示。

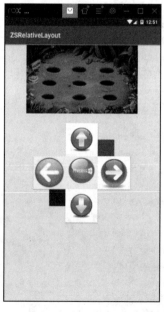

图 3.12　相对于控件定位的效果

微课视频

3.1.4 FrameLayout 布局

FrameLayout 又称为帧布局,是 Android 中最为简单的一种布局。采用帧布局设计界面时,所有控件都默认显示在屏幕的左上角,并按照先后放入的顺序重叠摆放,先放入的控件显示在最底层,后放入的控件显示在最顶层。

1. FrameLayout 案例

1) 案例分析

(1) 界面分析。布局界面中总体是 LinearLayout,自上而下的垂直方向,界面上方嵌套一个 FrameLayout 帧布局,头像和名字在帧布局的内部,下方的几个控件线性排列。

(2) 设计思路。整个布局界面可以通过布局嵌套达到用户需求,上方嵌套的帧布局中,人物图片和用户名默认在左上角重叠摆放,通过 android:layout_gravity 属性设置了对齐方式,通过 android:layout_margin 属性设定了控件周围的空白距离,达到用户的需求。

2) 实现步骤

(1) 创建一个新的工程,工程名为 ZSFrameLayout。

(2) 切换工程的 Project 项目结构。选择该模式下方的 app,依次展开,便看到工程的布局界面和工程的类文件,其中,activity_main.xml 是布局界面,MainActivity.java 为类文件。

(3) 修改布局界面。双击 layout 文件夹下的 activity_main.xml 文件,便可打开布局编辑器,修改布局类型为 LinearLayout,添加方向属性为垂直。依次添加嵌套的 FrameLayout 帧布局和其他控件,代码如下。

```xml
<?xml version = "1.0" encoding = "utf-8"?>
<LinearLayout xmlns:android = "http://schemas.android.com/apk/res/android"
    xmlns:app = "http://schemas.android.com/apk/res-auto"
    xmlns:tools = "http://schemas.android.com/tools"
    android:layout_width = "match_parent"
    android:layout_height = "match_parent"
    android:orientation = "vertical"
    tools:context = ".MainActivity">
    <FrameLayout
        android:layout_width = "match_parent"
        android:layout_height = "wrap_content"
        android:background = "@drawable/myinfo_login_bg"    >
        <ImageView
            android:layout_width = "80dp"
            android:layout_height = "80dp"
            android:src = "@drawable/default_icon"
            android:layout_gravity = "center_horizontal"
            android:layout_marginTop = "50dp"/>
        <TextView
            android:layout_width = "wrap_content"
            android:layout_height = "wrap_content"
            android:text = "admin"
```

```xml
            android:textSize = "30sp"
            android:textColor = "#ffffff"
            android:layout_gravity = "center_horizontal"
            android:layout_marginTop = "130dp"/>
</FrameLayout>
<TextView
    android:layout_width = "wrap_content"
    android:layout_height = "wrap_content"
    android:text = "个人中心"
    android:textSize = "30sp"
    android:textStyle = "bold"
    android:layout_gravity = "center"
    android:layout_marginBottom = "20dp"/>
<View
    android:layout_width = "match_parent"
    android:layout_height = "2dp"
    android:background = "#E91A1A"/>
<TextView
    android:layout_width = "wrap_content"
    android:layout_height = "wrap_content"
    android:text = "设置"
    android:textSize = "25sp"
    android:padding = "10dp"       />
<View
    android:layout_width = "match_parent"
    android:layout_height = "1dp"
    android:background = "#C5C2C2"/>
<TextView
    android:layout_width = "wrap_content"
    android:layout_height = "wrap_content"
    android:text = "修改密码"
    android:textSize = "25sp"
    android:padding = "10dp"       />
<View
    android:layout_width = "match_parent"
    android:layout_height = "1dp"
    android:background = "#C5C2C2"/>
<TextView
    android:layout_width = "wrap_content"
    android:layout_height = "wrap_content"
    android:text = "作品详情"
    android:textSize = "25sp"
    android:padding = "10dp"       />
<View
    android:layout_width = "match_parent"
    android:layout_height = "1dp"
    android:background = "#C5C2C2"/>
<TextView
    android:layout_width = "wrap_content"
```

```
                android:layout_height = "wrap_content"
                android:text = "注销"
                android:textSize = "25sp"
                android:padding = "10dp"           />
        <View
                android:layout_width = "match_parent"
                android:layout_height = "1dp"
                android:background = "#C5C2C2"/>
</LinearLayout>
```

上述代码中,线性布局内嵌套了帧布局。在帧布局中,使用 android:layout_gravity 属性让图片水平居中,使用 android:layout_marginTop 确定了用户头像图片的位置。采用同样的方法显示用户名,从而实现了多个控件重叠摆放的问题。另外,画线时使用的控件为 View,该控件的高度用来决定画线的粗细,同时使用 android:background 属性设置画线的颜色。

(4) 重新运行程序,效果如图 3.13 所示。

图 3.13 FrameLayout 布局效果图

3.2 Android 开发中的样式设计

Android 系统中包含很多样式(Style),这些样式用于定义界面上的布局风格,Style 是针对窗体元素级别的,改变指定控件或者 Layout 的样式。例如,TextView 或 Button 等控件,通过样式的设定,让布局合理而且美观,提升用户体验。

3.2.1 自定义控件样式

微课视频

样式是包含一种或多种控件的属性集合,可以指定控件的高度、宽度、字体大小及颜色等。Android 中的样式类似于网页中的 CSS 样式,可以让设计与内容分离。样式文件在 XML 资源文件中定义,并且可以继承、复用等,方便统一管理并减少代码量。

(1) 新建一个工程,工程名为 StyleTest。

(2) 切换工程的 Project 项目结构。选择该模式下方的 app,依次展开,便看到工程的布局界面和工程的类文件,其中,activity_main.xml 是布局界面,MainActivity.java 为类文件。

(3) 创建样式。在 app 目录结构下找到 res/values/style 目录下的 styles.xml 文件,双击打开,看到<resource>根标签和定义样式的<style>标签,它包含多个<item>来声明样式名称和属性,在其中编写两种 TextView 控件样式,代码如下。

```xml
<resources>
    <!-- Base application theme. -->
    <style name="AppTheme" parent="Theme.AppCompat.Light.DarkActionBar">
        <!-- Customize your theme here. -->
        <item name="colorPrimary">@color/colorPrimary</item>
        <item name="colorPrimaryDark">@color/colorPrimaryDark</item>
        <item name="colorAccent">@color/colorAccent</item>
    </style>
    <style name="textStyle_one">
        <item name="android:layout_width">match_parent</item>
        <item name="android:layout_height">wrap_content</item>
        <item name="android:textColor">#ff00ac</item>
        <item name="android:textSize">35sp</item>
    </style>
    <style name="textStyle_two" parent="@style/textStyle_one">
        <item name="android:textSize">25sp</item>
    </style>
</resources>
```

上述代码中,第 1 个<style>标签中的代码是系统自带的样式,其中,name 属性是样式名称,parent 属性表示继承某个样式,并且通过<item>标签以键值对的形式定义属性和属性值。textStyle_one 是自定义的样式,设置了控件的高、宽、字体颜色、字体大小四个属性。textStyle_two 样式继承了 textStyle_one,并在该属性中重新定义了 android:textSize 的属性。

(4) 设计布局界面。双击 layout 文件夹下的 activity_main.xml 文件,便可打开布局编辑器,修改布局类型为 LinearLayout,添加方向属性为垂直。依次添加两个 TextView 控件,代码如下。

```xml
<?xml version="1.0" encoding="utf-8"?>
<LinearLayout xmlns:android="http://schemas.android.com/apk/res/android"
    xmlns:app="http://schemas.android.com/apk/res-auto"
```

```xml
xmlns:tools = "http://schemas.android.com/tools"
android:layout_width = "match_parent"
android:layout_height = "match_parent"
android:orientation = "vertical"
tools:context = ".MainActivity">
<TextView
    style = "@style/textStyle_one"
    android:text = "TextView 样式一"             />
<TextView
    style = "@style/textStyle_two"
    android:text = "TextView 样式二"             />
</LinearLayout>
```

上述代码中,两个 TextView 只需要以 style="@style/xxx"这种方式就可以引用自定义样式中的所有属性,其中属性值对应自定义样式名称,运行程序,预览效果如图 3.14 所示。

图 3.14 TextView 样式

微课视频

3.2.2 自定义背景样式

在设计开发项目时,也可以自己创建样式文件,通过定义 XML 文件来实现不同的样式。下面以定义一个具有边框色和填充色的输入框 EditText 控件的样式为例,继续在 StyleTest 工程中完善内容。

（1）边框和填充样式文件创建。展开自定义样式1创建的工程文件 StyleTest，在 app 目录结构下找到 res/drawable，右击 drawable 文件夹，选择 New→Drawable resource file，输入样式名称和节点类型，单击 OK 按钮，如图 3.15 所示。

图 3.15　样式文件定义

（2）在 editstyle.xml 样式文件中输入如下代码。

```xml
<?xml version = "1.0" encoding = "utf-8"?>
<shape xmlns:android = "http://schemas.android.com/apk/res/android">
    <stroke android:width = "3dp" android:color = "#ff0000"/>
    <solid android:color = "#FFEB3B"/>
</shape>
```

上述代码中，stroke 节点用来定义边框，使用 android:width 定义边框宽度，使用 android:color 定义边框颜色。solid 节点用来定义填充，android:color 表示填充的颜色。

（3）设计布局界面。双击 layout 文件夹下的 activity_main.xml 文件，便可打开布局编辑器，添加 EditText 控件，引用定义的样式文件，代码如下。

```xml
<?xml version = "1.0" encoding = "utf-8"?>
<LinearLayout xmlns:android = "http://schemas.android.com/apk/res/android"
    xmlns:app = "http://schemas.android.com/apk/res-auto"
    xmlns:tools = "http://schemas.android.com/tools"
    android:layout_width = "match_parent"
    android:layout_height = "match_parent"
    android:orientation = "vertical"
    tools:context = ".MainActivity">
<TextView
        style = "@style/textStyle_one"
        android:text = "TextView 样式一"          />
<TextView
        style = "@style/textStyle_two"
        android:text = "TextView 样式二"          />
<EditText
        android:layout_width = "match_parent"
        android:layout_height = "wrap_content"
        android:hint = "请输入用户名"
        android:textSize = "40sp"
        android:background = "@drawable/editstyle"
        android:layout_margin = "10dp"/>
</LinearLayout>
```

上述代码中，使用 android:background 给 EditText 控件设定背景，其值为 @drawable/editstyle 表示引用 drawable 文件夹下的样式文件。运行程序，预览效果如图 3.16 所示。

按钮具有弹起和按下两种状态，使用背景选择器可以让按钮在正常情况下显示为紫色填充和黑色边框，按下时显示为蓝色填充和红色边框。继续在 StyleTest 工程中完善内容。

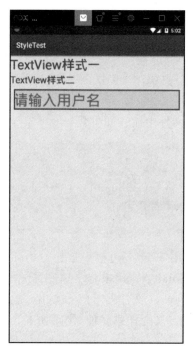

图 3.16　EditText 控件样式

（1）创建按钮正常情况下的样式。在 app 目录结构下找到 res/drawable，右击 drawable 文件夹，选择 New→Drawable resource file，输入样式名称和节点类型，单击 OK 按钮，如图 3.17 所示。

（2）在 btnstyleone.xml 样式文件中输入如下代码。

图 3.17　样式文件定义

```xml
<?xml version = "1.0" encoding = "utf - 8"?>
<shape xmlns:android = "http://schemas.android.com/apk/res/android">
    <stroke android:width = "8dp" android:color = "#090808"/>
    <solid android:color = "#C70AF0"/>
    <corners android:radius = "50dp"/>
</shape>
```

上述代码中，stroke 节点用来定义边框，solid 节点用来定义填充，corners 节点用来定义圆角半径。

（3）创建按钮按下时的样式。方法同上，新建样式名称为 btnstyletwo.xml，节点类型为 shape，在样式文件 btnstyletwo.xml 中输入如下代码。

```xml
<?xml version = "1.0" encoding = "utf - 8"?>
<shape xmlns:android = "http://schemas.android.com/apk/res/android">
    <stroke android:width = "8dp" android:color = "#E91E4D"/>
    <solid android:color = "#172CD5"/>
    <corners android:radius = "50dp"/>
</shape>
```

(4) 创建按钮选择器样式文件。在 app 目录结构下找到 res/drawable，右击 drawable 文件夹，选择 New → Drawable resource file，输入选择器名称和节点类型，单击 OK 按钮，如图 3.18 所示。

(5) 在 btnselector.xml 选择器样式文件中输入如下代码。

图 3.18 按钮选择器样式

```
<?xml version = "1.0" encoding = "utf - 8"?>
< selector xmlns:android = "http://schemas.android.com/apk/res/android">
    < item android:drawable = "@drawable/btnstyleone" android:state_pressed = "false"/>
    < item android:drawable = "@drawable/btnstyletwo" android:state_pressed = "true"/>
</selector>
```

上述代码中，selector 标签可以添加一个或多个 item 子标签，而相应的状态是在 item 标签中定义的。item 必须指定 android:drawable 属性，drawable 的属性值可以是一张图片，也可以是一个样式文件。android:state_pressed 设置是否按压状态，默认为 false 表示未单击按钮，属性值为 true 时表示单击按钮。

第一行 item 表示按钮正常情况下默认时的背景图片，第二行 item 表示按钮单击时的背景图片。

(6) 设计布局界面。双击 layout 文件夹下的 activity_main.xml 文件，便可打开布局编辑器，添加 Button 控件，引用定义的按钮选择器样式文件，代码如下。

```
<?xml version = "1.0" encoding = "utf - 8"?>
< LinearLayout xmlns:android = "http://schemas.android.com/apk/res/android"
    xmlns:app = "http://schemas.android.com/apk/res - auto"
    xmlns:tools = "http://schemas.android.com/tools"
    android:layout_width = "match_parent"
    android:layout_height = "match_parent"
    android:orientation = "vertical"
    tools:context = ".MainActivity">
    < TextView
        style = "@style/textStyle_one"
        android:text = "TextView 样式一" />
    < TextView
        style = "@style/textStyle_two"
        android:text = "TextView 样式二" />
    < EditText
        android:layout_width = "match_parent"
        android:layout_height = "wrap_content"
        android:hint = "请输入用户名"
        android:textSize = "40sp"
        android:background = "@drawable/editstyle"
        android:layout_margin = "10dp"/>
    < Button
        android:layout_width = "100dp"
        android:layout_height = "100dp"
```

```
            android:text = "美"
            android:textSize = "50sp"
            android:textColor = "#ffffff"
            android:padding = "5dp"
            android:background = "@drawable/btnselector"/>
</LinearLayout>
```

上述代码中,使用 android:background 给 Button 控件设定背景,其值为@drawable/btnselector 表示引用 drawable 文件夹下按钮选择器的样式文件。运行程序,预览效果如图 3.19 所示。

图 3.19 Button 控件选择器样式

3.3 Android 开发中的主题设计

主题(Theme)是应用到整个 Activity 和 Application 的样式,而不是只应用到单个视图,当设置好主题后,Activity 或整个程序中的视图都将使用主题中的属性。当主题和样式中的属性发生冲突时,样式的优先级要高于主题。

主题与样式在代码结构上是一样的,不同之处在于引用方式上,主题要在 AndroidManifest.xml 配置文件中引用,下面通过一个实例说明。

(1)新建一个工程,工程名为 ThemeTest。

(2)切换工程的 Project 项目结构。选择该模式下方的 app,依次展开,便看到工程的布局界面和工程的类文件,其中,activity_main.xml 是布局界面,MainActivity.java 为类文件。

（3）创建主题样式。在 app 目录结构下找到 res/values/style 目录下 styles.xml 文件，双击打开，在其内部编写主题样式代码，具体代码如下。

```xml
<resources>
    <!-- Base application theme. -->
    <style name="AppTheme" parent="Theme.AppCompat.Light.DarkActionBar">
        <!-- Customize your theme here. -->
        <item name="colorPrimary">@color/colorPrimary</item>
        <item name="colorPrimaryDark">@color/colorPrimaryDark</item>
        <item name="colorAccent">@color/colorAccent</item>
    </style>
    <style name="grayTheme" parent="Theme.AppCompat.Light.DarkActionBar">
        <item name="android:background">#999999</item>
    </style>
</resources>
```

上述代码中，定义了一个灰色的主题背景。需要注意的是，在定义主题时，需要用到 parent 属性去继承 Theme.AppCompat.Light.DarkActionBar 来保证它的兼容性，否则运行时会出现异常。接下来打开 AndroidManifest.xml 文件引用这个主题，在<activity>标签中添加 android:theme="@style/grayTheme"属性，具体代码如下。

```xml
<?xml version="1.0" encoding="utf-8"?>
<manifest xmlns:android="http://schemas.android.com/apk/res/android"
    package="com.example.themetest">
    <application
        android:allowBackup="true"
        android:icon="@mipmap/ic_launcher"
        android:label="@string/app_name"
        android:roundIcon="@mipmap/ic_launcher_round"
        android:supportsRtl="true"
        android:theme="@style/AppTheme">
        <activity android:name=".MainActivity"
            android:theme="@style/grayTheme">
            <intent-filter>
                <action android:name="android.intent.action.MAIN" />
                <category android:name="android.intent.category.LAUNCHER" />
            </intent-filter>
        </activity>
    </application>
</manifest>
```

在上述代码中，大家会发现在<application>标签中同样存在 android:theme 属性，此处是整个应用程序主题的样式，而<activity>标签中是改变当前界面的主题样式，这里要注意区分清楚。运行效果如图 3.20 所示。

图 3.20　主题样式运行结果

微课视频

3.4　实战案例——智能计算器

　　智能计算器开发主要涉及 LinearLayout、Style、EditText、Button 的使用,为 Android 入门基础内容。本节课将利用项目 3 所学过的知识,引导学生设计并实现智能计算器,系统具有良好的界面、必要的交互信息、简约的美观效果,使用人员能够快捷简单地进行操作,及时准确地获取需要的计算结果,充分降低了数字计算的难度,节约了时间成本。

3.4.1　界面分析

　　本案例中共有一个界面,布局总体为 LinearLayout,自上而下的垂直方向 vertical,通过线性布局的嵌套,实现水平方向按钮的摆放,通过自定义 Style 样式文件,让 EditText 和 Button 呈现较美的外观,界面运行效果如图 3.21 所示。

3.4.2　实现思路

　　上方的 EditText 控件用来显示参与运算的数据,同时也显示运算的结果。单击数字按钮接受 0~9 组合起来的数字显示到上方的 EditText 控件中,单击加、减、

图 3.21　智能计算器界面效果图

乘、除运算符按钮时标志着第一个运算数输入结束。此时获取 EditText 中的数值便得到了参与运算的第一个数。继续单击数字按钮 0~9 组合起来的数字，数字依然会显示到上方的 EditText 控件中，当单击等号时意味着第二个运算数输入结束，此时获取 EditText 中的数值便得到了参与运算的第二个数。根据单击的运算符号，对获得的两个运算数进行运算，便可得到结果。最后将运算结果显示到 EditText 编辑框中。

3.4.3 任务实施

【任务 3-1】 自定义 style 样式文件

（1）创建一个新的工程，工程名为 ZSCalculator。

（2）切换工程的 Project 项目结构。选择该模式下方的 app，依次展开，便看到工程的布局界面和工程的类文件，其中，activity_main.xml 是布局界面，MainActivity.java 为类文件。

（3）准备一张图片，图片名为 calculatorbg.jpg，将其粘贴到 app 目录结构中 res 下方的 drawable 文件夹下，图片作为页面布局的背景。

（4）EditText 样式文件创建。在 app 目录结构下找到 res/drawable，右击 drawable 文件夹，选择 New→Drawable resource file，输入样式名称 editstyle1 和节点类型 shape，单击 OK 按钮，在 editstyle1.xml 样式文件中输入如下代码。

```xml
<?xml version = "1.0" encoding = "utf-8"?>
<shape xmlns:android = "http://schemas.android.com/apk/res/android">
    <solid android:color = "#CDDC39"/>
    <stroke android:color = "#D6C9C9" android:width = "3dp"/>
</shape>
```

（5）Button 按钮未单击时样式文件创建。在 app 目录下找到 res/drawable，右击 drawable 文件夹，选择 New→Drawable resource file，输入样式名称 btnstyle3 和节点类型 shape，输入如下代码。

```xml
<?xml version = "1.0" encoding = "utf-8"?>
<shape xmlns:android = "http://schemas.android.com/apk/res/android">
    <solid android:color = "#D024FF"/>
    <stroke android:color = "#000000" android:width = "5dp"/>
    <corners android:radius = "1000dp"/>
</shape>
```

（6）Button 按钮被单击时样式文件创建。输入样式名称 btnstyle4 和节点类型 shape，输入如下代码。

```xml
<?xml version = "1.0" encoding = "utf-8"?>
<shape xmlns:android = "http://schemas.android.com/apk/res/android">
    <solid android:color = "#FF24AF"/>
    <stroke android:color = "#28B474" android:width = "5dp"/>
    <corners android:radius = "1000dp"/>
</shape>
```

（7）Button 按钮选择器样式文件创建。输入选择器名称 btn34 和节点类型 selector，输入如下代码。

```xml
<?xml version = "1.0" encoding = "utf-8"?>
<selector xmlns:android = "http://schemas.android.com/apk/res/android">
<item android:drawable = "@drawable/btnstyle3" android:state_pressed = "false"/>
<item android:drawable = "@drawable/btnstyle4" android:state_pressed = "true"/>
</selector>
```

【任务 3-2】 计算器 UI 界面设计

设计布局界面。双击 layout 文件夹下的 activity_main.xml 文件，便可打开布局编辑器，修改布局界面类型为 LinearLayout，方向为 vertical，依次添加所需控件，并引用刚刚创建的样式文件，代码如下。

```xml
<?xml version = "1.0" encoding = "utf-8"?>
<LinearLayout xmlns:android = "http://schemas.android.com/apk/res/android"
    xmlns:app = "http://schemas.android.com/apk/res-auto"
    xmlns:tools = "http://schemas.android.com/tools"
    android:layout_width = "match_parent"
    android:layout_height = "match_parent"
    android:orientation = "vertical"
    android:background = "@drawable/calculatorbg"
    tools:context = ".MainActivity">
    <EditText
        android:id = "@+id/txt_result"
        android:layout_width = "match_parent"
        android:layout_height = "wrap_content"
        android:hint = "0"
        android:gravity = "right"
        android:textSize = "40sp"
        android:textColor = "#000000"
        android:padding = "10dp"
        android:layout_margin = "10dp"
        android:background = "@drawable/editstyle1" />
    <LinearLayout
        android:layout_width = "match_parent"
        android:layout_height = "wrap_content"
        android:orientation = "horizontal"
        android:weightSum = "4">
        <Button
            android:id = "@+id/btn_7"
            android:layout_width = "wrap_content"
            android:layout_height = "wrap_content"
            android:text = "7"
            android:textSize = "60sp"
            android:textColor = "#ffffff"
            android:textStyle = "bold"
            android:background = "@drawable/btn34"
```

```xml
            android:padding = "5dp"
            android:layout_margin = "10dp"
            android:layout_weight = "1"/>
        <Button
            android:id = "@+id/btn_8"
            android:layout_width = "wrap_content"
            android:layout_height = "wrap_content"
            android:text = "8"
            android:textSize = "60sp"
            android:textColor = "#ffffff"
            android:textStyle = "bold"
            android:background = "@drawable/btn34"
            android:padding = "5dp"
            android:layout_margin = "10dp"
            android:layout_weight = "1"/>
        <Button
            android:id = "@+id/btn_9"
            android:layout_width = "wrap_content"
            android:layout_height = "wrap_content"
            android:text = "9"
            android:textSize = "60sp"
            android:textColor = "#ffffff"
            android:textStyle = "bold"
            android:background = "@drawable/btn34"
            android:padding = "5dp"
            android:layout_margin = "10dp"
            android:layout_weight = "1"/>
        <Button
            android:id = "@+id/btn_jia"
            android:layout_width = "wrap_content"
            android:layout_height = "wrap_content"
            android:text = " + "
            android:textSize = "60sp"
            android:textColor = "#ffffff"
            android:textStyle = "bold"
            android:background = "@drawable/btn34"
            android:padding = "5dp"
            android:layout_margin = "10dp"
            android:layout_weight = "1"/>
    </LinearLayout>
        ...
</LinearLayout>
```

上述代码中，第一行 EditText 控件用来存放输入的运算数及运算结果。下方依次是四个线性布局，按自左向右的水平方向放置。

第一行水平线性布局放置数字 7、8、9、＋，其 ID 分别为 btn_7、btn_8、btn_9、btn_jia；代码如上，因下方三个水平线性布局与第一行线性布局代码雷同，请自行复制将代码补充完整。

第二行水平线性布局放置数字 4、5、6、−，其 ID 分别为 btn_4、btn_5、btn_6、btn_jian。

第三行水平线性布局放置数字 1、2、3、*，其 ID 分别为 btn_1、btn_2、btn_3、btn_cheng。

最后一行水平线性布局放置数字 0、C、=、/，其 ID 分别为 btn_0、btn_qing、btn_deng、btn_chu。

【任务 3-3】 计算器功能设计

实现 Activity 页面的功能。打开 MainActivity.java 类文件，修改 MainActivity 的代码，如下。

```java
public class MainActivity extends AppCompatActivity {
    //定义对象
    private TextView txt_result;
    private Button btn_7;
    private Button btn_8;
    private Button btn_9;
    private Button btn_jia;
    private Button btn_4;
    private Button btn_5;
    private Button btn_6;
    private Button btn_jian;
    private Button btn_1;
    private Button btn_2;
    private Button btn_3;
    private Button btn_cheng;
    private Button btn_0;
    private Button btn_qing;
    private Button btn_deng;
    private Button btn_chu;
    private double num1 = 0, num2 = 0;           //声明两个参数,接收数据
    private double result = 0;                    //运算结果
    private Boolean isClickdeng = false;          //判断是否单击了 =
    private String op = " % ";                    //操作符 +- * /

    @Override
    protected void onCreate(Bundle savedInstanceState) {
        super.onCreate(savedInstanceState);
        setContentView(R.layout.activity_main);
    //绑定控件
        txt_result = findViewById(R.id.txt_result);
        btn_7 = findViewById(R.id.btn_7);
        btn_8 = findViewById(R.id.btn_8);
        btn_9 = (Button) findViewById(R.id.btn_9);
        btn_jia = (Button) findViewById(R.id.btn_jia);
        btn_4 = (Button) findViewById(R.id.btn_4);
        btn_5 = (Button) findViewById(R.id.btn_5);
        btn_6 = (Button) findViewById(R.id.btn_6);
```

```java
        btn_jian = (Button) findViewById(R.id.btn_jian);
        btn_1 = (Button) findViewById(R.id.btn_1);
        btn_2 = (Button) findViewById(R.id.btn_2);
        btn_3 = (Button) findViewById(R.id.btn_3);
        btn_cheng = (Button) findViewById(R.id.btn_cheng);
        btn_0 = (Button) findViewById(R.id.btn_0);
        btn_qing = (Button) findViewById(R.id.btn_qing);
        btn_deng = (Button) findViewById(R.id.btn_deng);
        btn_chu = (Button) findViewById(R.id.btn_chu);

//按钮的单击事件
//数字0~9按钮代码
        btn_7.setOnClickListener(new View.OnClickListener() {
            @Override
            public void onClick(View v) {
                //按钮单击逻辑
                if(isClickdeng){//说明刚单击了=,上一个运算刚结束
                    txt_result.setText(""); //重新计算,文本框清空
                    isClickdeng = false; //更改=按钮的状态
                }
                txt_result.setText(txt_result.getText().toString() + "7"); //第一种情况,单击7直接显示在文本控件里.第二种情况:先单击别的数字,再单击7
            }
        });
        btn_8.setOnClickListener(new View.OnClickListener() {
            @Override
            public void onClick(View v) {
                //按钮单击逻辑
                if(isClickdeng){//说明刚单击了=,上一个运算刚结束
                    txt_result.setText(""); //重新计算,文本框清空
                    isClickdeng = false; //更改=按钮的状态
                }
                txt_result.setText(txt_result.getText().toString() + "8"); //第一种情况,单击8直接显示在文本控件里.第二种情况:先单击别的数字,再单击8

            }
        });
        btn_9.setOnClickListener(new View.OnClickListener() {
            @Override
            public void onClick(View v) {
                //按钮单击逻辑
                if(isClickdeng){//说明刚单击了=,上一个运算刚结束
                    txt_result.setText(""); //重新计算,文本框清空
                    isClickdeng = false; //更改=按钮的状态
                }
                txt_result.setText(txt_result.getText().toString() + "9"); //第一种情况,单击9直接显示在文本控件里.第二种情况:先单击别的数字,再单击9

            }
```

```java
        });

            …        //此处省略了4,5,6,按钮代码,请自行补充完整
            …        //此处省略了1,2,3按钮代码,请自行补充完整
            …        //此处省略了0按钮代码,请自行补充完整
//运算符 +、-、*、/ 按钮单击事件
//运算符 + 按钮代码
        btn_jia.setOnClickListener(new View.OnClickListener() {
            @Override
            public void onClick(View v) {
                String st1 = txt_result.getText().toString();  //获取单击加号之前字符串类
                                                               //型的数据
                if(st1.equals("")){//判断获取的数据是否为空
                    return;  //返回,什么都不做
                }
                num1 = Double.parseDouble(st1);  //将获取的字符串类型的数据转换为Double小数类型,这
                                                 //是第一个数
                txt_result.setText("");  //清空文本控件中的第一个数
                op = " + ";  //表示进行加法计算
                isClickdeng = false;  //单击加号按钮时,等号按钮不起作用
            }
        });

            …//此处省略了运算符 - 按钮代码,请自行补充完整
            …//此处省略了运算符 * 按钮代码,请自行补充完整
            …//此处省略了运算符 / 按钮代码,请自行补充完整

//清除 c 按钮代码
        btn_qing.setOnClickListener(new View.OnClickListener() {
            @Override
            public void onClick(View v) {
                txt_result.setText("");  //清除文本框中的内容
            }
        });

// = 按钮代码
        btn_deng.setOnClickListener(new View.OnClickListener() {
            @Override
            public void onClick(View v) {
                String str2 = txt_result.getText().toString();  //获取文本框中的数据
                if(str2.equals("")){    //判断获取的数据是否为空
                    return;  //返回,什么都不做
                }
                num2 = Double.parseDouble(str2);  //将获取的字符串类型的数据转换为Double小数类型,这
                                                  //是第二个数据
                txt_result.setText("");  //清空文本控件中的第二个数
                switch (op){    //此时判断进行何种操作
                    case " + ":result = num1 + num2;break;  //op = +   加法
```

```
                    case "-":result = num1 - num2;break; //op = -    减法
                    case "*":result = num1 * num2;break; //op = *    乘法
                    case "/":result = num1/num2;break;   //op = /    除法
                    case "%":result = num2;break;        //op = %    不进行计算
                    default:result = 0.0;break;
                }
                txt_result.setText(result + "");  //将 +- * /运算的运算结果转换为字符串显
                                                  //示到结果文本框中

                op = "%";
                isClickdeng = true;
            }
        });


    }
}
```

上述代码中，0～9数字按钮代码雷同。以数字7按钮为例，单击数字7时，首先要判断等号的状态，如果 isClickdeng 状态为 true，说明上一个运算结果刚结束，此时要清空 EditText 中的运算结果，同时将等号的状态 isClickdeng 置为 false，这样就可以开始接收下一个运算数。如果直接单击数字7，则将数字7显示到 EditText 控件中，代码可表示为：txt_result.setText("7")。如果先单击别的数字之后又单击了数字7，则需要将先单击数字与7连接到一起，形成一个新的数字，显示到 EditText 控件中，因此便出现了此行代码：txt_result.setText(txt_result.getText().toString()+"7")。如果单击的是数字8按钮，那么此行代码就需要更改为：txt_result.setText(txt_result.getText().toString()+"8")。0～9这10个数字按钮单击事件内部代码相似，请根据已有代码自行补充完整。

+、-、*、/运算符号代码雷同，以运算符+为例，单击运算符+按钮时，表示第一个运算数输入结束，此时获取 EditText 控件的内容便得到了第一个运算数。判断所获得运算数，如果为空，说明用户并没有输入数据，return 表示程序不做任何运算，直接返回空。如果获得第一个运算数不为空，则将获取的字符串类型的数据转换为 double 类型，同时将 EditText 清空，操作运算符置为+，等号状态 isClickdeng 置为 false，开始准备接收第二个运算数。如果单击的是-按钮，则操作运算符 op="-"；单击乘号*按钮时将 op="*"；单击/按钮时将 op="/"。这四个运算符按钮单击事件内部代码相似，请根据已有代码自行补充完整。

"清除"按钮代码较为简单，即清除 EditText 中的输入运算数或者是清空计算结果。

=按钮代码较为复杂，首先单击=时，表示第二个运算符输入结束，此时获取 EditText 控件的内容便得到了第二个运算数。同理，判断所获得运算数是否为空，若为空，什么都不做，直接返回。如果获得第二个运算数不为空，则将获取的字符串类型的数据转换为 double 类型，然后判断操作运算符，当操作运算符为+时，将两个运算数进行加法操作；当操作运算符为-时，将两个运算数进行减法操作；当操作运算符为*时，将两个运算数进行乘法操作；当操作运算符为/时，将两个运算数进行除法操作，并将运算结果显示到 EditText 中。

项目小结

本项目内容紧紧围绕 Android 当中的 UI 布局和 Android 开发中的 Style 样式设计的内容来展开,通过三种常用布局的介绍,开阔学生眼界,提升学生设计布局界面的能力。通过 Style 的设计,对布局界面的复杂性和观赏性提出了更高的要求。最后通过智能计算器的案例,培养学生移动端 App 的设计与开发能力。

习题

1. 下面选项中,(　　)布局是线性布局。
   A. RelativeLayout　　B. LinearLayout　　C. FrameLayout　　D. TableLayout
2. 线性布局只有两种方向:水平和垂直。(　　)
   A. 正确　　　　　　B. 错误
3. 在定义输入文本框的样式时,选择的节点类型为(　　)。
   A. menu　　　　　B. selector　　　　C. shape　　　　D. anim
4. 在定义控件的样式时,(　　)属性表示描边。
   A. < solid android:color="#dddddd"/>
   B. < stroke android:color="#dfcdae" android:width="3dp"/>
   C. < corners android:radius="10dp"/>
   D. < padding android:top="20dp"/>
5. 下面关于相对布局,说法错误的是(　　)。
   A. 控件相对于父容器进行定位
   B. 控件相对于控件进行定位
   C. 所有的控件都从左上角开始定位
   D. 一个控件可以位于另一个控件的上方、下方、左侧、右侧
6. 下面属性中,(　　)属性表示线性布局的方向是垂直的。
   A. android:orientation="vertical"　　　　B. android:orientation="horizontal"
   C. android:layout_gravity="center"　　　D. android:layout_weight="1"
7. 所有的 View 都会放在左上角,并且后添加进去的 View 会覆盖之前放进去的 View。具有这种特点的布局是帧布局。(　　)
   A. 正确　　　　　　B. 错误
8. 在定义控件的样式时,(　　)属性表示圆角半径。
   A. < solid android:color="#dddddd"/>
   B. < stroke android:color="#dfcdae" android:width="3dp"/>
   C. < corners android:radius="10dp"/>
   D. < padding android:top="20dp"/>
9. 线性布局中可以嵌套帧布局。(　　)
   A. 正确　　　　　　B. 错误

项目 4

打地鼠小游戏

【教学导航】

学习目标	（1）掌握 Toast 消息提示的方法和使用。 （2）掌握 Dialog 对话框的方法和使用。 （3）掌握 Spinner 控件的方法和使用。 （4）了解 Android 中的适配器。 （5）理解 Menu 菜单的使用方法。 （6）具备简单游戏的设计开发能力。
教学方法	任务驱动法、理论实践一体化、探究学习法、分组讨论法
课时建议	6 课时

4.1 Android 的事件处理与交互实现程序设计

在 Android 中，事件是用户对图形界面的操作，而 Android 应用程序是通过事件和信息来实现人机交互的。Android 中的事件包括按下、弹起、滑动、双击等，针对不同的控件，存在着对应事件处理的相关方法。Android 中的信息包括信息框、对话框、通知等。本节介绍 Android 中与事件处理和交互相关的组件及知识，包括 Toast 提示、Dialog 对话框、Menu 菜单和 Spinner 下拉列表框等。

4.1.1 提示

Toast（消息提示框）是 Android 中的轻量级信息提醒机制，显示在应用程序界面的最上层。例如，在 App 的注册界面，通常需要用户发送手机验证码进行验证，当用户单击"发送验证码"后，通常在页面下方会出现一条信息"已发送，注意查收"。该信息出现的时间大约 3～5s，并且自动消失，像这样的信息就是 Toast 提示信息，它一般用于提示一些不那么引人注目，但是又希望用户看见的消息，无须用户维护。

微课视频

1. Toast 创建步骤

使用 Toast 显示信息提示框的步骤如下。

(1) 创建 Toast 对象。

通常有两种方法：一种是通过构造方法创建，另一种是通过 Toast 类的 makeText() 方法创建。

(2) 设置消息提示的样式。

Toast 类提供了很多方法用于设置消息提示框的样式，如 setGravity() 方法用来设置消息对齐方式、setMargin() 方法用来设置页边距等，该步骤根据需求选择性进行设置。

(3) 显示消息。

调用 Toast 类的 show() 方法显示消息提示框。

2. Toast 常用方法

Toast 常用方法如表 4.1 所示。

表 4.1 Toast 常用方法

方　　法	作　　用	参数说明
makeText(Context context, CharSequence text, int duration)	创建 Toast	context：当前的上下文环境 text：要显示的字符串 duration：显示的时间长短
show()	显示 Toast	无
setGravity(int gravity, int xOffset, int yOffset)	设置 Toast 的位置	gravity：Toast 在屏幕中显示的位置 xOffset：相对于 gravity 设置 Toast 位置的横向 X 轴的偏移量，正数向右偏移，负数向左偏移 yOffset：Y 轴偏移量，同上
setMargin(float horizontalMargin, float verticalMargin)	设置 Toast 的页边距	horizontalMargin：水平边距 verticalMargin：垂直边距
getView()	获取布局视图	无

3. Toast 案例

1) 案例分析

(1) 界面分析。本案例共有一个布局界面，在该布局界面中添加两个按钮控件，在页面创建时显示一条短的 Toast，单击第一个按钮显示一条长的 Toast，单击第二个按钮显示一条居中的带图片的 Toast。

(2) 设计思路。布局界面中的两个控件通过添加属性的方式实现需求；显示 Toast 通过按钮单击事件和 Toast 常用方法实现需求。

2) 实现步骤

(1) 创建一个新的工程，工程名为 ToastZhangsan。

(2) 切换工程的 Project 项目结构。选择该模式下方的 app，依次展开，便看到工程的布局界面和工程的类文件，其中，activity_main.xml 是布局界面，MainActivity.java 为类文件。

(3) 设计布局界面。双击 layout 文件夹下的 activity_main.xml 文件,便可打开布局编辑器,修改布局类型为 LinearLayout,添加方向属性为 vertical。依次添加两个控件,代码如下。

```xml
<?xml version = "1.0" encoding = "utf-8"?>
<LinearLayout xmlns:android = "http://schemas.android.com/apk/res/android"
    xmlns:app = "http://schemas.android.com/apk/res-auto"
    xmlns:tools = "http://schemas.android.com/tools"
    android:layout_width = "match_parent"
    android:layout_height = "match_parent"
    android:orientation = "vertical"
    tools:context = ".MainActivity">
    <Button
        android:id = "@+id/button_1"
        android:layout_width = "match_parent"
        android:layout_height = "wrap_content"
        android:text = "长的toast"
        android:textAllCaps = "false"
        android:textSize = "28sp"/>
    <Button
        android:id = "@+id/button_2"
        android:layout_width = "match_parent"
        android:layout_height = "wrap_content"
        android:text = "居中带图片的toast"
        android:textAllCaps = "false"
        android:textSize = "28sp"/>
</LinearLayout>
```

(4) 实现案例功能。可以在程序中不同位置添加创建和显示 Toast 的代码来实现显示消息,打开 MainActivity.java 类文件,修改 MainActivity 的代码如下。

```java
public class MainActivity extends AppCompatActivity {
    //第一步:定义对象
    Button btn_1;
    Button btn_2;
    @Override
    protected void onCreate(Bundle savedInstanceState) {
        super.onCreate(savedInstanceState);
        setContentView(R.layout.activity_main);
        //第二步:绑定控件
        btn_1 = findViewById(R.id.button_1);
        btn_2 = findViewById(R.id.button_2);
        //第三步:创建一条短的Toast并显示
        Toast.makeText(MainActivity.this,"这是一条短的toast",Toast.LENGTH_SHORT).show();
        //第四步:创建一条长的Toast并显示
        btn_1.setOnClickListener(new View.OnClickListener() {
            @Override
            public void onClick(View v) {
                Toast.makeText(MainActivity.this,"这是一条长的toast",Toast.LENGTH_LONG).show();
```

```
            }
        });
        //第五步:创建一条居中的带图片的 Toast 并显示
        btn_2.setOnClickListener(new View.OnClickListener() {
            @Override
            public void onClick(View v) {
                Toast toast = Toast.makeText(MainActivity.this,"这是一条居中的带图片的
toast",Toast.LENGTH_LONG);
                toast.setGravity(Gravity.CENTER,0,0); //让 Toast 居中,左右偏移量为 0
                ImageView image1 = new ImageView(MainActivity.this); //定义图片控件
                image1.setImageResource(R.mipmap.ic_launcher_round); //设置图片
                LinearLayout toastview = (LinearLayout)toast.getView(); //定义 Toast 布局视
图为线性布局
                toastview.addView(image1); //将图片加载到 Toast 布局中
                toast.show(); //显示 Toast
            }
        });
    }
}
```

上述代码中,第一步:定义了两个 Button 对象。第二步:绑定控件,通过 findViewById()方法分别获取到布局文件中的两个控件。第三步:实现在页面创建时显示一条短的 Toast,使用 Toast 类的 makeText()方法创建一条短 Toast,在此方法中,第一个参数设置为当前环境 MainActivity.this,第二个参数设置提示消息内容,第三个参数显示时间长短,设置为 Toast.LENGTH_SHORT,时间为 3s 左右,最后通过 show()方法显示消息。第四步:实现单击 btn_1 显示一条长的 Toast,通过在 onClick()方法中添加 makeText()和 show()方法创建并显示消息,与前一条 Toast 不同的是第三个参数,此时设置为 Toast.LENGTH_LONG,时间为 5s 左右。第五步:实现单击 btn_2 显示居中的带图片的 Toast,通过在 onClick()方法中添加代码,首先通过 makeText()方法创建消息,然后通过 setGravity()方法设置 Toast 居中,由于需要显示图片,所以需要定义一个图片控件并设置图片为 mipmap 文件夹下的 ic_launcher_round.jpg,再定义 Toast 布局视图为线性布局,这里通过 getView()方法获取布局视图,接着通过 addView()方法将图片加载到布局中,最后通过 show()方法显示。

(5) 运行程序,效果如图 4.1 所示。

微课视频

4.1.2 对话框

Dialog(对话框)是程序与用户交互的一种方式,通常用于显示当前程序提示信息以及相关说明,以小窗口形式展现。例如,在 App 界面中选择退出时,通常会弹出一个对话框,让用户确认是否确定要退出,像这样的提示框就是 Dialog,它通常需要用户做出决定后才会继续执行。Android 中的对话框有很多种,除了常见的普通对话框外,还有单选对话框、多选对话框、进度条对话框、编辑对话框等,另外,还支持用户自定义对话框。

1. Dialog 创建步骤

创建 Dialog 对话框的步骤如下。

图 4.1　Toast 运行效果

(1) 创建构造器 Builder。

Android 中创建对话框是通过 AlertDialog 类下的构造器 Builder 创建的,因此首先要创建一个 Builder 对象。

(2) 设置对话框的样式和功能。

Builder 提供了很多方法设置对话框不同类型的组件,如 setIcon()方法用来设置对话框的图标、setTitle()方法用来设置对话框的标题等。另外,对话框上通常会有按钮,Builder 也提供了不同方法设置按钮单击的功能,如 setPositiveButton()方法用来设置"确定"按钮单击的功能、setNegativeButton()方法用来设置"取消"按钮单击的功能等,该步骤根据需求选择性进行设置。

(3) 创建并显示对话框。

通过调用 AlertDialog 类的 create()和 show()方法创建并显示对话框。

2. Dialog 常用方法

Dialog 常用的方法如表 4.2 所示。前 6 个方法从属 AlertDialog.Builder,最后两个方法从属 AlertDialog。

表 4.2　Dialog 常用方法

方　　法	作　　用	参数说明
setIcon(int iconId)	设置 Dialog 的图标	iconID：Dialog 图标文件
setTitle(CharSequence title)	设置 Dialog 的标题	title：Dialog 标题
setMessage(CharSequence message)	设置 Dialog 的内容	message：Dialog 内容
setPositiveButton(CharSequence text, DialogInterface.OnClickListener listener)	设置"确定"按钮	text：按钮文本内容 listener：监听器
setNegativeButton(CharSequence text, DialogInterface.OnClickListener listener)	设置"取消"按钮	text：按钮文本内容 listener：监听器

续表

方　　法	作　　用	参数说明
setNeutralButton(CharSequence text, DialogInterface.OnClickListener listener)	设置"忽略"按钮	text：按钮文本内容 listener：监听器
setSingleChoiceItems（CharSequence[] items, intcheckedItem, android. content. DialogInterface. OnClickListener listener)	设置单选列表	items：单选框的数据集合。 checkedItem：默认选中的单选项 listener：监听器
setMultiChoiceItems(CharSequence[] items, boolean[] checkedItems, android. content. DialogInterface. OnMultiChoiceClickListener listener)	设置多选列表	items：多选框的数据集合。 checkedItem：默认选中的多选项 listener：监听器
create()	创建 Dialog	无
show()	显示 Dialog	无

3. Dialog 案例

1）案例分析

（1）界面分析。本案例共有一个布局界面，在该布局界面中添加两个按钮控件，单击第一个按钮弹出一个普通的对话框，单击第二个按钮弹出一个单选对话框。

（2）设计思路。布局界面中的两个控件通过添加属性的方式实现需求；普通的对话框可以通过对话框构造器完成；单选对话框中需要定义一个字符串类型的数组设置选项内容，然后调用 setSingleChoiceItems()方法设置对话框中的单选列表，其余部分与普通对话框方法相同。

2）实现步骤

（1）创建一个新的工程，工程名为 AlertDialogZhangsan。

（2）切换工程的 Project 项目结构。选择该模式下方的 app，依次展开，便看到工程的布局界面和工程的类文件，其中，activity_main.xml 是布局界面，MainActivity.java 为类文件。

（3）设计布局界面。双击 layout 文件夹下的 activity_main.xml 文件，便可打开布局编辑器，修改布局类型为 LinearLayout，添加方向属性为 vertical。依次添加两个控件，代码如下。

```xml
<?xml version = "1.0" encoding = "utf - 8"?>
<LinearLayout xmlns:android = "http://schemas.android.com/apk/res/android"
    xmlns:app = "http://schemas.android.com/apk/res - auto"
    xmlns:tools = "http://schemas.android.com/tools"
    android:layout_width = "match_parent"
    android:layout_height = "match_parent"
    android:orientation = "vertical"
    tools:context = ".MainActivity">
    <Button
        android:id = "@ + id/button_1"
```

```
            android:layout_width = "match_parent"
            android:layout_height = "wrap_content"
            android:text = "普通对话框"
            android:textAllCaps = "false"
            android:textSize = "28sp"/>
    <Button
            android:id = "@ + id/button_2"
            android:layout_width = "match_parent"
            android:layout_height = "wrap_content"
            android:text = "单选对话框"
            android:textAllCaps = "false"
            android:textSize = "28sp"/>
</LinearLayout>
```

(4)实现案例功能。可以在程序中不同位置添加创建和显示 Dialog 的代码来实现显示对话框,打开 MainActivity.java 类文件,修改 MainActivity 的代码如下。

```
public class MainActivity extends AppCompatActivity {
    //第一步:定义对象
    Button btn_1;
    Button btn_2;
    @Override
    protected void onCreate(Bundle savedInstanceState) {
        super.onCreate(savedInstanceState);
        setContentView(R.layout.activity_main);
        //第二步:绑定控件
        btn_1 = findViewById(R.id.button_1);
        btn_2 = findViewById(R.id.button_2);
        //第三步:创建一个普通 AlertDialog
        btn_1.setOnClickListener(new View.OnClickListener() {
            @Override
            public void onClick(View v) {
                //1.创建一个 AlertDialog 的构造器
                AlertDialog.Builder builder = new AlertDialog.Builder(MainActivity.this);
                //2.用构造器构建 AlertDialog 的每一个组成部分
                builder.setIcon(R.mipmap.ic_launcher_round); //设置图标
                builder.setTitle("温馨提示"); //设置标题
                builder.setMessage("确定要退出吗?"); //设置提示内容
                builder.setPositiveButton("确定", new DialogInterface.OnClickListener() {
                    @Override
                    public void onClick(DialogInterface dialog, int which) {
                        //添加"确定"按钮后的代码
                        finish(); //程序页面关闭
                    }
                });
                builder.setNegativeButton("取消", new DialogInterface.OnClickListener() {
                    @Override
                    public void onClick(DialogInterface dialog, int which) {
                        //添加"取消"按钮后的代码
```

```java
            }
        });
        //3.组合AlertDialog的各部分并显示
        builder.create().show();
    }
});
//第四步:创建一个单选AlertDialog
btn_2.setOnClickListener(new View.OnClickListener() {
    @Override
    public void onClick(View v) {
        //1.创建一个AlertDialog的构造器
        AlertDialog.Builder builder = new AlertDialog.Builder(MainActivity.this);
        //2.用构造器构建AlertDialog的每一个组成部分
        builder.setIcon(R.mipmap.ic_launcher_round); //设置图标
        builder.setTitle("请选择城市"); //设置标题
        final String[] cities = {"北京", "上海", "广州", "深圳"}; //设置选项内容
        builder.setSingleChoiceItems(cities, 0, new DialogInterface.OnClickListener() {
            @Override
            public void onClick(DialogInterface dialog, int which) {

            }
        });
        builder.setPositiveButton("确定", new DialogInterface.OnClickListener() {
            @Override
            public void onClick(DialogInterface dialog, int which) {
                //添加"确定"按钮后的代码
            }
        });
        builder.setNegativeButton("取消", new DialogInterface.OnClickListener() {
            @Override
            public void onClick(DialogInterface dialog, int which) {
                //添加"取消"按钮后的代码
            }
        });
        //3.组合AlertDialog的各部分并显示
        builder.create().show();
    }
});
    }
}
```

上述代码中,第一步:定义了两个Button对象。第二步:绑定控件,通过findViewById()方法分别获取到布局文件中的两个控件。第三步:实现单击btn_1弹出一个普通AlertDialog。在onClick()方法中,首先创建一个AlertDialog的构造器builder,然后利用builder的不同方法构建AlertDialog的每一个组成部分。setIcon()方法设置图标;setTitle()方法设置标题;setMessage()方法设置提示内容;setPositiveButton()方法设置"确定"按钮,其中,第一个参数设置按钮文本内容,第二个参数设置监听器。单击"确定"按钮后的代码在onClick()方法中书写,本案例中单击"确定"按钮后页面关闭。

setNegativeButton()方法设置"取消"按钮,本案例中单击"取消"按钮后什么也不操作。最后通过 create()和 show()方法组合 AlertDialog 的各部分并显示。第四步:实现单击 btn_2 弹出一个单选 AlertDialog。在 onClick()方法中,首先创建一个 AlertDialog 的构造器 builder,然后利用 builder 的不同方法构建 AlertDialog 的每一个组成部分,同时定义一个字符串数组 cities 放置单选列表的内容,接着利用 setSingleChoiceItems()方法设置单选列表,其中,第一个参数是单选列表数据集合,此处设置为字符串数组 cities,第二个参数设置默认选中的单选项,第三个参数设置监听器,单击选项后的代码在 onClick()方法中书写。后面的步骤和第三步一样,最后也是通过 create()和 show()方法组合 AlertDialog 的各部分并显示。

(5) 运行程序,效果如图 4.2 所示。

图 4.2　　Dialog 运行效果

4.1.3　菜单

Menu(菜单)在 Android 应用中很常见,通常用于对内容或事件进行分类及触发不同的操作。菜单通常有很多子菜单,入口通常是一个菜单图标,单击入口后出现子菜单,用户根据需求选择相应的菜单处理事件。Android 中的菜单主要分为三类:选项菜单、上下文菜单、弹出菜单。Android 中提供了两种创建菜单的方法,分别是通过 Java 代码和使用菜单资源文件创建菜单。下面针对后者进行详细介绍。

1. 菜单创建步骤

创建菜单的步骤如下。

(1) 创建菜单资源文件。

对于所有菜单类型,Android 提供了标准的 XML 格式来定义菜单项,因此需要先创建菜单资源文件,创建的位置是在项目的 res/menu/目录下,由于 menu 目录系统默认不创建,因此也需要手动创建该目录。

(2) 设计菜单选项。

创建好菜单资源文件后,需要设计菜单选项内容及样式。菜单资源文件的根元素默认是<menu></menu>标签,在此标签中,通过<item/>标签添加菜单项。在 item 标签中,可通过属性设置菜单项的标题样式,常用的属性如表 4.3 所示。该步骤根据需求选择性进行设置。

表 4.3 菜单常用属性

属　　性	功能描述	实　　例
android:id	设置菜单项 ID,即唯一标识	android:id="@+id/item_1" 菜单项 ID 为"item_1"
android:title	设置菜单项标题	android:title="添加" 菜单项标题为"添加"
android:alphabeticShortcut	设置菜单项字符快捷键	android:alphabeticShortcut="a" 菜单项字符快捷键为 a
android:numericShortcut	设置菜单项数字快捷键	android:numericShortcut="1" 菜单项数字快捷键为 1
android:icon	设置菜单项图标	android:icon="@drawable/ic_launcher_background" 菜单项图标为 ic_launcher_background
android:enabled	设置菜单项是否可用	android:enabled="true" 菜单项可用 android:enabled="false" 菜单项不可用
android:checkable	设置菜单项是否可选	android:checkable="true" 菜单项可选 android:checkable="false" 菜单项不可选
android:checked	设置菜单项是否选中	android:checked="true" 菜单项选中 android:checked="false" 菜单项未选中

(3) 创建菜单及设计菜单选项被选择时应做的处理。

在设计完菜单选项后,需要将其显示并添加响应菜单项选择的代码。实现这两步需要调用 Activity 类的方法,创建并显示菜单使用的是 onCreateOptionsMenu()方法,设计菜单选项被选择时应做的处理使用的是 onOptionsItemSelected()方法,具体如表 4.4 所示。

表 4.4 菜单常用方法

方　　法	作　　用	参数说明
onCreateOptionsMenu (Menu menu)	创建选项菜单	menu:Activity 的选项菜单对象 返回值:true 表示需要显示菜单,否则菜单不显示
onOptionsItemSelected (MenuItem item)	设计菜单选项被选择时应做的处理	item:被单击的菜单项 返回值: true 表示任务已经处理完 false 表示希望将菜单项单击的事件传递下去,继续触发其他处理 super.onOptionsItemSelected(item)表示系统默认返回 false

2. 菜单案例

1) 案例分析

（1）界面分析。本案例共有一个布局界面,一个菜单资源文件,菜单包括四个菜单选项："添加""删除""修改""退出"。单击前三个菜单选项时,分别显示一条 Toast 提示信息,单击最后一个菜单选项时,退出程序。

（2）设计思路。菜单通过在 res 目录下创建 menu 目录,再在 menu 目录下创建菜单资源文件,并通过属性设计菜单,同时调用 Activity 类的 onCreateOptionsMenu() 方法实现显示菜单需求；单击菜单选项的响应事件则通过调用 Activity 类的 onOptionsItemSelected() 方法实现需求。

2) 实现步骤

（1）创建一个新的工程,工程名为 MenuZhangsan。

（2）切换工程的 Project 项目结构。选择该模式下方的 app,依次展开,便看到工程的布局界面和工程的类文件,其中,activity_main.xml 是布局界面,MainActivity.java 为类文件。

（3）设计布局界面。双击 layout 文件夹下的 activity_main.xml 文件,便可打开布局编辑器,修改布局类型为 LinearLayout,添加方向属性为 vertical,本案例无须添加控件,代码如下。

```xml
<?xml version = "1.0" encoding = "utf-8"?>
<LinearLayout xmlns:android = "http://schemas.android.com/apk/res/android"
    xmlns:app = "http://schemas.android.com/apk/res-auto"
    xmlns:tools = "http://schemas.android.com/tools"
    android:layout_width = "match_parent"
    android:layout_height = "match_parent"
    android:orientation = "vertical"
    tools:context = ".MainActivity">

</LinearLayout>
```

（4）创建菜单资源文件。右击工程的 res 目录,在弹出的快捷菜单中依次选择 New→Directory 命令,弹出 New Directory 对话框,如图 4.3 所示。

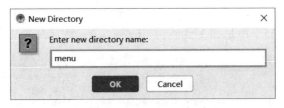

图 4.3 创建菜单目录

在 Enter new directory name 输入框中输入"menu",菜单目录创建完成,如图 4.4 所示。

右击 menu 目录,在弹出的快捷菜单中依次选择 New→Menu resource file 命令,弹出 New Menu Resource File 对话框,如图 4.5 所示。

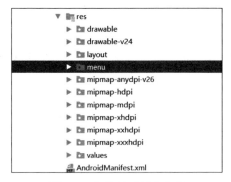

图 4.4 菜单目录创建完成

图 4.5 创建菜单资源文件

在 Enter a new file name 输入框中输入菜单资源名称"main",菜单资源文件创建完成,打开如图 4.6 所示窗口。

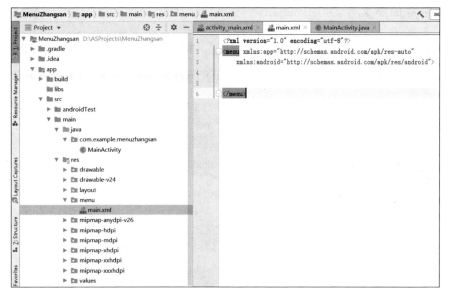

图 4.6 菜单资源文件创建完成

(5) 设计菜单选项。

创建好菜单资源文件后,需要设计菜单选项内容及样式。打开 main.xml 文件,在 <menu></menu>标签中添加如下代码。

```xml
<?xml version = "1.0" encoding = "utf-8"?>
<menu xmlns:app = "http://schemas.android.com/apk/res-auto"
    xmlns:android = "http://schemas.android.com/apk/res/android">
    <item
        android:id = "@+id/add_item"
        android:title = "添加" />
    <item
        android:id = "@+id/delete_item"
        android:title = "删除" />
```

```xml
    <item
        android:id = "@ + id/modify_item"
        android:title = "修改" />
    <item
        android:id = "@ + id/exit_item"
        android:title = "退出" />
</menu>
```

上述代码中,< menu > </menu >标签是自动生成的。在此标签中,通过< item/>标签添加菜单项,在此添加了四个菜单项,通过 android:title 属性设置菜单项的标题,android:id 属性设置菜单项的 ID。

(6) 创建菜单及设计菜单选项被选择时应做的处理。

在设计完菜单选项后,运行程序,会发现页面上没有菜单,这是因为菜单在类文件中还没有被创建和显示。打开 MainActivity.java 类文件,修改 MainActivity 的代码如下。

```java
public class MainActivity extends AppCompatActivity {
    @Override
    protected void onCreate(Bundle savedInstanceState) {
        super.onCreate(savedInstanceState);
        setContentView(R.layout.activity_main);
    }
    //第一步:添加选项菜单,让创建的菜单显示出来,重写 onCreateOptionsMenu()方法
    @Override
    public boolean onCreateOptionsMenu(Menu menu) {
        getMenuInflater().inflate(R.menu.main,menu); //获取当前菜单对象,加载菜单布局文件
        return true;//显示菜单
    }
    //第二步:给菜单项添加选中时的响应事件,重写 onOptionsItemSelected()方法
    @Override
    public boolean onOptionsItemSelected(@NonNull MenuItem item) {
        switch (item.getItemId()){
            case R.id.add_item:
                Toast.makeText(MainActivity.this,"您单击了添加菜单",Toast.LENGTH_LONG).show();
                break;
            case R.id.delete_item:
                Toast.makeText(MainActivity.this, "您单击了删除菜单", Toast.LENGTH_LONG).show();
                break;
            case R.id.modify_item:
                Toast.makeText(MainActivity.this, "您单击了修改菜单", Toast.LENGTH_LONG).show();
                break;
            case R.id.exit_item:
                finish();
                break;
            default:
        }
```

```
            return true;
      }
}
```

上述代码中,实现了创建并显示菜单以及响应菜单项的操作。第一步:创建并显示菜单是通过重写 onCreateOptionsMenu()方法实现的,在此,通过 getMenuInflater()方法获取当前菜单对象,通过 inflate()方法加载菜单布局文件 menu.main,返回值 true 表示将菜单显示出来。第二步:给菜单项添加选中时的响应事件是通过重写 onOptionsItemSelected()方法实现,在此,使用 switch 语句实现对不同菜单项的响应,选中"添加""删除""修改"菜单项时,显示一条 Toast,选中"退出"菜单项时,退出程序。

(7) 运行程序,效果如图 4.7 所示。

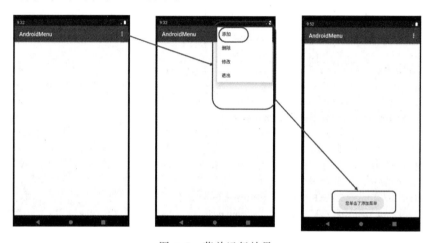

图 4.7 菜单运行效果

微课视频

4.1.4 下拉列表框

Spinner(下拉列表框)用于提供一系列可以选择的列表项供用户进行选择。下拉列表框通常在右侧会有一个小三角图标,单击该图标后出现下拉列表项,用户根据需求选择相应的列表项,选择完成后,被选中的列表项将出现在列表框中,同时下拉列表消失。

1. Spinner 下拉列表框步骤

创建 Spinner 下拉列表框的步骤如下。

(1) 添加 Spinner 控件。

Android 中提供了下拉列表框控件,在布局文件中通过<Spinner/>标记添加即可。Spinner 继承自 View,除了 View 的属性外,它自己也有些常用属性,如表 4.5 所示。

表 4.5 Spinner 常用属性

属 性	功能描述	实 例
android:entries	设置下拉列表选项	android:entries="@array/数组名称"
android:dropDownWidth	设置下拉列表框宽度	android:dropDownWidth="250dp"
android:popupBackground	设置菜单项标题	android:popupBackground="#F1D0D0"

其中，android:entries 用于指定列表项。如果在布局文件中不指定该属性，可以在 Java 代码中通过为其定义适配器的方式指定。本教材介绍第二种方法。

（2）创建适配器。

适配器（Adapter）是连接控件与数据的桥梁，它将各种数据以合适的形式显示在 View 中给用户查看。常用的适配器如下。

① ArrayAdapter：数组适配器，是最简单的适配器，适用于列表项中单一文本的情况。

② SimpleAdapter：简单适配器，适用于每一个列表项中含有多种子控件的情况。

③ SimpleCursorAdapter：简单游标适配器，适用于数据源为数据库的情况。

适配器提供了很多方法，由于下拉列表框中的数据通常是单一的文本，因此这里只对 ArrayAdapter 进行介绍。创建 ArrayAdapter 使用的是 ArrayAdapter(Context context, int textViewResourceId, T[] objects)构造方法，其中，第一个参数是上下文环境；第二个参数是设置数据的显示方式（布局）；第三个参数是要显示的数据，即适配器需要连接的数组数据。示例如下。

```
String[] arr1 = {"全部","纪录片","漫画","音乐","舞蹈","游戏"};
ArrayAdapter<String> adapter1 = new ArrayAdapter<String>
                    (MainActivity.this, android.R.layout.simple_spinner_item, arr1);
```

（3）显示 Spinner 数据及设计选项被选择时应做的处理。

这一步骤需要使用 Spinner 的常用方法，如 setAdapter()方法是设置 Spinner 使用的适配器，如表 4.6 所示。

表 4.6 ArrayAdapter 常用方法

方　　法	作　　用	参数说明
setAdapter(SpinnerAdapter adapter)	设置 Spinner 使用的适配器	adapter：适配器
setOnItemSelectedListener（AdapterView.OnItemSelectedListener listener）	监听 Spinner 选项被选中的事件	listener：监听器
onItemSelected(AdapterView<?> arg0，View arg1，int arg2，long arg3)	设置 Spinner 下拉选项被选中的处理	arg0：Spinner 控件 arg1：选中的列表项 arg2：选中项的位置 arg3：选中项 ID
onNothingSelected(AdapterView<?> parent)	设置 Spinner 下拉选项未被选中的处理	parent：没有任何选中条目的 AdapterView

2. Spinner 案例

1）案例分析

（1）界面分析。本案例共有一个布局界面，在该布局界面中添加一个下拉列表框控件，单击各下拉列表项，页面下方显示相应的 Toast。

（2）设计思路。布局界面中的下拉列表框控件通过添加属性的方式实现需求；下拉列表框的内容通过创建数组适配器实现需求，显示 Toast 通过 Spinner 常用方法实现需求。

2) 实现步骤

(1) 创建一个新的工程,工程名为 SpinnerZhangsan。

(2) 切换工程的 Project 项目结构。选择该模式下方的 app,依次展开,便看到工程的布局界面和工程的类文件,其中,activity_main.xml 是布局界面,MainActivity.java 为类文件。

(3) 设计布局界面。双击 layout 文件夹下的 activity_main.xml 文件,便可打开布局编辑器,修改布局类型为 LinearLayout,添加一个 Spinner 控件,代码如下。

```xml
<?xml version = "1.0" encoding = "utf-8"?>
<LinearLayout xmlns:android = "http://schemas.android.com/apk/res/android"
    xmlns:app = "http://schemas.android.com/apk/res-auto"
    xmlns:tools = "http://schemas.android.com/tools"
    android:layout_width = "match_parent"
    android:layout_height = "match_parent"
    tools:context = ".MainActivity">
    <Spinner
        android:id = "@+id/spinner_1"
        android:layout_width = "match_parent"
        android:layout_height = "wrap_content" />
</LinearLayout>
```

(4) 实现案例功能。通过创建数组适配器为下拉列表框添加数据,通过 Spinner 常用方法实现显示 Toast,打开 MainActivity.java 类文件,修改 MainActivity 的代码如下。

```java
public class MainActivity extends AppCompatActivity {
    //第一步:定义对象
    Spinner spin1;
    @Override
    protected void onCreate(Bundle savedInstanceState) {
        super.onCreate(savedInstanceState);
        setContentView(R.layout.activity_main);
        //第二步:绑定控件
        spin1 = findViewById(R.id.spinner_1);
        //第三步:设置控件 Spinner 的下拉列表项
        String[] arr1 = {"全部","纪录片","漫画","音乐","舞蹈","游戏"};
        //第四步:定义适配器(连接控件与数据的桥梁),告诉控件以什么样的方式显示数据
        ArrayAdapter<String> adapter1 = new ArrayAdapter<String>(MainActivity.this,
android.R.layout.simple_spinner_item,arr1);
        //第五步:让控件 Spinner 显示数据
        spin1.setAdapter(adapter1);
        //第六步:添加选择下拉选项后触发的事件
        spin1.setOnItemSelectedListener(new AdapterView.OnItemSelectedListener() {
            @Override
            public void onItemSelected(AdapterView<?> arg0, View arg1, int agr2, long agr3) {
                //选项被选中时执行的方法
                TextView txt1 = (TextView) arg1; //定义选中项
                String str1 = txt1.getText().toString(); //获取选中项的文本内容
```

```
                Toast.makeText(MainActivity.this,str1,Toast.LENGTH_LONG).show();  //显示 Toast
            }

            @Override
            public void onNothingSelected(AdapterView<?> parent) {
                //选项未被选中时执行的方法
            }
        });
    }
}
```

上述代码中，第一步：定义了 Spinner 对象。第二步：绑定控件，通过 findViewById()方法获取到 Spinner 控件。第三步：定义一个数组放置 Spinner 的下拉列表项。第四步，通过 ArrayAdapter()方法定义一个数组适配器并将第三步定义的数组与之相连，其中，第二个参数为 Android 自带布局 simple_spinner_item。第五步：通过 setAdapter()方法将适配器应用到 Spinner 中，显示下拉列表数据。第六步：通过 setOnItemSelectedListener()方法监听 Spinner 选项被选中的事件，其中，onItemSelected()方法设置 Spinner 下拉选项被选中的处理，在此方法中有四个参数，agr0 为 Spinner 控件，arg1 为选中的列表项，arg2 为选中项的位置，arg3 为选中项 ID，单击选项后通过 makeText()方法将选项的内容显示成 Toast，而未被选中时什么都不处理。

（5）运行程序，效果如图 4.8 所示。

图 4.8　Spinner 运行效果

微课视频

微课视频

4.2 实战案例——打地鼠小游戏

大家对于打地鼠游戏都不陌生,无论是在游戏机上还是在手机上,都可能玩过这样的游戏。在打地鼠游戏中,地鼠从地洞中出现,需要打到地鼠才能得分,打到地鼠后它会消失,同时在未知的地方再次出现,乐趣就在于在用时短的情况下打到尽可能多的地鼠。本案例就是利用项目4学过的知识,设计一个简单的打地鼠游戏,完成 Android 学习的第一款游戏设计。

4.2.1 界面分析

本案例包含一个界面,布局总体为 FrameLayout,布局上添加游戏背景图片和地鼠图片,运行效果如图4.9所示。

图4.9 打地鼠游戏界面效果图

4.2.2 实现思路

首先,要让地鼠从相应的洞穴中出现,需要创建一个二维数组,用于保存每个洞穴的位置坐标。然后,要控制地鼠随机出现,需要用到 Thread 线程和 Handler 消息处理机制。最后,记录打到地鼠的个数并显示,需要使用到 Toast 消息提示。

4.2.3 任务实施

【任务 4-1】 打地鼠游戏 UI 设计

(1) 创建一个新的工程,工程名为 MouseZhangsan。

(2) 切换工程的 Project 项目结构。选择该模式下方的 app,依次展开,便看到工程的布局界面和工程的类文件,其中,activity_main.xml 是布局界面,MainActivity.java 为类文件。

(3) 准备背景图片 background.png 和老鼠图片 mouse.png,将其复制粘贴到 app 目录结构中 res 目录下方的 drawable 文件夹下。

(4) 设计布局界面。双击 layout 文件夹下的 activity_main.xml 文件,便可打开布局编辑器,为方便定位,修改布局类型为 FrameLayout,设置背景图片为 background.png,并添加一个 ImageView 控件,设置图片为 mouse.png,代码如下。

```xml
<?xml version = "1.0" encoding = "utf-8"?>
<FrameLayout xmlns:android = "http://schemas.android.com/apk/res/android"
    xmlns:app = "http://schemas.android.com/apk/res-auto"
    xmlns:tools = "http://schemas.android.com/tools"
    android:layout_width = "match_parent"
    android:layout_height = "match_parent"
    android:background = "@drawable/background"
    tools:context = ".MainActivity">
    <ImageView
    android:id = "@+id/imageView_1"
    android:layout_width = "72dp"
    android:layout_height = "72dp"
    android:src = "@drawable/mouse"/>
</FrameLayout>
```

【任务 4-2】 打地鼠游戏功能实现

实现游戏功能。打开 MainActivity.java 类文件，修改 MainActivity 的代码如下。

```java
public class MainActivity extends AppCompatActivity {
/************第一步:定义变量、对象、洞穴坐标******************/
    private int i = 0; //记录打到的地鼠个数
    private ImageView mouse; //定义 mouse 对象
    private Handler handler; //声明一个 Handler 对象
    public int[][] position = new int[][]{
            {360, 250}, {750, 250}, {1200, 250},
            {300, 450}, {750, 450}, {1250, 430},
            {280, 650}, {750, 650}, {1300, 650}
    }; //创建一个表示地鼠位置的数组

@Override
    protected void onCreate(Bundle savedInstanceState) {
      super.onCreate(savedInstanceState);
      setContentView(R.layout.activity_main);

      /************第二步:绑定控件*****************/
      mouse = (ImageView) findViewById(R.id.imageView_1);

      /************第三步:实现地鼠随机出现*****************/
      //创建 Handler 消息处理机制
      handler = new Handler() {
          @Override
          public void handleMessage(@NonNull Message msg) {
              //需要处理的消息
              int index;
              if (msg.what == 0x101) {
                  index = msg.arg1; //获取位置索引值
                  mouse.setX(position[index][0]); //设置 X 轴坐标
                  mouse.setY(position[index][1]); //设置 Y 轴坐标(不包括程序名称栏)
                  mouse.setVisibility(View.VISIBLE); //设置地鼠显示
```

```java
            }
            super.handleMessage(msg);
        }
    };
    //创建线程
    Thread t = new Thread(new Runnable() {
        @Override
        public void run() {
            int index = 0; //定义一个记录地鼠位置的索引值
            while (!Thread.currentThread().isInterrupted()) {
                index = new Random().nextInt(position.length); //产生一个随机整数
                Message m = handler.obtainMessage(); //创建消息对象
                m.what = 0x101; //设置消息标志
                m.arg1 = index; //保存地鼠标位置的索引值<>
                handler.sendMessage(m); //发送消息通知 Handler 处理

                try {
                    Thread.sleep(new Random().nextInt(500) + 500);//休眠一段时间
                } catch (InterruptedException e) {
                    e.printStackTrace();
                }
            }
        }
    });
    t.start();
    /******** 第四步:实现单击地鼠后的事件:让地鼠不显示&显示消息 **********/
    //添加触摸 mouse 后的事件
    mouse.setOnTouchListener(new View.OnTouchListener() {
        @Override
        public boolean onTouch(View v, MotionEvent event) {
            v.setVisibility(View.INVISIBLE); //设置地鼠不显示
            i++;
            Toast.makeText(MainActivity.this, "打到[ " + i + " ]只地鼠!",
                    Toast.LENGTH_SHORT).show(); //显示消息提示框
            return false;
        }
    });

    }
}
```

上述代码中,第一步:定义变量 i 记录打到的地鼠个数,定义 mouse 对象和 handler 对象,定义洞穴坐标,坐标值是提前获取的。第二步:绑定控件,通过 findViewById()方法获取到 ImageView 控件。第三步:通过线程和消息处理机制实现地鼠随机出现,Android 中子线程不允许操作主线程中的组件,因此需要通过 Handler 消息处理机制来更新,在线程中定义一个记录地鼠位置的索引值 index,通过 Random().nextInt(position.length)产生一个随机数(0≤index<数组长度),再通过 Handler 的 obtainMessage()方法创建消息对象,设置消息标志 what 字段便于 Handler 判断,将地鼠位置的索引值 index 保存在 arg1

字段,接着通过 sendMessage()方法发送消息通知 Handler 处理。在 Handler 中,需要处理的消息写在 handleMessage()方法中,在此获取线程发送的索引值,通过 setX()和 setY()方法设置地鼠的随机坐标,随机是通过下标 index 控制的,同时设置地鼠可见。另外,还需调用 handleMessage()处理该消息。最后,启动线程,实现地鼠随机出现在洞穴中。第四步:实现单击地鼠后的事件,即让地鼠不显示并且显示消息,通过 setOnTouchListener()方法设置触摸地鼠后的事件,代码写在 onTouch()方法中,通过 setVisibility()方法设置地鼠的可见性,利用 Toast 显示打到的地鼠个数。

图 4.10 第一次运行打地鼠游戏效果图

【任务 4-3】 游戏运行与调试

运行程序,如图 4.10 所示。虽然游戏可以进行,但还不够完美,存在如下问题:首先,由于打地鼠游戏通常是在横屏模式下操作,因此如果游戏打开时默认切换为横屏会更合理;其次,游戏上方的项目名栏目和系统时间栏目的显示让游戏体验较差,我们希望游戏是全屏的;最后,地鼠没有从相应的洞穴中出来。针对这几个问题,需要一一调试并解决。

(1) 游戏启动时默认切换为横屏。

若希望游戏启动时默认切换为横屏,使用 setRequestedOrientation()方法,参数设置为横屏即可,在类文件的 onCreate()方法中添加如下代码。

```
setRequestedOrientation(ActivityInfo.SCREEN_ORIENTATION_LANDSCAPE);    //设置横屏模式
```

运行程序,将发现程序默认切换为横屏。

(2) 隐藏顶部项目名栏目和系统时间栏目。

要隐藏项目名栏目,需要打开 app/src/main/res/values 目录下的 styles 文件,将 style 标签下的 parent 属性由默认的 Theme.AppCompat.Light.DarkActionBar 修改为 Theme.AppCompat.Light.NoActionBar,表示继承无项目名栏目的主题,如下。

```
< style name = "AppTheme" parent = "Theme.AppCompat.Light.NoActionBar">
```

而隐藏系统时间栏目,则需要使用 getWindow()和 setFlags()方法,参数设置为全屏。操作如下:在类文件中 onCreate()方法中添加如下代码。

```
getWindow().setFlags(WindowManager.LayoutParams.FLAG_FULLSCREEN, WindowManager.LayoutParams.FLAG_FULLSCREEN);    //设置不显示顶部栏
```

(3) 地鼠从相应的洞穴中出来。

不同的模拟器屏幕大小不一致,因此案例中二维数组的数值不是唯一的,要想确定

当前模拟器中洞穴的位置,可以通过定义一个辅助控件 TextView,通过触摸它并将洞穴坐标输出到 Logcat() 中查看,操作如下。在布局文件中添加一个 TextView 控件,代码如下。

```xml
<?xml version="1.0" encoding="utf-8"?>
<FrameLayout xmlns:android="http://schemas.android.com/apk/res/android"
    xmlns:app="http://schemas.android.com/apk/res-auto"
    xmlns:tools="http://schemas.android.com/tools"
    android:layout_width="match_parent"
    android:layout_height="match_parent"
    android:background="@drawable/background"
    tools:context=".MainActivity">
    <ImageView
    android:id="@+id/imageView_1"
    android:layout_width="72dp"
    android:layout_height="72dp"
    android:src="@drawable/mouse"/>
<TextView
    android:id="@+id/info"
    android:layout_width="fill_parent"
    android:layout_height="fill_parent" />
</FrameLayout>
```

上述代码中,让该控件填充满整个屏幕,这样才能保证能在任何位置触摸到它。

接着,在类文件的主程序中添加如下代码。

```
/************ 获取洞穴位置 ***************/
//通过 Logcat 查看【注】:getRawY():触摸点距离屏幕上方的长度(此长度包括程序项目名栏的长度)

info1 = findViewById(R.id.info);
info1.setOnTouchListener(new View.OnTouchListener() {
    @Override
    public boolean onTouch(View v, MotionEvent event) {
        switch (event.getAction()) {
            case MotionEvent.ACTION_DOWN:
                float x = event.getRawX();
                float y = event.getRawY();
                Log.i("x:" + x, "y:" + y);
                break;
            default:
                break;
        }
        return false;
    }
});
```

上述代码中,将 info 控件与定义的 info1 绑定,通过 setOnTouchListener() 设置触摸 info 控件后的事件,代码写在 onTouch() 方法中,当动作为 ACTION_DOWN 即触摸时,通过 getRawX() 和 getRawY() 获取到触摸点的坐标,并输出到 Logcat 中。这里要注意的是,触摸时需要考虑到地鼠是有一定高度和宽度的,此坐标以其左上角的坐标值计算;另外,getRawY()计算的是触摸点距离屏幕上方的长度,如有项目名栏,此长度是包括程序项目名称栏高度的,而 setY()方法是不包括程序名称栏高度的,为防止坐标无法对齐的情况,建议尽量保证项目名栏是隐藏的。

最终 MainActivity.java 类文件的完整代码如下。

```java
public class MainActivity extends AppCompatActivity {
/************第一步:定义变量、对象、洞穴坐标******************/
    private int i = 0; //记录打到的地鼠个数
    private ImageView mouse; //定义 mouse 对象
    private TextView info1; //定义 info1 对象(用于查看洞穴坐标)
    private Handler handler; //声明一个 Handler 对象
    public int[][] position = new int[][]{
            {360, 250}, {750, 250}, {1200, 250},
            {300, 450}, {750, 450}, {1250, 430},
            {280, 650}, {750, 650}, {1300, 650}
    }; //创建一个表示地鼠位置的数组

    @Override
    protected void onCreate(Bundle savedInstanceState) {
      super.onCreate(savedInstanceState);
      setContentView(R.layout.activity_main);
      getWindow().setFlags(WindowManager.LayoutParams.FLAG_FULLSCREEN, WindowManager.LayoutParams.FLAG_FULLSCREEN); //设置不显示顶部栏
        setRequestedOrientation(ActivityInfo.SCREEN_ORIENTATION_LANDSCAPE); //设置横屏模式

        /************第二步:绑定控件*****************/
        mouse = (ImageView) findViewById(R.id.imageView_1);
        info1 = findViewById(R.id.info);

        /************获取洞穴位置*****************/
        info1.setOnTouchListener(new View.OnTouchListener() {
            @Override
            public boolean onTouch(View v, MotionEvent event) {
                switch (event.getAction()) {
                    case MotionEvent.ACTION_DOWN:
                        float x = event.getRawX();
                        float y = event.getRawY();
                        Log.i("x:" + x, "y:" + y);
                        break;
                    default:
                        break;
                }
                return false;
            }
        });

        /************第三步:实现地鼠随机出现*****************/
        //创建 Handler 消息处理机制
```

```java
            handler = new Handler() {
                @Override
                public void handleMessage(@NonNull Message msg) {
                    //需要处理的消息
                    int index;
                    if (msg.what == 0x101) {
                        index = msg.arg1; //获取位置索引值
                        mouse.setX(position[index][0]); //设置 X 轴坐标
                        mouse.setY(position[index][1]); //设置 Y 轴坐标(不包括程序名称栏)
                        mouse.setVisibility(View.VISIBLE); //设置地鼠显示
                    }
                    super.handleMessage(msg);
                }
            };
            //创建线程
            Thread t = new Thread(new Runnable() {
                @Override
                public void run() {
                    int index = 0; //定义一个记录地鼠位置的索引值
                    while (!Thread.currentThread().isInterrupted()) {
                        index = new Random().nextInt(position.length); //产生一个随机整数
                        Message m = handler.obtainMessage(); //创建消息对象
                        m.what = 0x101; //设置消息标志
                        m.arg1 = index; //保存地鼠标位置的索引值
                        handler.sendMessage(m); //发送消息通知 Handler 处理

                        try {
                            Thread.sleep(new Random().nextInt(500) + 500); //休眠一段时间
                        } catch (InterruptedException e) {
                            e.printStackTrace();
                        }
                    }
                }
            });
            t.start();
            /********* 第四步:实现单击地鼠后的事件:让地鼠不显示 & 显示消息 ************/
            //添加触摸 mouse 后的事件
            mouse.setOnTouchListener(new View.OnTouchListener() {
                @Override
                public boolean onTouch(View v, MotionEvent event) {
                    v.setVisibility(View.INVISIBLE); //设置地鼠不显示
                    i++;
Toast.makeText(MainActivity.this, "打到[ " + i + " ]只地鼠!", Toast.LENGTH_SHORT).show(); //显示消息提示框
                    return false;
                }
            });
        }
    }
```

至此，程序的问题调试完毕，此时运行程序，会发现问题都解决了，可以开始玩自己设计的游戏了。

项目小结

本项目主要讲解了 Android 的事件处理与交互实现程序设计以及打地鼠小游戏的开发。通过 Toast 和 Dialog 的学习，读者理解了 Android 中的交互实现机制；通过 Menu 和 Spinner 的学习，读者掌握了 Android 中更多常用控件的使用；最后，通过打地鼠小游戏，读者了解了线程消息处理技术在实际中的应用。本项目的完成，为前面所学的 Android 常用控件、布局、事件处理及消息提示等知识的巩固提供了帮助。

习题

1. 创建 Toast 的方法是（   ）。
   A．makeText()　　　B．show()　　　C．makeToast()　　　D．showToast()
2. 设置 Toast 位置的方法是（   ）。
   A．makeText()　　　B．show()　　　C．getGravity()　　　D．setGravity()
3. 设定 Dialog 内容使用的方法是（   ）。
   A．setIcon()　　　B．setTitle()　　　C．setMessage()　　　D．SetButton()
4. AlertDialog.Builder setPositiveButton()用来设定（   ）按钮。
   A．取消　　　B．关闭　　　C．确定　　　D．忽略
5. Dialog 中设定单选按钮使用的方法是（   ）。
   A．setMultiChoiceItems()　　　B．setSingleChoiceItems()
   C．setNegativeButton()　　　D．setNeutralButton()
6. 添加选项菜单，让创建的菜单显示出来需要重写的方法是（   ）。
   A．onCreateOptionsMenu()　　　B．onOptionsItemSelected()
   C．onItemSelected()　　　D．onCreatMenu()
7. 获取当前菜单对象的方法是（   ）。
   A．inflate()　　　B．getMenuInflater()
   C．getMenu()　　　D．getMenuInflate()
8. 显示 Spinner 需要使用到的适配器是（   ）。
   A．ArrayAdapter　　　B．SimpleAdapter
   C．SimpleCursorAdapter　　　D．Adapter
9. 添加 Spinner 选择下拉选项后触发事件的方法是（   ）。
   A．onItemSelected()　　　B．setOnItemSelectedListener()
   C．onOptionsItemSelected()　　　D．setOnItemSelected ()
10. 在 Android 的 Handler 实例中，下列表示发送 Handler 消息的方法是（   ）。
    A．message()　　　B．HandleMessage()
    C．dispatchMessage()　　　D．sendMessage()

项目 5

记忆的仓库——备忘录

【教学导航】

学习目标	（1）了解 SharedPreferences 实现简单的数据存储方法。 （2）理解 Android 中的文件存储方法。 （3）掌握摄像头调用和相册获取图片的方法。 （4）了解 Android 运行时权限设置。 （5）具备较复杂案例的设计开发能力。
教学方法	任务驱动法、理论实践一体化、探究学习法、分组讨论法
课时建议	8 课时

5.1 SharedPreferences 存储

Android 中常用的三大存储方式分别是 SharedPreferences 存储、文件存储和数据库存储。其中，SharedPreferences 存储在程序中有少量的数据需要保存时使用，它存储数据的格式很简单，采用键值对的方式来存储数据。文件存储主要用来存储资源文件，如一篇日记、一首歌等。数据库存储常用于当程序中有大量数据需要存储、访问时，就需要借助于数据库，Android 内置了 SQLite 数据库。本节将针对 SharedPreferences 存储进行讲解。

微课视频

5.1.1 SharedPreferences 简介

SharedPreferences 是 Android 平台上一个轻量级的存储类，常用于存储一些应用程序的配置参数，例如用户名、密码、是否打开同步、是否打开音响、自定义参数的设置等。SharedPreferences 中存储的数据是以 key-value（键值）对的形式保存在 XML 文件中，需要注意的是，SharedPreferences 中的 value 值只能是 float、int、long、boolean、String、StringSet 类型数据。

1. SharedPreferences 常用方法

SharedPreferences 提供了一系列的方法用于获取应用程序中的数据,具体如表 5.1 所示。

表 5.1 SharedPreferences 常用方法

属性	功能描述	实例
getSharedPreferences(String name, String mode)	获取 SharedPreference 的实例对象	SharedPreference sp = this.getSharedPreferences("myinfo",0)
getString(String key, String defValue)	获取 SharedPreferences 中指定的 key 对应的 String 值	getSharedPreferences("myinfo",0).getString("name","");
getInt(String key, int defValue)	获取 SharedPreferences 中指定的 key 对应的 int 值	getSharedPreferences("myinfo",0).getInt("age",0);
getBoolean(String key, boolean defValue)	获取 SharedPreferences 中指定的 key 对应的 boolean 值	getSharedPreferences("info",0).getBoolean("married",false);
getFloat(String key, float defValue)	获取 SharedPreferences 中指定的 key 对应的 float 值	getSharedPreferences("info",0).getFloat("weight",0);
getLong(String key, long defValue)	获取 SharedPreferences 中指定的 key 对应的 long 值	getSharedPreferences("info",0).getLong("birthdistance",0);

表 5.1 中,使用 SharedPreferences 类存储数据时,首先需要通过 context.getSharedPreferences(String name, int mode) 获取 SharedPreferences 的实例对象,其中 context 表示上下文环境,在 Activity 中可以直接使用 this 代表上下文,如果不是在 Activity 中则需要传入一个 Context 对象获取上下文。

getSharedPreferences(String name, String mode) 方法中,第一个参数 name 表示文件名,如果指定的文件不存在则会创建一个新文件,该文件存放在 data/data/<package name>/shared_prefs 目录下。第二个参数 mode 用于指定文件操作模式,目前只有 MODE_PRIVATE 这一种模式可选,指定该 SharedPreferences 数据只能被本应用程序读写,它是默认的操作模式,和直接传入 0 效果是相同的,其他几种模式从 Android 4.2 版本开始均已废弃。

getXxx(String, xxx value) 获取 SharedPreferences 数据里指定 key 对应的 value。如果该 key 不存在,则返回默认值 value。其中,xxx 可以是 boolean、float、int、long、String 等多种基本类型。

2. SharedPreferences.Editor 常用方法

SharedPreferences 对象本身只能获取数据,并不支持数据的存储和修改,数据的存储和修改需要通过 SharedPreferences.Editor 对象来实现。要想获取 Edit 实例对象,需要调用 SharedPreferences.Editor 的 edit() 方法,Editor 提供一系列方法来向 SharedPreferences 写入数据,如表 5.2 所示。

表 5.2 SharedPreferences.Editor 常用方法

方法	功能描述	实例
SharedPreferences.Editor edit()	创建一个 Editor 对象	SharedPreferences.Editor editor=this.getSharedPreferences("myinfo",0).edit();
SharedPreferences.Editor putString(String key, String value)	向 SharedPreferences 中存入指定 key 对应的 String 值	editor.putString("name","张三");
SharedPreferences.Editor putInt(String key, int value)	向 SharedPreferences 中存入指定 key 对应的 int 值	editor.putInt("age",20);
SharedPreferences.Editor putBoolean(String key, boolean value)	向 SharedPreferences 中存入指定 key 对应的 boolean 值	editor.putBoolean("married",false);
SharedPreferences.Editor putFloat(String key,float value)	向 SharedPreferences 中存入指定 key 对应的 float 值	editor.putFloat("weight",76.5f);（其中,f 表示 float）
SharedPreferences.Editor putLong(String key,long value)	向 SharedPreferences 中存入指定 key 对应的 long 值	editor.putLong("birthdistance",1546729204391);（其中,l 表示 long）
SharedPreferences.Editor remove(String key)	删除 SharedPreferences 中指定 key 对应的数据	editor.remove("weight");
boolean commit()	提交数据,返回 boolean 表明提交是否成功	editor.commit();
SharedPreferences.Editor apply()	提交数据,没有返回值	editor.apply();

SharedPreferences.Editor clear()	清除 SharedPreferences 中的所有数据	SharedPreferences.Editor editor1=getSharedPreferences("myinfo",0).edit();
		editor1.clear();
		editor1.commit();

微课视频

5.1.2 SharedPreferences 的使用

1. 将数据存储到 SharedPreferences 中

使用 SharedPreferences 存储数据时,需要先获取 SharedPreferences 对象,通过该对象的 edit()方法获取到 Editor 对象,然后通过 Editor 对象的相关方法存储数据,具体代码如下。

```
SharedPreferences.Editor editor = getSharedPreferences("myinfo",0).edit();   //获取编辑器 editor
editor.putString("name","张三");              //存入 String 类型数据
editor.putInt("age",20);                      //存入 int 类型数
editor.putBoolean("married",false);           //存入 boolean 类型数据
editor.putFloat("weight",76.8f);              //存入 float 类型数据
editor.putLong("birthdistance",1546729204391);//存入 long 类型数据
editor.commit();                              //提交数据,保存成功
```

2. 从 SharedPreferences 中读取数据

SharedPreferences 获取数据比较简单，只需要创建 SharedPreferences 对象，然后使用该对象获取相应 key 的值即可。具体代码如下。

```
String myname = getSharedPreferences("myinfo",0).getString("name","");   //获取用户名
int myage = getSharedPreferences("myinfo",0).getInt("age",0);            //获取年龄
float weight = getSharedPreferences("info",0).getFloat("weight",0);      //获取体重
boolean mymarried = getSharedPreferences("info",0).getBoolean("married",false);
                                                                         //获取婚姻状态
long mybirthdistance = getSharedPreferences("info",0).getLong("birthdistance",0);
                                                                         //获取出生到现在的时长
```

3. 删除 SharedPreferences 中的数据

SharedPreferences 删除数据与存储数据类似，同样需要先获取到 Editor 对象，然后通过该对象删除数据，最后提交，具体代码如下。

```
SharedPreferences.Editor editor1 = getSharedPreferences("myinfo",0).edit();
                                              //获取编辑器对象,取名为 editor1
editor1.remove("name");      //删除一条数据
editor1.clear();             //删除所有数据
editor1.commit();            //提交数据,保存成功
```

SharedPreferences 使用很简单，但一定要注意以下两点。

（1）存入数据和删除数据时，一定要在最后使用 editor.commit()方法提交数据。

（2）读取数据的 key 值与存入数据的 key 值的数据类型要一致，否则查找不到数据。

4. SharedPreferences 案例

1）案例分析

（1）界面分析。布局界面中有 5 个控件，其中两个 EditText 控件用于输入用户名和密码，复选框 CheckBox 供用户选择是否需要记住密码，"登录"按钮保存信息并跳转到第二个页面。

（2）设计思路。布局界面中的控件通过添加属性的方式达到用户需求，在登录按钮单击事件内部实现用户名、密码、复选框状态等数据的存储，下次启动 App 时，根据复选框的状态决定是否显示存储的用户名和密码。

2）实现步骤

（1）创建一个新的工程，工程名为 ZSSharedPreferences。

（2）切换工程的 Project 项目结构。选择该模式下方的 App，依次展开，便看到工程的布局界面和工程的类文件，其中，activity_main.xml 是布局界面，MainActivity.java 为类文件。

（3）在工程中添加一个新的页面。右击 com.example.zssharedpreferences 包→New→Activity→Empty Activity，会弹出一个创建活动的对话框，将活动命名为 SecondActivity，默认勾选 Generate Layout File 关联布局界面，布局界面名称为 activity_second，但不要勾选 Launcher Activity。单击 Finish 按钮，便完成了第二个页面的添加。

(4) 准备一张图片,图片名为 bgtwo.jpg,将其粘贴到 app 目录结构中 res 下方的 drawable 文件夹下,作为登录页面的背景图片。

(5) 修改布局界面。在 app 目录下的结构中,双击 layout 文件夹下的 activity_main.xml 文件,便可打开布局编辑器,切换到 Text 选项卡,输入代码如下。

```xml
<?xml version="1.0" encoding="utf-8"?>
<LinearLayout xmlns:android="http://schemas.android.com/apk/res/android"
    xmlns:app="http://schemas.android.com/apk/res-auto"
    xmlns:tools="http://schemas.android.com/tools"
    android:layout_width="match_parent"
    android:layout_height="match_parent"
    android:orientation="vertical"
    android:background="@drawable/bgtwo"
    tools:context=".MainActivity">
    <ImageView
        android:layout_width="wrap_content"
        android:layout_height="wrap_content"
        android:src="@mipmap/ic_launcher_round"
        android:layout_marginTop="150dp"
        android:layout_marginBottom="50dp"
        android:layout_gravity="center"/>
    <EditText
        android:id="@+id/editText_inputname"
        android:layout_width="match_parent"
        android:layout_height="wrap_content"
        android:hint="请输入用户名"
        android:textColor="#000000"
        android:textSize="25sp"
        android:textStyle="bold"/>
    <EditText
        android:id="@+id/editText_inputpwd"
        android:layout_width="match_parent"
        android:layout_height="wrap_content"
        android:hint="请输入密码"
        android:inputType="textPassword"
        android:textColor="#000000"
        android:textSize="25sp"
        android:textStyle="bold"/>
    <CheckBox
        android:id="@+id/checkBox_reme"
        android:layout_width="wrap_content"
        android:layout_height="wrap_content"
        android:text="记住密码"
        android:layout_gravity="right"
        android:layout_marginRight="10dp"
        android:textColor="#000000"
        android:textSize="20sp"
```

```xml
        android:textStyle = "bold"
        android:layout_marginTop = "10dp"
        android:layout_marginBottom = "10dp"/>
    <Button
        android:id = "@+id/button_login"
        android:layout_width = "match_parent"
        android:layout_height = "wrap_content"
        android:text = "登录"
        android:textColor = "#000000"
        android:textSize = "25sp"
        android:textStyle = "bold"/>
</LinearLayout>
```

上述代码中，根节点为线性布局，方向为 vertical，其内部放置了一个图片控件、输入用户名的 EditText 控件和输入密码的 EditText 控件，以及复选框 checkBox 控件，方便用户选择是否记住密码。最下方是 Button 控件，用来实现数据存储和跳转功能。

（6）实现第一个 Activity 页面的功能。打开 MainActivity.java 类文件，修改 MainActivity 的代码如下。

```java
public class MainActivity extends AppCompatActivity {
    //第一步:定义对象
    EditText edit_name,edit_pwd;
    CheckBox check_reme;
    Button btn_login;
    @Override
    protected void onCreate(Bundle savedInstanceState) {
        super.onCreate(savedInstanceState);
        setContentView(R.layout.activity_main);
        //第二步:绑定控件
        edit_name = findViewById(R.id.editText_inputname);
        edit_pwd = findViewById(R.id.editText_inputpwd);
        check_reme = findViewById(R.id.checkBox_reme);
        btn_login = findViewById(R.id.button_login);
        //第三步:按钮单击事件
        btn_login.setOnClickListener(new View.OnClickListener() {
            @Override
            public void onClick(View view) {
                //单击按钮将输入的用户名、密码、复选框的状态保存起来
                SharedPreferences.Editor editor = getSharedPreferences("myfile",0).edit();
                editor.putString("name", edit_name.getText().toString());
                editor.putString("pwd",edit_pwd.getText().toString());
                editor.putBoolean("st",check_reme.isChecked());
                editor.commit();
                Intent intent = new Intent(MainActivity.this, SecondActivity.class);
                startActivity(intent);
                finish();
            }
        });
```

```
            //第四步:如果选中了"记住密码"复选框,下一次启动,能够获取用户名和密码并显示出来
            String myname = getSharedPreferences("myfile",0).getString("name","");
            String mypwd = getSharedPreferences("myfile",0).getString("pwd","");
            Boolean myst = getSharedPreferences("myfile",0).getBoolean("st",false);
        if(myst == true){
            edit_name.setText(myname);
            edit_pwd.setText(mypwd);
            check_reme.setChecked(true);
        }else{
            edit_name.setText("");
            edit_pwd.setText("");
            check_reme.setChecked(false);
        }
    }
}
```

上述代码中,在按钮单击事件内部,首先使用 SharedPreferences 将输入的用户名、密码、复选框的状态存储起来,然后实现页面的跳转,即从当前页面跳转到第二个页面。下次启动时,从 SharedPreferences 中获取用户名、密码和复选框的状态等数据信息,根据获取的复选框的状态进行判断,如果为真,说明用户选中了"记住密码",此时将获取的用户名和密码显示到 EditText 控件上,并将复选框设置为选中。如果为假,说明用户没有选中"记住密码",EditText 控件内容设置为空,并将复选框设置为未选中。

(7) 运行程序,效果如图 5.1(a)所示,在界面中输入用户名和密码,选中"记住密码"复选框,单击"登录"按钮,将用户名和密码保存并跳转到第二个页面,再次运行程序,效果如图 5.1(b)所示。

(a)　　　　　　　　　　(b)

图 5.1　程序运行效果

（8）查看 SharedPreferences 中保存的数据。在百度中下载 RE 文件管理器，把下载好的 RE 文件管理器软件拖放到模拟器中，模拟器便会开始安装 RE 文件管理器，大约 5s 即可安装成功。在模拟器中打开 RE 文件管理器，挂载为可读写模式，根据文件的存储路径依次操作：data/data/<package name>/shared_prefs/文件名，单击 data 文件夹，在打开的模拟器页面中再次单击 data 文件夹，在列表中找到工程的包名 com.example.zssharedpreferences，在打开的页面中双击 Shared_prefs 将其打开，看到了保存的文件名 myfile.xml，双击查看保存的用户名和密码信息。

图 5.2　在模拟器中查看 SharedPreferences 中保存信息

5.2　Android 的文件存储

文件存储是 Android 中最基本的一种数据存储方式，它不会对存储的内容进行任何的格式化处理，它与 Java 中的文件存储类似，都是通过 I/O 流的形式把数据原封不动地存储到文档中。不同的是，Android 中的文件存储分为内部存储和外部存储，本节分别针对这两种存储方式进行详细的讲解。

5.2.1　内部存储

内部存储不是内存，它是手机中的一块存储区域，是系统本身和系统应用程序主要的数据存储位置。内部存储是指将应用程序中的数据以文件方式存储到设备的内部存储空间中，与 SharedPrefences 存储位置类似，文件存储的位置也位于 data/data/<package name>/files/目录下。内部存储方式所存储的文件被其所创建的应用程序私有，如果其他应用程序要操作本应用程序中的文件，需要设置权限。当创建的应用程序被卸载时，其内部存储文件也随之删除。

1. 文件内部存储常用方法

Android 提供了 openFileInput()和 openFileOutput()两种方法，向应用程序"读取"和"写入"数据流。按照流向分可以分为输入流和输出流，输入流只能从中读取数据，不能写入数据；输出流只能向其中写入数据，不能读取数据。写入和读取数据时，按照操作的数据单元分为字节流和字符流，字节流操作的数据单元是 8 位的字节，字符流操作的数据单元是 16 位的字节。

系统提供了文件内部存储的方法，如表 5.3 所示。

表 5.3 文件内部存储方法

方　　法	功能描述	实　　例
将数据存储到文件中的方法		
openFileOutput（String name，int model）	得到一个文件输出流对象 FileOutputStream 类向文件中写入数据	FileOutputStream fout = openFileOutput("zs.txt",0)
write(int b) throws IOException	一次写一个字节,b 表示要写入的字节	fout.write(edit_input.getText().toString().getBytes())
write（byte[ ] b）throws IOException；	一次写一个字节数组	
write(byte[ ] b, int off, int len) throws IOException；	一次写一部分字节数组	
close()throws IOException；	关闭此文件输出流并释放与此流有关的所有系统资源	fout.close()
从文件中读取数据的方法		
openFileInput（String name）	得到一个文件输入流对象 FileInputStream 类 从某个文件中读取字节	FileInputStream fin = openFileInput("zs.txt");
public int read() throws IOException	从此输入流中读取一个数据字节	
public int read（byte[ ] b）throws IOException	从此输入流中将最多 b.length 个字节的数据读入一个字节数组中	byte[ ] arr = new byte[fin.available()]; fin.read(arr);
public int read(byte[ ] b, int off, int len) throws IOException	从此输入流中将最多 len 个字节的数据读入一个字节数组中	
public voidclose() throws IOException	关闭此文件输入流并释放与此流有关的所有系统资源	fout.close();
删除文件的方法		
deleteFile(filename)	可以根据文件名,删除文件	deleteFile("zs.txt");

上述方法中,openFileOutput()得到一个文件输出流对象 FileOutputStream,用于打开应用程序中对应的输出流,将数据存储到指定的文件中。其中,参数 name 表示文件名；mode 表示文件的操作模式,有两种模式可选：MODE_PRIVATE 和 MODE_APPEND,其中,MODE_PRIVATE 是默认的操作模式,表示当指定同名文件的时候,所写入的内容会覆盖原文件中的内容,而 MODE_APPEND 是追加模式,表示如果该文件已存在,就往文件里面追加内容,不存在就创建新的文件。openFileInput()得到一个文件输入流对象 FileInputStream,用于打开应用程序中对应的输入流,读取文件中的内容并显示出来。

Android 中文件读写的原理：首先,所有文件的存储都是字节的存储。其次,在磁盘上保留的并不是文件的字符,而是先把字符编码转换成字节,再存储这些字节到磁盘。最后,在读取文件特别是文本文件时,也是一个字节一个字节地读取以形成字节序列。因此,在存储数据时,需要将字符串类型的数据转换成二进制字节流进行保存。在读取数据时,则需将二进制字节流转换为字符串再显示出来。

2. 文件内部存储案例

1) 案例分析

（1）界面分析。布局界面中有三个控件：TextView 控件用来显示评论人，EditText 控件用来输入评论内容，Button 控件用来提交评论，提交后评论内容以文件方式保存到内存中。

（2）设计思路。布局界面中的控件通过添加属性的方式达到用户需求，在按钮单击事件中，调用文件内部存储的方法，将文件保存到内存中，下次运行程序时，读取内存中的文件内容并显示出来，用户可输入新的评论内容并提交，达到用户的需求。

2) 实现步骤

（1）创建一个新的工程，工程名为 ZSFileSave。

（2）切换工程的 Project 项目结构。选择该模式下方的 app，依次展开，便可看到工程的布局界面和工程的类文件，其中，activity_main.xml 是布局界面，MainActivity.java 为类文件。

（3）准备一张图片，图片名为 filesavebg.jpg，将其粘贴到 app 目录结构中 res 下方的 drawable 文件夹下，图片作为评论页面的背景图片。

（4）修改布局界面。在 app 目录下的结构中，双击 layout 文件夹下的 activity_main.xml 文件，便可打开布局编辑器，切换到 Text 选项卡，输入代码如下。

```xml
<?xml version = "1.0" encoding = "utf-8"?>
<LinearLayout xmlns:android = "http://schemas.android.com/apk/res/android"
    xmlns:app = "http://schemas.android.com/apk/res-auto"
    xmlns:tools = "http://schemas.android.com/tools"
    android:layout_width = "match_parent"
    android:layout_height = "match_parent"
    android:orientation = "vertical"
    android:background = "@drawable/filesavebg"
    tools:context = ".MainActivity">
    <TextView
        android:layout_width = "wrap_content"
        android:layout_height = "wrap_content"
        android:text = "贝壳评论:"
        android:textSize = "40sp"
        android:layout_marginBottom = "70dp"
        android:layout_marginTop = "20dp"/>
    <EditText
        android:id = "@+id/editText_input1"
        android:layout_width = "match_parent"
        android:layout_height = "wrap_content"
        android:layout_margin = "5dp"
        android:layout_weight = "1"
        android:background = "#A4E49C"
        android:gravity = "top"
        android:hint = "请在此留下你的内容"
        android:textSize = "20sp"
        android:textColor = "#000000"            />
```

```xml
<Button
    android:id = "@+id/button_sava1"
    android:layout_width = "match_parent"
    android:layout_height = "wrap_content"
    android:text = "提交评论"
    android:textSize = "30sp"
    android:textStyle = "bold"
    android:background = "#009688"
    android:textColor = "#ffffff"/>
</LinearLayout>
```

上述代码中，根节点为线性布局，方向为 vertical，其内部放置了一个 TextView 控件用来显示发表评论的用户名，使用 EditText 控件输入评论内容，最下方是 Button 控件，用来实现文件存储功能。

（5）实现提交评论内容的功能。打开 MainActivity.java 类文件，修改 MainActivity 的代码如下。

```java
public class MainActivity extends AppCompatActivity {
    //第一步:定义对象
    private EditText edit_input;
    private Button btn_save;
    @Override
    protected void onCreate(Bundle savedInstanceState) {
        super.onCreate(savedInstanceState);
        setContentView(R.layout.activity_main);
        //绑定控件
        initView();
        //按钮单击事件,把输入的内容保存到文件里面
        onbuttonClick();
        //下次打开页面时,读取文件内容,显示出来
        readFile();
        //完善工程,自动显示当前时间
        displaytime();
    }
    //第二步:绑定控件
    private void initView() {
        edit_input = findViewById(R.id.editText_input1);
        btn_save = findViewById(R.id.button_sava1);
    }
    //第三步:按钮单击事件
    private void onbuttonClick() {
        btn_save.setOnClickListener(new View.OnClickListener() {
            @Override
            public void onClick(View view) {
                //把输入的内容保存到文件里面
                try {
                    //Context 类中提供了一个方法 openFileOutput 可以实现向文件中写入数据
                    //参数:("文件名", 文件的操作模式:覆盖/追加),路径:data/data/包名/files/
```

```java
                FileOutputStream fout = openFileOutput("zs.txt",0);
                fout.write(edit_input.getText().toString().getBytes());
                                                        //数据写入文件中
                fout.close(); //关闭文件输出流
Toast.makeText(MainActivity.this,"提交成功",Toast.LENGTH_SHORT).show();
                                                        //弹出提示,提交成功
            } catch (FileNotFoundException e) {
                e.printStackTrace();
            } catch (IOException e) {
                e.printStackTrace();
            }
        }
    });
}
//第四步:下次打开页面时,读取文件内容,显示出来
private void readFile() {
    try {
        //调用 Context 类的 openFileInput 方法得到一个文件输入流对象
        FileInputStream fin = openFileInput("zs.txt");
        byte[] arr = new byte[fin.available()];
                        //创建一个数组缓冲区,用来存放读取的很多字节数据
        fin.read(arr); //数据的读取
        fin.close(); //关闭输入流
        edit_input.setText(new String(arr)); //数据显示出来,byte 类型的数据转为字符串
    } catch (FileNotFoundException e) {
        e.printStackTrace();
    } catch (IOException e) {
        e.printStackTrace();
    }
}
//第五步:完善工程:自动显示当前时间
private void displaytime() {
    Time time = new Time(); //创建一个时间类,注意导入包 import android.text.format.Time
    time.setToNow(); //获取系统当前时间
    edit_input.append("\n\n" + time.year + "年" + (time.month + 1) + "月" + time.monthDay
+ "日" + "\n");//把获取的年月日显示出来
    edit_input.setSelection(edit_input.getText().length());    //将光标移动至文字末尾
    }
}
```

上述代码中,考虑到程序越来越复杂,onCreate()方法中的代码量会越来越多,为了避免代码混乱,根据程序实现的功能自定义几个方法,这种方式不仅可以让 onCreate()内部的代码结构清晰,而且独立的方法有利于程序的复用。方法命名规范一般由两个或多个单词组成,从第二个单词开始首字母大写。

从功能实现角度,在按钮单击事件中,通过 openFileOutput 文件输出流来保存数据到 zs.txt 文件中,由于从 EditText 控件获取用户输入的内容为字符串类型,而所有文件的存储都是采用字节存储的,因此采用 getBytes()方法将字符串转换为二进制字节流,再调用 write()方法将一个一个字节写入文件进行存储。下次运行程序,用户希望原来存储的数据

可以直接显示,此时需要调用 openFileInput 文件输入流从 zs.txt 文件中读取内容。首先建立数据存储空间即字节类型的数组 arr,然后调用 read()方法开始读取数据,并将读取的数据存放到数组 arr 中,最后将数组中的字节流转换为字符串并显示在 EditText 控件上。为了提升用户体验,在重新运行 App 时,自动显示系统时间。

(6) 运行程序,效果如图 5.3 所示。

(7) 查看保存到内存中的评论内容。在模拟器中打开 RE 文件管理器,挂载为可读写模式,根据文件的存储路径依次操作:data/data/< package name >/files/文件名,单击 data,在打开的模拟器页面中再次单击 data,在列表中找到工程的包名 com.example.zsfilesave,在打开的页面中显示了刚刚保存的文件 zs.txt,如图 5.4 所示。双击 zs.txt 将其打开,看到了保存的评论内容,如图 5.5 所示。

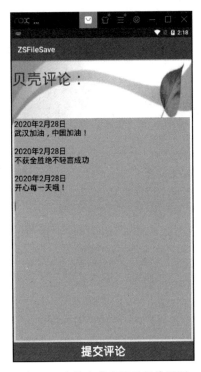

图 5.3 文件内部存储运行效果图

图 5.4 查看内部存储文件的保存位置

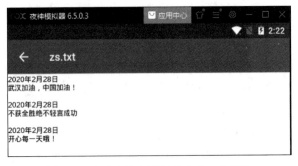

图 5.5 查看文件内部存储中保存的信息

5.2.2 外部存储

微课视频

手机的内部存储通常不会很大,一旦手机的内部存储容量被用完,可能会出现手机无法使用的情形。对于开发者来说,不宜存储视频等大文件,而适合存储一些小文件。例如,常用的 SharedPreferences 和 SQLite 数据库都是存储在内部存储中的,不会占用太大的空间。

1. 外部存储简介

外部存储是指将文件存储到一些外围设备上,如 SD 卡或者设备内嵌的存储卡,属于永久性的存储方式。外部存储的文件可以被其他应用程序所共享,当将外围存储设备连接到计算机时,这些文件可以被浏览、修改和删除,因此这种方式不安全。

由于外围存储设备可能被移除、丢失或者处于其他状态,因此在使用外围设备之前,建议用户使用 Environment.getExternalStorageState()方法来确认外围设备是否可用,当外围设备可用并且具有读写权限时,就可以通过 FileInputStream、FileOutputStream 或者 FileReader、FileWrite 对象来读写外围设备中的文件。

2. 外部存储调用的方法

访问外部存储的方法如表 5.4 所示。

表 5.4 获取外部存储的方法

获取路径的方法	功能描述	实例和存储路径
getExternalStorageDirectory()	主要的外部存储目录/mnt/sdcar 或者/storage/emulated/0 或者/storage/sdcard0	String path=Environment.getExternalStorageDirectory()+ "/"+"zs.txt"; /storage/emulated/0/zs.txt(存储路径)
context.getExternalCacheDir()	应用在外部存储上的缓存目录	Stringpath = MainActivity.this.getExternalCacheDir()+ "/"+"zs.txt"; /storage/emulated/0/Android/data/com.example.filesavetest/cache/zs.txt(存储路径)
context.getExternalFilesDir(Environment.DIRECTORY_MOVIES)	应用在外部存储上的目录,其中,movies 可更换为任意目录	Stringpath = MainActivity.this.getExternalFilesDir(Environment.DIRECTORY_MOVIES)+ "/"+ "zs.txt"; /storage/emulated/0/Android/data/com.example.filesavetest/files/Movies/ zs.txt(存储路径,文件夹可改为 MUSIC、DCIM、PICTURES 等)
context.getFilesDir()	获取当前程序路径应用在内存上的目录	Stringpath = MainActivity.this.getFilesDir()+ "/"+"zs.txt"; /data/data/com.example.filesavetest/files/zs.txt(存储路径)
context.getCacheDir()	应用在内存上的缓存目录	Stringpath = MainActivity.this.getExternalCacheDir()+ "/"+"zs.txt"; /storage/emulated/0/Android/data/com.example.filesavetest/cache/zs.txt(存储路径)

在表 5.4 中,前三行列举了获取外部存储路径调用的方法,后两行列举了获取当前程序

路径应用在内存上目录调用的方法。在 Android 开发中，用户可根据程序需求自行将文件存放到需要的位置。

3. 文件外部存储案例

1）案例分析

（1）界面分析。布局界面中有四个控件，TextView 控件用来显示标题，EditText 控件用来输入内容，两个 Button 控件用来保存内容和删除内容，保存后内容以文件方式保存到外存 SD 卡中。

（2）设计思路。布局界面中的控件通过添加属性的方式达到用户需求，在"保存"按钮单击事件内部，调用文件外部存储的方法，将文件保存到 SD 卡中，下次运行程序时，读取 SD 卡中的文件内容并显示出来。单击"删除"按钮，删除文件内容，达到用户的需求。

2）实现步骤

（1）创建一个新的工程，工程名为 ZSFileSaveTest。

（2）切换工程的 Project 项目结构。选择该模式下方的 app，依次展开，便看到工程的布局界面和工程的类文件，其中，activity_main.xml 是布局界面，MainActivity.java 为类文件。

（3）准备一张图片，图片名为 filesavebgt.jpg，将其粘贴到 app 目录结构中 res 下方的 drawable 文件夹下，图片作为页面的背景图片。

（4）修改布局界面。在 app 目录下的结构中，双击 layout 文件夹下的 activity_main.xml 文件，便可打开布局编辑器，切换到 Text 选项卡，输入代码如下。

```xml
<?xml version = "1.0" encoding = "utf-8"?>
<LinearLayout xmlns:android = "http://schemas.android.com/apk/res/android"
    xmlns:app = "http://schemas.android.com/apk/res-auto"
    xmlns:tools = "http://schemas.android.com/tools"
    android:layout_width = "match_parent"
    android:layout_height = "match_parent"
    android:orientation = "vertical"
    android:background = "@drawable/filesavebgt"
    tools:context = ".MainActivity">
    <TextView
        android:layout_width = "match_parent"
        android:layout_height = "wrap_content"
        android:text = "随手记事"
        android:textSize = "30sp"
        android:textColor = "#000000"
        android:textStyle = "bold"
        android:gravity = "center"
        android:layout_marginBottom = "10dp"
        android:layout_marginTop = "80dp"/>
    <View
        android:layout_width = "match_parent"
        android:layout_height = "5dp"
        android:background = "#000000"/>
    <EditText
        android:id = "@+id/editText_input1"
```

```xml
        android:layout_width = "match_parent"
        android:layout_height = "wrap_content"
        android:layout_margin = "5dp"
        android:layout_weight = "1"
        android:background = "#DF9CE4"
        android:gravity = "top"
        android:hint = "请在此留下你的内容"
        android:textSize = "18sp"
        android:textColor = "#000000"            />
    <View
        android:layout_width = "match_parent"
        android:layout_height = "8dp"
        android:background = "#000000"/>
    <LinearLayout
        android:layout_width = "match_parent"
        android:layout_height = "wrap_content"
        android:orientation = "horizontal"
        android:layout_marginTop = "3dp">
        <Button
            android:id = "@+id/button_sava1"
            android:layout_width = "wrap_content"
            android:layout_height = "wrap_content"
            android:text = "保存"
            android:textSize = "30sp"
            android:textStyle = "bold"
            android:background = "#FF5722"
            android:textColor = "#ffffff"
            android:layout_weight = "1"/>
        <Button
            android:id = "@+id/button_delete"
            android:layout_width = "wrap_content"
            android:layout_height = "wrap_content"
            android:text = "删除"
            android:textSize = "30sp"
            android:textStyle = "bold"
            android:background = "#E91E63"
            android:textColor = "#ffffff"
            android:layout_weight = "1"
            android:layout_marginBottom = "2dp"/>
    </LinearLayout>
</LinearLayout>
```

（5）实现外部文件存储与读取功能。重点关注向外围设备（SD卡）中存储数据和从外围设备（SD卡）中读取数据的示例代码，具体如下。

```java
public class MainActivity extends AppCompatActivity {
    private static final String TAG = "MainActivity";
    //第一步:定义对象
    private EditText edit_input1;
    private Button btn_save1;
```

```java
    private String path;
    private File file;
    private String filename = "note.txt";
    @Override
    protected void onCreate(Bundle savedInstanceState) {
        super.onCreate(savedInstanceState);
        setContentView(R.layout.activity_main);
        //绑定控件
        initView();
        //按钮单击事件,把输入的内容保存到文件里面
        onbuttonClick();
        //下次打开页面时,读取文件内容,显示出来
        readFile();
        //完善工程,自动显示当前时间
        displaytime();
    }
    //第二步:绑定控件
    private void initView() {
        edit_input1 = findViewById(R.id.editText_input1);
        btn_save1 = findViewById(R.id.button_sava1);
    }
    //第三步:按钮单击事件
    private void onbuttonClick() {
        btn_save1.setOnClickListener(new View.OnClickListener() {
            @Override
            public void onClick(View view) {
                //把输入的内容保存到文件里面
                try {
                    //利用外部 getExternalStorageDirectory 获取根目录,直接在后面加上想创建的文件
                    path = Environment.getExternalStorageDirectory() + "/" + filename; //路径是:/storage/emulated/0/ note.txt
                    file = new File(path); //创建文件
                    if(!file.exists()){ //文件若不存在则调用 file.createNewFile()
                        file.createNewFile();
                    }
                    FileOutputStream fout = openFileOutput(filename,0); //向文件中写内容,自然要创建文件
                                                                        //输出流的操作
                    fout.write(edit_input1.getText().toString().getBytes()); //数据写入文件中,调用输出
                                                                              //流的 write 方法
                    fout.close(); //关闭文件输出流
                    Toast.makeText(MainActivity.this,"提交成功",Toast.LENGTH_SHORT).show(); //弹出提示,
                                                                                              //提交成功

                } catch (FileNotFoundException e) {
                    e.printStackTrace();
                } catch (IOException e) {
                    e.printStackTrace();
                }
            }
        });
```

```java
        btn_delete.setOnClickListener(new View.OnClickListener() {
            @Override
            public void onClick(View view) {
                deleteFile(filename); //删除文件
    Toast.makeText(MainActivity.this,"删除成功",Toast.LENGTH_SHORT).show(); //弹出提示,
                                                                           //删除成功
            }
        });
    }
    //第四步:下次打开页面时,读取文件内容,显示出来
    private void readFile() {
        //检查 SD 卡的状态,Environment.MEDIA_MOUNTED 表示 SD 卡在手机上正常使用状态
        if(Environment.getExternalStorageState().equals(Environment.MEDIA_MOUNTED)){
            //SD 卡可用,可以开始读取数据
            try {
FileInputStream fin = openFileInput(filename); //调用 Context 类 openFileInput 方法得到一个
                                                //文件输入流对象
byte[] arr = new byte[fin.available()]; //创建一个数组缓冲区,用来存放读取的数据
                fin.read(arr); //数据的读取
                fin.close(); //关闭输入流
                edit_input1.setText(new String(arr));//数据显示出来,byte 类型的数据转
                                                     //为字符串
            } catch (FileNotFoundException e) {
                e.printStackTrace();
            } catch (IOException e) {
                e.printStackTrace();
            }
        }else{
            //SD 卡不可用
            Toast.makeText(MainActivity.this,"SD 卡不可用",Toast.LENGTH_SHORT).show();
        }
    }
    //第五步:完善工程:自动显示当前时间
    private void displaytime() {
        Time time = new Time(); //创建一个时间类,导入包 import android.text.format.Time;
        time.setToNow(); //获取系统当前时间
        edit_input1.append("\n\n" + time.year + "年" + (time.month + 1) + "月" + time.monthDay
+ "日" + "\n");//把获取的年月日显示出来
        edit_input1.setSelection(edit_input.getText().length()); //将光标移动至文字末尾
    }
}
```

上述代码中,向外围设备(SD 卡)中存储数据时,使用到了 Environment.getExternalStorageDirectory()方法,该方法用于获取 SD 卡的路径。针对不同手机厂商 SD 卡根目录也不完全一致,用这种方法可以避免把路径写死而找不到 SD 卡。在该路径下创建文件,运行时文件若不存在则调用 file.createNewFile()方法创建文件,之后创建文件输出流对象,调用 write()方法向文件中写内容。

从外围设备(SD 卡)中读取数据时,需要判断外围设备是否可用,Environment.

getExternalStorageState()方法用来确认外围设备是否可用,Environment. MEDIA_MOUNTED 表示 SD 卡处于正常使用状态,当外围设备可用并且具有读写权限时,通过 FileInputStream 对象读取指定目录下的文件。

(6) 声明权限。Android 系统为了保证应用程序的安全性做了相应规定,如果程序需要访问系统的一些关键信息,如读写 SD 卡、访问 Internet、打开相机、访问相册、访问通讯录等,必须在工程的配置文件 AndroidManifest. xml 中声明权限才可使用,否则程序运行时会直接崩溃。这里操作 SD 卡的信息就是系统中比较关键的信息,因此需要在清单文件的< manifest >节点下配置权限信息,切换工程的 Project 项目结构。选择该模式下方的 app→src→main→AndroidManifest. xml 依次展开,双击打开 AndroidManifest. xml 配置文件,具体代码如下。

```
< manifest xmlns:android = "http://schemas.android.com/apk/res/android"
    xmlns:tools = "http://schemas.android.com/tools"
    package = "com.example.filesavetest">
< uses - permission android:name = "android.permission.WRITE_EXTERNAL_STORAGE"/>
< uses - permission android:name = "android.permission.READ_EXTERNAL_STORAGE"/>
< application
    ...
```

上述代码指定了当前 App 程序具有读写外部 SD 卡的权限,因此应用程序便可以操作 SD 卡中的数据。

(7) 运行程序,效果如图 5.6 所示。

(8) 查看保存到外存中的随手记事内容。在模拟器找到文件管理器,双击打开文件管理器,根据文件的存储路径依次操作:/storage/emulated/0/note.txt 文件名,单击 storage 文件夹,打开的模拟器页面中单击 emulated 文件夹,单击 0 文件夹,在打开的页面中显示了刚刚保存的文件 note.txt,如图 5.7 所示。双击 note.txt 将其打开,便可查看随手记事的文件内容。

图 5.6　文件外部存储运行效果图

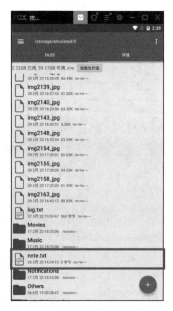
图 5.7　查看外部存储文件的保存位置

5.3 调用摄像头和相册

在使用 QQ 或者微信的时候经常要和别人分享图片,这些图片可以是用手机摄像头拍的,也可以是从相册中选取的。类似这样的功能太常见了,那么本节就学习一下调用摄像头和相册方面的知识。

微课视频

5.3.1 调用摄像头拍照

现在很多应用都会要求用户上传一张图片来作为头像,这时打开摄像头拍张照片是最简单快捷的,下面通过一个例子来学习一下,如何才能在应用程序中调用手机的摄像头进行拍照。

微课视频

1. 案例分析

(1) 界面分析。布局界面中有两个控件,其中,Button 控件用来打开相机拍照,ImageView 控件用来把拍摄的照片显示出来。

(2) 设计思路。布局界面中的控件通过添加属性的方式达到用户需求,单击拍照按钮时利用隐式跳转打开系统相机,并将拍摄的照片存放到指定的路径下,拍照结束后通过系统回调将拍摄的照片显示出来,达到用户的需求。

2. 实现步骤

(1) 创建一个新的工程,工程名为 ZSPhoto。

(2) 切换工程的 Project 项目结构,选择该模式下方的 app,依次展开,便看到工程的布局界面和工程的类文件,其中,activity_main.xml 是布局界面,MainActivity.java 为类文件。

(3) 准备一张图片,图片名为 zsphotobg.jpg,将其粘贴到 app 目录结构中 res 下方的 drawable 文件夹下,作为页面的背景图片。

(4) 设计布局界面。双击 layout 文件夹下的 activity_main.xml 文件,便可打开布局编辑器,修改布局类型为 LinearLayout,添加方向属性为 vertical。依次添加两个控件,代码如下。

```xml
<?xml version = "1.0" encoding = "utf-8"?>
<LinearLayout xmlns:android = "http://schemas.android.com/apk/res/android"
    xmlns:app = "http://schemas.android.com/apk/res-auto"
    xmlns:tools = "http://schemas.android.com/tools"
    android:layout_width = "match_parent"
    android:layout_height = "match_parent"
    android:orientation = "vertical"
    android:background = "@drawable/zsphotobg"
    tools:context = ".MainActivity">
    <ImageView
        android:id = "@+id/imageView_cam"
        android:layout_width = "match_parent"
        android:layout_height = "200dp"
```

```xml
            android:src = "@mipmap/ic_launcher_round"
            android:layout_gravity = "center"          />
    <Button
        android:id = "@+id/button_cam"
        android:layout_width = "match_parent"
        android:layout_height = "wrap_content"
        android:textColor = "#000000"
        android:text = "拍照"
        android:textSize = "30sp"/>
</LinearLayout>
```

上述代码中，Button 控件用来打开系统相机，ImageView 控件用来显示拍摄的图片。

(5) 按钮单击事件处理。打开 MainActivity.java 类文件，修改 MainActivity 的代码，具体如下。

```java
public class MainActivity extends AppCompatActivity {
    private static final String TAG = "MainActivity";
    //定义对象
    private ImageView img_camer;
    private Button btn_camer;
    //定义一个路径：①拍照的临时路径；②显示照片的最终路径
    String tmp_path,disp_path;
    @Override
    protected void onCreate(Bundle savedInstanceState) {
        super.onCreate(savedInstanceState);
        setContentView(R.layout.activity_main);
        //绑定控件
        initView();
        //"拍照"按钮单击事件
        btnCamOnclick();
        //拍完照之后，接收照片并显示，系统回调
    }
    //第一步:绑定控件
    private void initView() {
        img_camer = findViewById(R.id.imageView_cam);
        btn_camer = findViewById(R.id.button_cam);
    }
    //第二步:拍照按钮单击事件
    private void btnCamOnclick() {
        btn_camer.setOnClickListener(new View.OnClickListener() {
            @Override
            public void onClick(View view) {
                Log.d(TAG, "onClick:1 " + "单击了按钮");
                //拍照代码
                //前期准备:路径
                tmp_path = Environment.getExternalStorageDirectory() + "/img_" + randFileName() + ".jpg";
                File imgfile = new File(tmp_path);//创建保存照片的文件,照片都保存到本路径下
```

```java
                try {
                    if(imgfile.exists()){
                        imgfile.delete();
                    }
                    imgfile.createNewFile();
                } catch (IOException e) {
                    e.printStackTrace();
                }
        Intent intent = new Intent("android.media.action.IMAGE_CAPTURE");
                                                        //1.创建一个打开相机的Intent
        intent.putExtra(MediaStore.EXTRA_OUTPUT, Uri.fromFile(imgfile));
                                                        //2.告诉相机图片的保存位置
        startActivityForResult(intent,11); //3.打开相机
            }
        });
    }
    //第三步:拍完照之后,接收照片并显示,系统回调
    @Override
    protected void onActivityResult(int requestCode, int resultCode, Intent data) {
        switch (requestCode){
            case 11:
                if (resultCode == RESULT_OK){
                    //根据照片的真实路径来显示照片
                    disp_path = tmp_path; //确定图片的路径
                    Glide.with (MainActivity.this).load(disp_path).into(img_camer);
                                                        //显示照片
                }
                break;
            default:
                break;
        }
    }
//时间戳,为了保证每次拍的照片都能以不同的文件名保存到SD卡上
    private String randFileName(){
        Time t = new Time();
        t.setToNow(); //取得系统时间
        String strtime = t.year + "" + (t.month + 1) + "" + t.monthDay + "" + t.hour + "" + t.minute + "" + t.second + "";
        return strtime;
    }
}
```

上述代码中,在onCreate()方法内部自定义两个方法:initView()方法内部放绑定控件的代码,btnCamOnclick()方法内部放按钮单击事件的代码。这种方式不仅可以让onCreate()内部的代码结构清晰,而且独立的方法有利于程序的复用。

在按钮单击事件内部书写拍照代码,拍照之前先准备好照片的存放路径,调用Environment.getExternalStorageDirectory()方法将拍摄的照片存放到外存的SD卡上。

为了保证拍摄的照片能够全部保存到 SD 卡上,采用时间戳为图片文件名的一部分,这样可以确保文件名的唯一。路径确定好之后,调用 createNewFile()方法在该路径下创建保存照片的文件,一切前期工作准备就绪,就可以通过隐式跳转的方式打开系统相机,同时告诉相机拍完之后图片的保存位置,通过 startActivityForResult(intent,11)启动活动,方法中传递了期望在活动销毁时能够返回一个结果给上一个活动的请求码 11,用于在之后的回调中判断数据的来源。因此,当相机拍照结束以后,会回调上一个活动的 onActivityResult()方法,因此需要在 MainActivity 类文件中重写这个方法来得到返回的数据。

onActivityResult()方法中有三个参数:第一个参数 requestCode 是在启动相机活动时传入的请求码,第二个参数 resultCode 是返回数据时传入的处理结果,第三个参数 data 是携带者返回数据的 Intent。首先通过检查 requestCode 判断数据的来源,再通过 resultCode 的值判断处理结果是否成功,最后通过 Glide 图片加载库根据图片的路径将图片显示到 ImageView 控件中。

(6) Glide 图片加载库。Glide 是一个被 Google 所推荐的图片加载库,作者是 bumptech。这个库被广泛运用在 Google 的开源项目中。Glide 是滑行的意思,可以看出这个库的主旨就在于让图片加载变得流畅。使用 Glide 时需要在 Android Studio 上添加依赖,切换工程的 Project 项目结构。选择该模式下方的 app→src→ build.gradle 依次展开,双击打开 build.gradle 文件,添加 Glide 依赖,只需添加一行代码即可,代码如下。

```
apply plugin: 'com.android.application'
…
dependencies {
    implementation fileTree(dir: 'libs', include: ['*.jar'])
    …
    androidTestImplementation 'com.android.support.test.espresso:espresso-core:3.0.2'
    implementation 'com.github.bumptech.glide:glide:4.9.0'
}
```

(7) 声明权限。因为要访问系统相机,因此需要在清单文件的<manifest>节点下配置权限信息,切换工程的 Project 项目结构。选择该模式下方的 app→src→main→AndroidManifest.xml 依次展开,双击打开 AndroidManifest.xml 配置文件,具体代码如下。

```
<manifest xmlns:android = "http://schemas.android.com/apk/res/android"
    xmlns:tools = "http://schemas.android.com/tools"
    package = "com.example.filesavetest">
        <uses-permission android:name = "android.permission.CAMERA"/>
    <uses-permission android:name = "android.permission.WRITE_EXTERNAL_STORAGE"/>
    <uses-permission android:name = "android.permission.READ_EXTERNAL_STORAGE"/>

    <application
    …
```

上述代码指定了当前 SD 卡具有可写可读的权限,同时指定了本程序具有访问系统相机的权限,因此应用程序不仅可以打开系统相机拍照,而且还可以操作 SD 卡中的图片文件。

（8）运行程序，效果如图5.8所示。

图 5.8　打开摄像头拍照和拍照的最终效果

5.3.2　从相册中选择照片

虽然调用摄像头拍照既方便又快捷，但并不是每次都需要去当场拍照，因为每个人的手机相册里应该会有很多图片，直接从相册里选择一张现有的照片会比打开相机拍摄一张照片更加常用。一个优秀的应用程序应该将这两种选择方式都提供给用户，由用户决定使用哪一种。下面就来看一下如何才能实现从相册中选择照片。

微课视频

1. 案例分析

（1）界面分析。布局界面中有两个控件，其中，Button控件用来打开相册，ImageView控件用来把选择的照片显示出来。

（2）设计思路。布局界面中的控件通过添加属性的方式达到用户需求，单击Button打开相册时利用隐式跳转打开手机相册，照片选择结束后通过系统回调将选择照片显示出来，达到用户的需求。

2. 实现步骤

（1）无须创建新工程，在拍照案例工程ZSPhoto中继续完善程序即可。

（2）完善设计布局界面。双击layout文件夹下的activity_main.xml文件，便可打开布局编辑器，在布局界面上拍照案例的两个控件保持不动，继续添加两个控件，代码如下。

```xml
<?xml version = "1.0" encoding = "utf-8"?>
<LinearLayout xmlns:android = "http://schemas.android.com/apk/res/android"
    xmlns:app = "http://schemas.android.com/apk/res-auto"
    xmlns:tools = "http://schemas.android.com/tools"
    android:layout_width = "match_parent"
    android:layout_height = "match_parent"
    android:orientation = "vertical"
    android:background = "@drawable/zsphotobg"
    tools:context = ".MainActivity">

    …

    <Button
        android:id = "@+id/button_photo"
        android:layout_width = "match_parent"
        android:layout_height = "wrap_content"
        android:textColor = "#000000"
        android:text = "从相册选择"
        android:textSize = "30sp"/>
    <ImageView
        android:id = "@+id/imageView_photo"
        android:layout_width = "match_parent"
        android:layout_height = "200dp"
        android:src = "@drawable/ic_launcher_background"
        android:layout_gravity = "center"              />
</LinearLayout>
```

上述代码中，Button 控件用来打开系统相册，ImageView 控件用来显示选择的图片。

(3)"选择相册"按钮功能。打开 MainActivity.java 类文件，完善从相册中选择图片的功能，代码如下。

```java
public class MainActivity extends AppCompatActivity {
    private static final String TAG = "MainActivity";
    //定义对象
    private ImageView img_camer, img_photo;
    private Button btn_camer, btn_photo;
    //定义一个路径：①拍照的临时路径；②显示照片的最终路径
    String tmp_path, disp_path;
    @Override
    protected void onCreate(Bundle savedInstanceState) {
        super.onCreate(savedInstanceState);
        setContentView(R.layout.activity_main);
        //手动开启程序权限之后，拍照仍闪退，解决拍照闪退问题
        if (Build.VERSION.SDK_INT >= Build.VERSION_CODES.N) {
            StrictMode.VmPolicy.Builder builder = new StrictMode.VmPolicy.Builder();
            StrictMode.setVmPolicy(builder.build());
        }
```

```java
        //绑定控件
        initView();
        //拍照按钮单击事件
        btnCamOnclick();
//拍完照之后,接收照片并显示,系统回调
        //"打开相册"按钮单击事件
        btnPhoOnclick();
    }
    //第一步:绑定控件
    private void initView() {
        img_camer = findViewById(R.id.imageView_cam);
        btn_camer = findViewById(R.id.button_cam);
        img_photo = findViewById(R.id.imageView_photo);
        btn_photo = findViewById(R.id.button_photo);
    }
    //第二步:"拍照"按钮单击事件
    private void btnCamOnclick() {
        btn_camer.setOnClickListener(new View.OnClickListener() {
            @Override
            public void onClick(View view) {
                //打开相机拍照按钮,拍照案例已有,这里代码省略
                ...
            }
        });
    }
    //第四步:"打开相册"按钮单击事件
    private void btnPhoOnclick() {
        btn_photo.setOnClickListener(new View.OnClickListener() {
            @Override
            public void onClick(View view) {
                //打开相册
                Intent intent = new Intent("android.intent.action.GET_CONTENT");
                                    //创建打开相册的Intent
                intent.setType("image/*"); //打开页面的类型
                startActivityForResult(intent,22); //打开相册
            }
        });
    }
    //第三步:拍完照之后,接收照片并显示,系统回调
    @Override
    protected void onActivityResult(int requestCode, int resultCode, @Nullable Intent data) {
        switch (requestCode){
            case 11:
                if (resultCode == RESULT_OK){
                    //根据照片的真实路径来显示照片
                    disp_path = tmp_path; //确定图片的路径
                    Glide.with(MainActivity.this).load(disp_path).into(img_camer);
                                                                //显示照片
```

```
                    }
                    break;
                case 22:
                    if(resultCode == RESULT_OK){
                        //根据选择照片的路径,显示照片
                        Uri imageuri = data.getData();  //这是一个缩略图
                        if (imageuri == null){
                            return;
                        }
                        disp_path = UriUtils.uri2File(imageuri).getPath();
                                                //将缩略图转换为真实的图片路径
                        Glide.with(MainActivity.this).load( disp_path ).into(img_photo);
                    }
                    break;
                default:
                    break;
            }
        }
    }
    //时间戳,为了保证每次拍的照片都能以不同的文件名保存到SD卡上
    private String randFileName(){
        Time t = new Time();
        t.setToNow();  //取得系统时间
        String strtime = t.year + "" + (t.month + 1) + "" + t.monthDay + "" + t.hour + "" + t.minute + "" + t.second + "";
        return strtime;
    }
}
```

上述代码中,利用隐式跳转 android.intent.action.GET_CONTENT 打开系统页面,调用 setType("image/*")方法设定打开页面的类型为图片,系统回调中,通过返回的数据 data.getData()得到图片的缩略图路径,之后调用 UriUtils.uri2File(imageuri).getPath()方法将缩略图转换为真实的图片路径,最后通过 Glide 图片加载库根据图片的真实路径将图片显示到 ImageView 控件中。

(4) 将图片的缩略图路径转换为真实路径的方法。UriUtils 是一个依赖,需要在 Android Studio 上添加依赖,切换工程的 Project 项目结构,选择该模式下方的 app→src→build.gradle 依次展开,双击打开 build.gradle 文件,添加 UriUtils 依赖,代码如下。

```
dependencies {
    implementation fileTree(dir: 'libs', include: ['*.jar'])
    ...
    implementation 'com.github.bumptech.glide:glide:4.9.0'
    api 'com.blankj:utilcode:1.23.7'
}
```

(5) 运行程序,效果如图 5.9 所示。

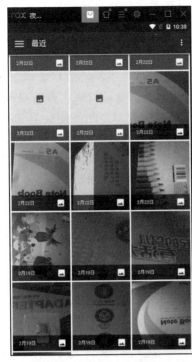 

图 5.9　打开手机相册和选择照片的最终效果

5.4　Android 运行时权限设置

Android 的权限机制并不是什么新鲜事物,从系统的第一个版本开始就已经存在了,但其实之前 Android 的权限机制在保护用户安全和隐私等方面起到的作用非常有限,为了更好地保护用户的安全和隐私,安卓系统的权限管理机制从 API 23 即 Android 6.0 之后发生了比较大的改变,在一些比较危险的权限上要求必须申请动态权限。那么本节就来详细学习一下 Android 6.0 系统中引入的新特性。

5.4.1　Android 权限机制介绍

1. 动态权限需求原因

Android 6.0 之前,权限在应用安装过程中只询问一次,以列表的形式展现给用户,然而大多数用户并不会注意到这些,直接就下一步了,应用安装成功后就会被赋予清单文件中的所有权限,应用就可以在用户不知情的情况下进行非法操作,如偷偷地上传用户数据等。其次,有些常用软件普遍存在着滥用权限的情况,如微信所申请的权限列表:存储、您的位置、电话、相机、短信、通讯录、身体传感器、麦克风等,其中的短信权限与微信关系不大,但用户若不认可微信所申请的所有权限,那么微信则会安装失败。

Android 开发团队也意识到了这个问题,于是在 Android 6.0 系统中加入了运行时动态权限申请,也就是说,用户不需要在安装软件的时候一次性授权所有申请的权限,而是可以在软件使用的过程中再对某一项权限申请进行授权。例如,一款相机应用在运行时申请了地理位置定位权限,即使用户拒绝了这个权限,仍然可以使用这个应用的其他功能,而不是像之前那样直接无法安装软件。

2. 需要动态申请的权限

并不是所有权限都需要在运行时申请,对于用户来说,不停地授权也很烦琐。Android现在将所有的权限归成了两类:一类是普通权限,一类是危险权限。普通权限指的是那些不会直接威胁到用户的安全和隐私的权限,对于这部分权限申请系统会自动授权,无须用户手动操作,例如,查看网络连接权限和开机启动权限。危险权限则是表示那些可能会触及用户隐私或者对设备安全性造成影响的权限,如获取设备联系人信息、定位设备的地址位置等,对于这部分权限申请,必须要由用户手动授权才可以,否则程序就无法使用相应的功能。

但是 Android 中一共有上百种权限,怎么区分哪些是普通权限、哪些是危险权限呢?其实并不难,因为危险权限总共就那么几个,除了危险权限之外,剩余的都是普通权限。Android 中所有的危险权限如表 5.5 所示。

表 5.5 需要动态申请的权限

权 限 名	功能描述	实 例
WRITE_EXTERNAL_STORAGE READ_EXTERNAL_STORAGE	外部存储权限	< uses-permission android: name = "android. permission. WRITE_EXTERNAL_STORAGE"/>
CAMERA	调用相机权限	
RECORD_AUDIO	录音权限	< uses-permission android: name = "android. permission. RECORD_AUDIO"/>
BODY_SENSORS	传感器权限	< uses-permission android: name = "android. permission. BODY_SENSORS"/>
WRITE_CALENDAR READ_CALENDAR	读写日历权限	< uses-permission android: name = "android. permission. WRITE_CALENDAR"/>
READ_CONTACTS WRITE_CONTACTS GET_ACCOUNTS	通讯录权限	< uses-permission android: name = "android. permission. READ_CONTACTS" />
SEND_SMS RECEIVE_SMS READ_SMS RECEIVE_WAP_PUSH RECEIVE_MMS USE _ SIP PROCESS _ OUTGOING _CALLS	消息权限	< uses-permission android: name = "android. permission. SEND_SMS"/>
CALL_PHONE READ_PHONE_STATE READ_CALL_LOG WRITE_CALL_LOG ADD_VOICEMAIL	手机状态相关	< uses-permission android: name = "android. permission. CALL_PHONE" />
ACCESS_FINE_LOCATION ACCESS_COARSE_LOCATION	定位权限	< uses-permission android: name = "android. permission. ACCESS_FINE_LOCATION"/>

表 5.5 中的权限共 9 大类,24 个权限,使用时从表中查看即可,如果是属于这张表中的权限,不仅需要在 AndroidManifest.xml 文件中添加权限声明,而且还要在程序代码中进行权限动态申请,如果所需的权限不在这张表中,那么只需要在 AndroidManifest.xml 文件中添加权限声明就可以了。

5.4.2 在程序运行时申请权限

1. 判断 Android 系统版本

在官方文档中,可以看到低于 API23 是不需要使用动态权限申请的,如果是 Android 6.0 以上的系统,需要进行判断,代码如下。

```
if (Build.VERSION.SDK_INT >= 23) {
        //此处做动态权限申请
    }
else {
        //低于 23 不需要特殊处理
    }
}
```

2. 运行时申请权限案例

1) 案例分析

(1) 界面分析。布局界面中有两个控件,其中,TextView 控件用来显示一段文字,Button 控件用来打电话。

(2) 设计思路。布局界面中的控件通过添加属性的方式达到用户需求,单击 Button 时,弹出"需要使用电话权限,是否允许?"的提示,如果禁止则程序关闭,如果允许则开始拨打电话,达到用户的需求。

2) 实现步骤

(1) 创建一个新的工程,工程名为 ZSRuntimePermission。

(2) 切换工程的 Project 项目结构,选择该模式下方的 app,依次展开,便看到工程的布局界面和工程的类文件,其中,activity_main.xml 是布局界面,MainActivity.java 为类文件。

(3) 准备一张图片,图片名为 runtimepbg.jpg,将其粘贴到 app 目录结构中 res 下方的 drawable 文件夹下,作为页面的背景图片。

(4) 修改布局界面。在 app 目录下的结构中,双击 layout 文件夹下的 activity_main.xml 文件,便可打开布局编辑器,切换到 Text 选项卡,输入如下代码。

```xml
<?xml version = "1.0" encoding = "utf-8"?>
<LinearLayout xmlns:android = "http://schemas.android.com/apk/res/android"
    xmlns:app = "http://schemas.android.com/apk/res-auto"
    xmlns:tools = "http://schemas.android.com/tools"
    android:layout_width = "match_parent"
    android:layout_height = "match_parent"
    android:orientation = "vertical"
```

```xml
    android:background = "@drawable/runtimepbg"
    tools:context = ".MainActivity">
    <TextView
        android:layout_width = "match_parent"
        android:layout_height = "wrap_content"
        android:text = "动态申请权限"
        android:textStyle = "bold"
        android:textSize = "30sp"
        android:textColor = "#0000ff"
        android:gravity = "center"
        android:layout_marginTop = "100dp"/>
    <Button
        android:id = "@+id/button_dadianhua"
        android:layout_width = "match_parent"
        android:layout_height = "wrap_content"
        android:text = "拨打电话"
        android:textStyle = "bold"
        android:textSize = "40sp"
        android:textColor = "#ff0000"
        android:gravity = "center"
        android:layout_marginTop = "50dp"/>
</LinearLayout>
```

上述代码中，根节点为线性布局，方向为 vertical，其内部放置了一个 TextView 控件用来显示提示文字，下方是 Button 控件，单击按钮可拨打电话。

（5）实现打电话功能。打开 MainActivity.java 类文件，修改 MainActivity 的代码，如下。

```java
public class MainActivity extends AppCompatActivity {
    //定义对象
    private Button btn_dadianhua;
    @Override
    protected void onCreate(Bundle savedInstanceState) {
        super.onCreate(savedInstanceState);
        setContentView(R.layout.activity_main);
        //绑定控件
        initView();
        //按钮单击事件
        btnOnClick();
    }
    //绑定控件
    private void initView() {
        btn_dadianhua = findViewById(R.id.button_dadianhua);
    }
    //按钮单击事件
    private void btnOnClick() {
        btn_dadianhua.setOnClickListener(new View.OnClickListener() {
            @Override
```

```java
                    public void onClick(View view) {
//检查权限:判断Android版本是否大于23
//参数:上下文Context和权限的名称.PERMISSION_GRANTED表示存在权限
                        if (Build.VERSION.SDK_INT > 23) {
if (ContextCompat.checkSelfPermission(MainActivity.this, Manifest.permission.CALL_PHONE)!=
PackageManager.PERMISSION_GRANTED ) {
                //申请权限
ActivityCompat.requestPermissions(MainActivity.this, new String[]{Manifest.permission.CALL
_PHONE },11);
                            }else {
                                //权限同意,不需要处理,调用打电话的方法
                                callPhone();
                            }
                        } else {
                            //低于23,不需要特殊处理
                            callPhone();
                        }
                    }
                });
    }
    /** 注册权限申请回调
     * @param requestCode 申请码:在申请权限的时候使用的唯一的申请码
     * @param permissions 申请的权限:String[] permission 则是权限列表,一般用不到
     * @param grantResults 结果:int[] grantResults 是用户的操作响应,包含该权限则请求成功
     * 由于在权限申请的时候,我们就申请了一个权限,所以此处的数组的长度都是1
     */
    @Override
    public void onRequestPermissionsResult(int requestCode, String[] permissions, int[]
grantResults)
    {
        switch (requestCode) //当权限较多时,建议使用Switch
        {
            case 11:
                if (grantResults[0] == PackageManager.PERMISSION_GRANTED ){
                    //权限申请成功
                    callPhone();
                }else {
                    //权限申请失败
Toast.makeText (MainActivity.this, "权限申请失败,用户拒绝权限", Toast.LENGTH_SHORT ).show();
                }
                break;
            default:
                break;
        }
    }
    /**
     * 拨号方法
```

```
    */
    private void callPhone()
    {
        Intent intent = new Intent();
        intent.setAction(Intent.ACTION_CALL);
        intent.setData(Uri.parse("tel:10086"));
        startActivity(intent);
    }
}
```

以上代码中,低于 API23 是不需要使用动态权限申请的,所以首先检测 Android 系统版本,如果版本高于 23,使用 ContextCompat.CheckSlefPermission 方法检查一下有没有权限,该方法有两个参数:上下文 Context 和权限的名称。返回值 PERMISSION_GRANTED 表示存在权限,PERMISSION_DENIED 表示不存在权限,使用方法的返回值和 PackageManager.PERMISSION_GRANTED 做比较,相等就说明用户已经授权,不等就表示用户没有授权。

如果已授权,直接去执行拨打电话的逻辑操作就可以了。这里把拨打电话的逻辑封装到了 callPhone()方法当中。如果没有授权,则需要调用 ActivityCompat.requestPermissions()方法向用户授权。

requestPermissions 方法包含三个参数:参数 1 是当前上下文;参数 2 是一个权限数组,把权限名放到数组中即可;参数 3 是请求码。使用权限数组,说明可以一次申请多个权限。请求码只要是唯一值就可以了,取值要大于 0,小于 65535,这里传入 11。由于这个权限请求是异步操作的,所以用户判断权限后需要回调函数,就用到这个请求码。

调用完 requestPermissions()方法之后,系统会弹出一个权限申请的对话框,用户可以选择同意或拒绝权限申请,不论是哪种结果,最终都会回调到 onRequestPermissionsResult()方法中。

回调函数的处理。由于程序是异步操作,在用户完成了操作后,需要调用回调函数,而此回调函数则是 onRequestPermissionsResult()方法,该方法有 3 个参数:requestCode 是申请权限的时候使用的唯一的申请码;String[] permission 则是权限列表,一般用不到;int[] grantResults 中封装了授权的结果。由于在权限申请的时候,申请了一个权限,所以此处的数组的长度都是 1,这里需要判断一下最后的授权结果,如果用户已同意授权就调用 callPhone()方法拨打电话,如果用户拒绝的话程序放弃操作,并且弹出一条失败的提示。

(6)声明权限。因为打电话会涉及手机的资费问题,因而被列为危险权限,因此必须在 AndroidManifes.xml 中加入权限声明,否则运行时程序就会崩溃。切换工程的 Project 项目结构。选择该模式下方的 app→src→main→AndroidManifest.xml 依次展开,双击打开 AndroidManifest.xml 配置文件,具体代码如下。

```
<manifest xmlns:android="http://schemas.android.com/apk/res/android"
    xmlns:tools="http://schemas.android.com/tools"
    package="com.example.zsruntimepermission">
```

```
<uses-permission android:name="android.permission.CALL_PHONE"/>

    <application
        ...
```

(7) 运行程序,效果如图 5.10 所示。

图 5.10 运行时权限申请效果

如果用户在低于 6.0 系统的设备上安装该程序,用户可以在应用管理界面查看任意一个程序的权限申请情况。在 Android 5.0 系统模拟器中查看权限方法为:"设置"→"应用管理"→ZSRuntimePermission,这样就可以非常清楚地知晓程序一共申请了哪些权限。同样,采用相同的步骤在 Android 6.0 系统真机上也可以查看该程序的权限,如图 5.11 所示。

图 5.11 Android 5.0 系统和 Android 6.0 系统权限展示

微课视频

5.5 实战案例——记忆的仓库:备忘录

时间如流水,几度夕阳红,回首往事,当迫不及待想记录一些事情的时候,有些人会选择用最原始的笔记本记录,但随着社会的发展,出现了新型的记录方式,经常伴随我们左右的智能手机和笔记本电脑也给用户随手记录提供了便利。本节将利用本章所学习的知识,设计和实现一款私人定制的手机备忘录 App,该备忘录功能简洁实用、界面美观、安全性高、易学易用,让我们一起动手开始私人订制的备忘录之旅吧!

5.5.1 界面分析

微课视频

本案例中包含三个界面:登录界面、主界面和信息添加界面,效果如图 5.12 所示。

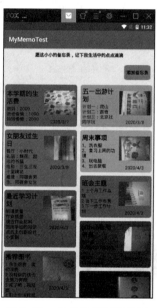

图 5.12 备忘录界面效果图

微课视频

5.5.2 实现思路

第一个界面是登录界面,用来保护备忘录信息的安全,采用 SharedPreference 键值对存储技术来保存用户名和密码;第二个界面是主界面,以滚动列表方式显示所有的备忘录信息,采用 RecyclerView 滚动控件和 SQLite 数据库实现;第三个界面是信息添加界面,用来添加新的备忘录内容,输入的标题和内容及选择的心情图片都会保存到 SQLite 数据库中。

微课视频

5.5.3 任务实施

【任务 5-1】 使用 SharedPreferences 保存输入的用户名和密码
(1)创建一个新的工程,工程名为 MyMemoTest。
(2)切换工程的 Project 项目结构,选择该模式下方的 app,依次展开,便看到工程的布局界面和工程的类文件,其中,activity_main.xml 是布局界面,MainActivity.java 为类文件。

(3) 在工程中添加新的页面。右击 com.example.zsintent 包→New→Activity→Empty Activity,会弹出一个创建活动的对话框,将活动命名为 LoginActivity,默认勾选 Generate Layout File 关联布局界面,布局界面名称为 activity_login 但不要勾选 Launcher Activity。单击 Finish 按钮,便可在工程中完成第二个页面的添加。同理,添加第三个页面,类文件为 AddInfoActivity,布局界面名称为 activity_add_info。

(4) 准备两张图片,图片名为 bgone.png 和 bgthree.jpg,将其粘贴到 app 目录结构中 res 下方的 drawable 文件夹下,作为登录页面和备忘录添加页面的背景图片。

(5) 准备三张小图片,将其粘贴到 app 目录结构中 res 下方的 drawable 文件夹下。其中,buttonbg.png 和 savebg.png 两张小图片作为"添加备忘录"按钮和"保存"按钮的背景图片,sunshine.jpg 图片显示到备忘录添加页面的图片控件中。

(6) 设计第一个 Activity 的布局界面。双击 layout 文件夹下的 activity_login.xml 文件,便可打开布局编辑器,修改布局类型为 LinearLayout,添加方向属性为 vertical。依次添加控件,代码如下:

```xml
<?xml version = "1.0" encoding = "utf-8"?>
<LinearLayout xmlns:android = "http://schemas.android.com/apk/res/android"
    xmlns:app = "http://schemas.android.com/apk/res-auto"
    xmlns:tools = "http://schemas.android.com/tools"
    android:layout_width = "match_parent"
    android:layout_height = "match_parent"
    android:orientation = "vertical"
    android:background = "@drawable/bgone"
    tools:context = ".LoginActivity">
    <TextView
        android:layout_width = "match_parent"
        android:layout_height = "wrap_content"
        android:text = "用户登录"
        android:textStyle = "bold"
        android:textColor = "#000000"
        android:textSize = "40sp"
        android:layout_margin = "100dp"
        android:gravity = "center"/>
    <LinearLayout
        android:layout_width = "match_parent"
        android:layout_height = "wrap_content"
        android:orientation = "horizontal"
        android:layout_margin = "10dp">
        <TextView
            android:layout_width = "wrap_content"
            android:layout_height = "wrap_content"
            android:text = "用户名:"
            android:textColor = "#000000"
            android:textSize = "30sp"            />
```

```xml
<EditText
    android:id="@+id/editText_inputname"
    android:layout_width="match_parent"
    android:layout_height="wrap_content"
    android:hint="请输入用户名"
    android:textColor="#000000"
    android:textSize="30sp"            />
</LinearLayout>
<LinearLayout
    android:layout_width="match_parent"
    android:layout_height="wrap_content"
    android:orientation="horizontal"
    android:layout_margin="10dp">
    <TextView
        android:layout_width="wrap_content"
        android:layout_height="wrap_content"
        android:text="密 码:"
        android:textColor="#000000"
        android:textSize="30sp"            />
    <EditText
        android:id="@+id/editText_inputpwd"
        android:layout_width="match_parent"
        android:layout_height="wrap_content"
        android:hint="请输入密码"
        android:textColor="#000000"
        android:inputType="textPassword"
        android:textSize="30sp"            />
</LinearLayout>
<CheckBox
    android:id="@+id/checkBox_reme"
    android:layout_width="wrap_content"
    android:layout_height="wrap_content"
    android:text="记住密码"
    android:layout_gravity="right"
    android:layout_margin="10dp"/>
<LinearLayout
    android:layout_width="match_parent"
    android:layout_height="wrap_content"
    android:orientation="horizontal"
    android:layout_margin="10dp">
    <Button
        android:layout_width="wrap_content"
        android:layout_height="wrap_content"
        android:text="取消"
        android:textColor="#000000"
        android:textSize="30sp"
        android:layout_weight="1"/>
    <Button
        android:id="@+id/button_login"
```

```xml
                    android:layout_width = "wrap_content"
                    android:layout_height = "wrap_content"
                    android:text = "登录"
                    android:textColor = "#000000"
                    android:textSize = "30sp"
                    android:layout_weight = "1"/>
    </LinearLayout>
</LinearLayout>
```

(7) 实现第一个 Activity 页面的功能。获取 EditText 控件输入的用户名和密码,采用 getSharedPreferences 键值对方式存储起来,根据复选框的状态进行判断,如果选中"记住密码"复选框则获取保存的用户名和密码并显示,如果没有选中"记住密码"复选框,则清空 EditText 控件的内容。打开 LoginActivity.java 类文件,修改 LoginActivity 的代码如下。

```java
public class LoginActivity extends AppCompatActivity {
//定义对象
    private EditText edit_inputname,edit_inputpwd;
    private CheckBox check_reme;
    private Button btn_login;
    @Override
    protected void onCreate(Bundle savedInstanceState) {
        super.onCreate(savedInstanceState);
        setContentView(R.layout.activity_login);
        //1.绑定控件
        initView();
        //2.单击"登录"按钮,将用户名和密码保存起来
        btnloginonClick();
        //3.下次启动,直接显示用户名和密码
        displayinfo();
    }
    //1. 绑定控件
    private void initView() {
        edit_inputname = findViewById(R.id.editText_inputname);
        edit_inputpwd = findViewById(R.id.editText_inputpwd);
        check_reme = findViewById(R.id.checkBox_reme);
        btn_login = findViewById(R.id.button_login);
    }
    //2.单击"登录"按钮,将用户名和密码保存起来
    private void btnloginonClick() {
        btn_login.setOnClickListener(new View.OnClickListener() {
            @Override
            public void onClick(View view) {
                //保存用户名和密码
                SharedPreferences.Editor editor = getSharedPreferences("myinfo",0).edit();
                editor.putString("name",edit_inputname.getText().toString());
                editor.putString("pwd",edit_inputpwd.getText().toString());
                editor.putBoolean("st",check_reme.isChecked());
                editor.commit();
```

```java
                    //跳转到第二页
                    Intent intent = new Intent(LoginActivity.this,MainActivity.class);
                    startActivity(intent);
                }
            });
        }
        //3.下次启动,直接显示用户名和密码
        private void displayinfo() {
            String strname = getSharedPreferences("myinfo",0).getString("name","");
            String strpwd = getSharedPreferences("myinfo",0).getString("pwd","");
            Boolean status = getSharedPreferences("myinfo",0).getBoolean("st",false);
            if(status == true){
                edit_inputname.setText(strname);
                edit_inputpwd.setText(strpwd);
                check_reme.setChecked(true);
            }else{
                edit_inputname.setText("");
                edit_inputpwd.setText("");
                check_reme.setChecked(false);
            }
        }
    }
```

上述代码中,在按钮单击事件的内部,先将用户名和密码保存起来,再跳转到下一个页面。当用户第二次运行程序时会根据复选框是否被选中的状态进行判断,从而确定是否显示用户名和密码。

(8) 修改 AndroidManifest.xml 配置文件,使程序运行时从第一个界面启动,代码如下。

```xml
<activity android:name=".AddInfoActivity"></activity>
<activity android:name=".LoginActivity" >
    <intent-filter>
        <action android:name="android.intent.action.MAIN" />
        <category android:name="android.intent.category.LAUNCHER" />
    </intent-filter>
</activity>
<activity android:name=".MainActivity">
</activity>
…
```

微课视频

(9) 运行程序,效果如图 5.13 所示。

【任务 5-2】 使用 RecyclerView 设计主界面

(1) 设计第二个 Activity 的布局界面。因在第二个布局界面中用到了 Android 新增的控件 RecyclerView,为了让 RecyclerView 在所有的 Android 版本上都能使用,Android 团队将 RecyclerView 定义在了 Support 库当中,如果想使用 RecyclerView 这个控件,只需在 Project 项目结构中 app→build.gradle 文件中添加相应的依赖库才行。打开 app/build.gradle 文件,在 dependencies 闭包中添加如下内容。

项目5 记忆的仓库——备忘录 | 183

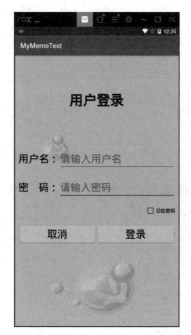

图 5.13 登录界面效果图

```
dependencies {
    implementation fileTree(dir: 'libs', include: ['*.jar'])
    implementation 'com.android.support:appcompat-v7:+'
    …
    implementation 'com.android.support:recyclerview-v7:+'
}
```

需要注意的是,recyclerview 版本号要与第三行代码 support 的版本号保持一致。添加完成后,单击本页面右上角的 Sync Now 来进行同步。

(2) 修改主界面的布局界面。双击 layout 文件夹下的 activity_main.xml 文件,便可打开布局编辑器,修改布局类型为 LinearLayout,添加方向属性为 vertical。依次添加所需控件,代码如下。

```
<?xml version = "1.0" encoding = "utf-8"?>
<LinearLayout xmlns:android = "http://schemas.android.com/apk/res/android"
    xmlns:app = "http://schemas.android.com/apk/res-auto"
    xmlns:tools = "http://schemas.android.com/tools"
    android:layout_width = "match_parent"
    android:layout_height = "match_parent"
    android:orientation = "vertical"
    tools:context = ".MainActivity">
    <TextView
        android:layout_width = "wrap_content"
        android:layout_height = "wrap_content"
        android:text = "愿这小小的备忘录,记下我生活中的点点滴滴"
        android:textStyle = "bold"
```

```xml
        android:textColor = "#000000"
        android:layout_gravity = "center"
        android:layout_margin = "10dp"              />
    <Button
        android:id = "@+id/button_add"
        android:layout_width = "wrap_content"
        android:layout_height = "wrap_content"
        android:text = "添加备忘录"
        android:textStyle = "bold"
        android:textColor = "#000000"
        android:layout_gravity = "right"
        android:background = "@drawable/buttonbg"
        android:layout_margin = "10dp"              />
    <android.support.v7.widget.RecyclerView
        android:id = "@+id/recy_view"
        android:layout_width = "match_parent"
        android:layout_height = "wrap_content"/>
</LinearLayout>
```

上述代码设计的布局界面非常简单,整体为线性布局,自上而下的垂直方向,TextView 控件用来显示一段文字,Button 控件用来跳转到下个页面,RecyclerView 控件用来显示用户添加的备忘录信息。

(3) 实现第二个 Activity 页面的功能。打开 MainActivity.java 类文件,修改 MainActivity 的代码,如下。

```java
public class MainActivity extends AppCompatActivity {
//定义对象
    private Button btn_add;
    private RecyclerView recy_view;
    @Override
    protected void onCreate(Bundle savedInstanceState) {
        super.onCreate(savedInstanceState);
        setContentView(R.layout.activity_main);
        //1.绑定控件
        initView();
        //2.对按钮添加单击事件
        btnonclicknext();
    }
//1.绑定控件
private void initView() {
    btn_add = findViewById(R.id.button_add);
    recy_view = findViewById(R.id.recy_view);
}
//2.对按钮添加单击事件
private void btnonclicknext() {
        btn_add.setOnClickListener(new View.OnClickListener() {
            @Override
            public void onClick(View view) {
```

```
                //单击后跳转到一下页
                Intent intent = new Intent(MainActivity.this,AddInfoActivity.class);
                startActivity(intent);
                finish();
            }
        });
    }
}
```

上述代码是主界面的一部分代码，由于用户还未添加备忘录信息，因此 RecyclerView 控件并未显示任何内容，在完成备忘录添加功能之后，再来完善本页面的功能代码。

（4）运行程序，效果如图 5.14 所示。

图 5.14　主界面设计效果图

【任务 5-3】 使用 SQLite 数据库实现信息添加

（1）设计第三个 Activity 的布局界面。双击 layout 文件夹下的 activity_add_info.xml 文件，便可打开布局编辑器，修改布局类型为 LinearLayout，添加方向属性为 vertical。依次添加所需控件，代码如下。

```xml
<?xml version = "1.0" encoding = "utf-8"?>
<LinearLayout xmlns:android = "http://schemas.android.com/apk/res/android"
    xmlns:app = "http://schemas.android.com/apk/res-auto"
    xmlns:tools = "http://schemas.android.com/tools"
    android:layout_width = "match_parent"
    android:layout_height = "match_parent"
    android:orientation = "vertical"
    android:background = "@drawable/bgthree"
    tools:context = ".AddInfoActivity">
```

```xml
<EditText
    android:id="@+id/editText_title"
    android:layout_width="match_parent"
    android:layout_height="wrap_content"
    android:hint="备忘录的标题"
    android:textStyle="bold"
    android:textColor="#000000"
    android:textSize="30sp"/>
<View
    android:layout_width="match_parent"
    android:layout_height="2dp"
    android:background="#009688"
    android:layout_marginTop="-10dp"/>
<EditText
    android:id="@+id/editText_content"
    android:layout_width="match_parent"
    android:layout_height="200dp"
    android:hint="备忘录的内容"
    android:textStyle="bold"
    android:textColor="#000000"
    android:textSize="20sp"
    android:gravity="top"/>
<View
    android:layout_width="match_parent"
    android:layout_height="2dp"
    android:background="#009688"
    android:layout_marginTop="-10dp"/>
<TextView
    android:layout_width="match_parent"
    android:layout_height="wrap_content"
    android:text="请选用下面的任一种方式,添加一张心情图片:"
    android:textStyle="bold"
    android:textColor="#000000"
    android:textSize="15sp"
    android:gravity="top"
    android:layout_margin="10dp"/>
<LinearLayout
    android:layout_width="match_parent"
    android:layout_height="wrap_content"
    android:orientation="horizontal"
    android:layout_margin="10dp">
    <Button
        android:id="@+id/button_camera"
        android:layout_width="wrap_content"
        android:layout_height="wrap_content"
        android:text="拍照"
        android:textStyle="bold"
        android:textColor="#000000"
        android:textSize="15sp"
        android:layout_margin="10dp"/>
```

```xml
    <Button
        android:id = "@+id/button_photo"
        android:layout_width = "wrap_content"
        android:layout_height = "wrap_content"
        android:text = "从图库中选择"
        android:textStyle = "bold"
        android:textColor = "#000000"
        android:textSize = "15sp"
        android:layout_margin = "10dp"/>
</LinearLayout>
<ImageView
    android:id = "@+id/imageView_preview"
    android:layout_width = "wrap_content"
    android:layout_height = "200dp"
    android:src = "@drawable/sunshine"
    android:layout_marginBottom = "20dp"
    android:layout_gravity = "center"/>
<Button
    android:id = "@+id/button_save"
    android:layout_width = "match_parent"
    android:layout_height = "wrap_content"
    android:text = "保存"
    android:textStyle = "bold"
    android:textColor = "#000000"
    android:textSize = "30sp"
    android:background = "@drawable/savebg"
    android:layout_margin = "10dp"/>
</LinearLayout>
```

上述代码中,View 控件用来定义一条线,使用 layout_height 属性设置线的粗细,使用 background 属性设置线的颜色,layout_marginTop 属性取值为 -10 的目的是想让这条线与 EditText 控件输入框底边线重合,增强页面的美观效果。

(2) 创建数据库类文件。因考虑到用户会存放很多的备忘录信息,而且每条备忘录信息包含文字和图片两种不同类型的数据,因此采用 SQLite 数据库存放备忘录信息。为了使程序文件条理清晰,创建 db 文件夹来存放数据库文件。方法为:右击 com.example.mymemotest→New→Package→输入名称"db"→OK,便可完成 db 文件夹的创建。然后创建数据库类文件,右击 db→New→Java Class→在 Name 中输入类名"MyDbHelper"→Superclass 中输入父类"SQLiteOpenHelper",会自动出现全称 android.database.sqlite.SQLiteOpenHelper→OK。在数据库类文件中输入如下代码。

```java
public class MyDbHelper extends SQLiteOpenHelper {
    //定义数据库名和数据库的版本号
    private static String DBNAME = "zsmemo.db";
    private static int VERSION = 1;
    //构造方法
    public MyDbHelper( Context context) {
        super(context, DBNAME, null, VERSION );
```

```java
        }
        //创建数据库
        @Override
        public void onCreate(SQLiteDatabase db) {
            //创建数据表
            db.execSQL("create table tb_memory(_id Integer primary key,title String (200),
content String (2000),imgpath String (200),mtime String (50))");
        }
        //升级数据库
        @Override
        public void onUpgrade(SQLiteDatabase sqLiteDatabase, int i, int i1) {
        }
    }
```

上述代码中,SQLiteOpenHelper 是专门管理数据库的帮助类,借助于这个类就可以非常简单地对数据库进行创建和升级。SQLiteOpenHelper 是一个抽象类,使用时需要创建一个自定义的帮助类去继承它。

SQLiteOpenHelper 中有两个抽象方法,分别是 onCreate()和 onUpgrade()方法,用户必须在自定义的帮助类里面重写这两个方法,然后分别在这两个方法中去实现创建、升级数据库的逻辑。SQLiteOpenHelper 中还有两个非常重要的实例方法:getWritableDatabase()和 getReadableDatabase(),这两个方法用于创建或打开一个现有的数据库,并返回一个可对数据库进行读写操作的对象 SQLiteDatabase。

SQLiteOpenHelper 的构造方法中接收四个参数:第一个参数是上下文环境 Context;第二个参数是数据库名;第三个参数是查询数据的时候返回一个自定义的 Cursor,一般传入 null;第四个参数是当前数据库的版本号,版本号用于对数据库进行升级操作。SQLiteDatabase 的 execSQL()方法创建数据表 tb_memory。

(3) 添加图片加载库 Glide 依赖,同时添加图片缩略图路径与真实路径转换的依赖,方便第三个页面中图片控件中显示拍照图片或者显示从相册中选择的图片。打开 Project/app/build.gradle 文件,在 dependencies 闭包中添加如下内容。

```
dependencies {
    implementation fileTree(dir: 'libs', include: ['*.jar'])
    implementation 'com.android.support:appcompat-v7:+'
    …
    implementation 'com.android.support:recyclerview-v7:+'
    implementation 'com.github.bumptech.glide:glide:4.9.0'
    api 'com.blankj:utilcode:1.23.7'
}
```

(4) 实现第三个 Activity 页面的功能。打开 AddInfoActivity.java 类文件,修改 AddInfoActivity 的代码,具体如下。

```java
public class AddInfoActivity extends AppCompatActivity {
    //定义对象
    private EditText edit_title,edit_content;
```

```java
    private Button btn_camera,btn_photo,btn_save;
    private ImageView img_preview;
    private String tmp_path,disp_path;
    private MyDbHelper mhelper;
    private SQLiteDatabase db;
    @Override
    protected void onCreate(Bundle savedInstanceState) {
        super.onCreate(savedInstanceState);
        setContentView(R.layout.activity_add_info);
        //1.绑定控件
        initView();
        //2.单击按钮、拍照、从图库中选择照片
        btnOnClick();
        //3.接受拍好照片、接受从图库当中选择的照片  ------ 方法:系统回调
        //4.把信息保存到数据库中
        btnSave();
    }
    //1.绑定控件
    private void initView() {
        edit_title = findViewById(R.id.editText_title);
        edit_content = findViewById(R.id.editText_content);
        btn_camera = findViewById(R.id.button_camera);
        btn_photo = findViewById(R.id.button_photo);
        img_preview = findViewById(R.id.imageView_preview);
        btn_save = findViewById(R.id.button_save);
        mhelper = new MyDbHelper(AddInfoActivity.this); //
        db = mhelper.getWritableDatabase();
    }
    //2.单击按钮、拍照
    private void btnOnClick() {
        btn_camera.setOnClickListener(new View.OnClickListener() {
            @Override
            public void onClick(View view) {
                //拍照
                Time time = new Time();
                time.setToNow();
String randtime = time.year + (time.month + 1) + time.monthDay + time.hour + time.minute + time.second + "";
                tmp_path = Environment.getExternalStorageDirectory() + "/image" + randtime + ".jpg";
                File imgfile = new File(tmp_path);
                try {
                    imgfile.createNewFile();
                } catch (IOException e) {
                    e.printStackTrace();
                }
                Intent intent = new Intent("android.media.action.IMAGE_CAPTURE");
                intent.putExtra(MediaStore.EXTRA_OUTPUT, Uri.fromFile(imgfile));
                startActivityForResult(intent,11);
            }
```

```java
        });
        //从相册中选择图片
        btn_photo.setOnClickListener(new View.OnClickListener() {
            @Override
            public void onClick(View view) {
                //选择照片
                Intent intent = new Intent("android.intent.action.GET_CONTENT");
                intent.setType("image/*");
                startActivityForResult(intent,22);
            }
        });
    }
    //3.接受拍好照片、接受从图库当中选择的照片  ------ 方法:系统回调
    @Override
    protected void onActivityResult(int requestCode, int resultCode, @Nullable Intent data) {
        switch (requestCode){
            case 11:
                if(resultCode == RESULT_OK){
                    disp_path = tmp_path;
                    Glide.with(AddInfoActivity.this).load(disp_path).into(img_preview);
                }
                break;
            case 22:
                Uri imageuri = data.getData();
                if (imageuri == null)
                {
                    return;
                }
                disp_path = UriUtils.uri2File(imageuri).getPath();
                Glide.with(AddInfoActivity.this).load(disp_path).into(img_preview);
                break;
            default:
                break;
        }
    }
    //4.把信息保存到数据库中
    private void btnSave() {
        btn_save.setOnClickListener(new View.OnClickListener() {
            @Override
            public void onClick(View view) {
                //保存信息到数据库代码
                Time time = new Time();
                time.setToNow();
                ContentValues contentValues = new ContentValues(); //1 行
                contentValues.put("title",edit_title.getText().toString());
                                                                    //1 行——title 列
                contentValues.put("content",edit_content.getText().toString());
                                                                    //1 行——content 列
```

```
                contentValues.put("imgpath",disp_path); //1 行——imgpath 列
                contentValues.put("mtime",time.year + "/" + (time.month + 1) + "/" + time.
monthDay); // mtime 列
                db.insert("tb_memory",null,contentValues);//将该行数据保存到数据表中
                Toast.makeText(AddInfoActivity.this,"保存成功",Toast.LENGTH_SHORT).show();
                //跳转到主界面
                Intent intent = new Intent(AddInfoActivity.this,MainActivity.class);
                startActivity(intent);
                finish();
            }
        });
    }
}
```

上述代码中,调用摄像头拍照和从相册中选择图片的代码在5.3节中已学习过,这里不再重复讲解,需要说明的是,本页面是如何把数据保存到数据库中的呢？首先定义数据库帮助类 MyDbHelper 和数据库类 SQLiteDatabase；其次调用 new MyDbHelper(AddInfoActivity.this)语句实例化数据库帮助类,对象名为 mhelper,调用数据库帮助类的 getWritableDatabase()方法获得 SQLiteDatabase 数据库对象,对象名为 db,借助于 SQLiteDatabase 对象就可以对数据库中的数据表进行增删改查操作了。

本页面主要实现了向数据库中 tb_memory 数据表添加数据,首先调用 new ContentValues()方法构建一行数据,其次采用 put()方法将用户输入的备忘录标题放到 title 列、备忘录内容放到 content 列、心情图片路径放到 imgpath 列、添加备忘录的时间放到 mtime 列,最后调用数据库的 insert()方法将本行数据保存到数据表中。数据保存结束后跳转到主界面,及时查看添加的备忘录信息。

(5) 声明权限。双击打开 AndroidManifest.xml 配置文件,具体代码如下。

```
<manifest xmlns:android = "http://schemas.android.com/apk/res/android"
    xmlns:tools = "http://schemas.android.com/tools"
    package = "com.example.filesavetest">
    <uses-permission android:name = "android.permission.CAMERA"/>
    <uses-permission android:name = "android.permission.WRITE_EXTERNAL_STORAGE"/>
    <uses-permission android:name = "android.permission.READ_EXTERNAL_STORAGE"/>

    <application
    ...
```

上述代码指定了当前 SD 卡具有可写可读的权限,同时指定了本程序具有访问系统相机的权限,因此应用程序不仅可以打开系统相机拍照,而且还可以操作 SD 卡中的图片文件。

(6) 信息添加页面完成后,运行程序,在页面中输入标题、内容、图片等数据内容,单击"保存"按钮完成数据的添加。效果如图 5.15(a)所示。

(7) 查看保存在数据库中的信息。在模拟器中打开 RE 文件管理器,挂载为可读写模式,根据存储路径依次操作：data/data/com.example.mymemotest/databases/zsmemo.db/tb_memory,双击 tb_memory 数据表将其打开,看到了保存的数据信息,效果如图 5.15

(b)所示。用户可以将模拟器切换到横屏模式查看完整的数据表字段。

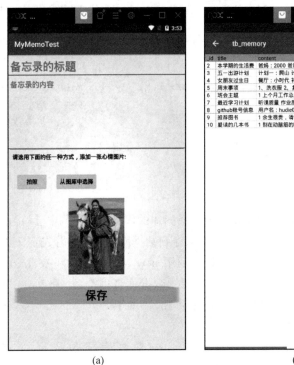

图 5.15 信息添加界面及查看存储的数据

【任务 5-4】 读取 SQLite 数据库中数据显示到备忘录主界面

RecyclerView 控件显示数据时,需要提前准备好数据、item 子布局、适配器、布局管理器。①数据:由于 RecyclerView 是滚动列表,可显示大量数据,因此一般会用 List 来存放 RecyclerView 控件所需要的数据。②item 子布局:RecyclerView 显示数据时,每一行都是一个 item,该行中的文字和图片排放位置如何,大小间距如何,都需要通过 item 子布局来设定每行数据显示的外观。③适配器:要将 List 中的数据以 item 子布局的外观样式显示出来,那么就需要在数据和子布局之间架起一座桥梁,这个桥梁即为适配器,通过适配器来控制数据以子布局设定的外观样式展示给用户。④布局管理器:RecyclerView 中的数据既可以横向滚动,又可以纵向滚动,用户可根据需要挑选 RecyclerView 控件的布局管理器。

(1)新建数据类 MemoBean。为了让 RecyclerView 能够显示很多数据,定义 List 动态数组列表来存放数据,List 列表支持单一数据类型,不能把文字和图片两种不同类型的数据单独放到里面,此时需要把备忘录标题、备忘录内容、心情图片三个数据封装成一个类,以 List<MemoBean> arr1 方式定义列表。这种数据类 MemoBean 有一个特殊的统称,被称为标准的 JavaBean。为了使程序条理清晰,创建 bean 文件夹来存放 JavaBean 类文件。方法为:右击 com.example.mymemotest→New→Package→输入名称"bean"→OK,便可完成 bean 文件夹的创建。然后创建 JavaBean 类文件,右击 bean→New→Java Class→在 Name 中输入类名"MemoBean"→OK。双击 MemoBean 类文件将其打开,输入如下代码。

```java
public class MemoBean {
    //定义四个属性
    private String title;
    private String content;
    private String imgpath;
    private String time;
//按 Alt + Insert 组合键,选择 Contructor 构造方法,全选四个属性,单击 OK 按钮
    public MemoBean(String title, String content, String imgpath, String time) {
        this.title = title;
        this.content = content;
        this.imgpath = imgpath;
        this.time = time;
    }
//按 Alt + Insert 组合键,选择 Getter and Setter 方法,全选四个属性,单击 OK 按钮
    public String getTitle() {
        return title;
    }
    public void setTitle(String title) {
        this.title = title;
    }
    public String getContent() {
        return content;
    }
    public void setContent(String content) {
        this.content = content;
    }
    public String getImgpath() {
        return imgpath;
    }
    public void setImgpath(String imgpath) {
        this.imgpath = imgpath;
    }
    public String getTime() {
        return time;
    }
    public void setTime(String time) {
        this.time = time;
    }
}
```

上述代码中,在 MemoBean 类文件中定义了四个属性,分别是 title、content、imgpath、time,这四个属性正好是 List 动态数组中要存放的四个数据,也是主页面 RecyclerView 控件中需要显示的内容,同时还是数据库 tb_memory 数据表里的四列字段名。

（2）新建 RecyclerView 子布局。在子布局中明确标题、内容和图片的位置关系。将工程依次展开为 app→src→main→res→layout,右击 layout→new→Layout resource file→file name 输入"recy_item"→OK,双击打开 recy_item.xml 布局文件,修改布局代码如下。

```xml
<?xml version = "1.0" encoding = "utf-8"?>
<LinearLayout
    xmlns:android = "http://schemas.android.com/apk/res/android"
    android:layout_width = "match_parent"
    android:layout_height = "wrap_content"
    android:orientation = "horizontal"
    android:background = "#7AECCC"
    android:id = "@+id/item_layout"
    android:layout_margin = "5dp" >
    <LinearLayout
        android:layout_width = "wrap_content"
        android:layout_height = "wrap_content"
        android:orientation = "vertical"
        android:layout_gravity = "center"
        android:layout_weight = "1"
        android:layout_margin = "5dp" >
        <TextView
            android:id = "@+id/item_title"
            android:layout_width = "match_parent"
            android:layout_height = "wrap_content"
            android:text = "标题"
            android:textSize = "20sp"
            android:textStyle = "bold"
            android:textColor = "#000000"/>
        <TextView
            android:id = "@+id/item_content"
            android:layout_width = "wrap_content"
            android:layout_height = "wrap_content"
            android:text = "内容"
            android:textColor = "#000000"/>
    </LinearLayout>
    <LinearLayout
        android:layout_width = "wrap_content"
        android:layout_height = "wrap_content"
        android:orientation = "vertical"
        android:layout_margin = "5dp">
        <ImageView
            android:id = "@+id/item_image"
            android:layout_width = "100dp"
            android:layout_height = "100dp"
            android:src = "@mipmap/ic_launcher_round"/>
        <TextView
            android:id = "@+id/item_time"
            android:layout_width = "wrap_content"
            android:layout_height = "wrap_content"
            android:text = "时间"
            android:textColor = "#000000"
            android:layout_gravity = "center"/>
    </LinearLayout>
</LinearLayout>
```

上述代码中,子布局总体采用水平方向线性布局,左侧嵌套垂直方向线性布局用来显示标题和内容,右侧嵌套垂直方向的线性布局,用来显示图片和日期,如图 5.16 所示。

图 5.16 RecyclerView 控件的子布局 item

(3) 建立 RecyclerView 适配器 MemoAdapter。为了让 JavaBean 中数据能够以 recy_item 子布局的外观样式显示到 RecyclerView 控件上,需要在数据与子布局之间搭建一个桥梁,即适配器。为了使程序文件条理清晰,创建 adapter 文件夹来存放适配器类文件。方法为:右击 com.example.mymemotest→New→Package→输入名称"adapter"→OK,便可完成 adapter 文件夹的创建。然后创建 JavaBean 类文件,右击 bean→New→Java Class→Name 中输入类名"MemoAdapter"→Superclass 中输入父类"RecyclerView.Adapter",会自动出现全称 android.support.v7.widget.RecyclerView.Adapter→OK。在适配器类文件中输入如下代码。

```
//1.类文件后面添加泛型
//2.鼠标定位类文件行红色波浪线处,按 Alt + Enter 组合键:添加未实现的方法
//3.鼠标定位类文件行 ViewHolder 处,按 Alt + Enter 组合键:添加内部类
//4.鼠标定位界面最下方内部类 ViewHolder 处,添加 extends RecyclerView.ViewHolder
//5.鼠标定位界面最下方内部类 ViewHolder 红色波浪线处,按 Alt + Enter 组合键:添加构造方法
//6.定义两个对象:上下文环境和数组
//7.定义两个对象下方的空白处:Alt + Insert 键,添加适配器的构造方法
public class MemoAdapter extends RecyclerView.Adapter<MemoAdapter.ViewHolder> {
    private Context mcontext;
    private List<MemoBean> arr1;
    public MemoAdapter(Context mcontext, List<MemoBean> arr1) {
        this.mcontext = mcontext;
        this.arr1 = arr1;
    }
//负责加载 item 布局
    @NonNull
    @Override
    public MemoAdapter.ViewHolder onCreateViewHolder(@NonNull ViewGroup parent, int i) {
        View view = LayoutInflater.from(mcontext).inflate(R.layout.recy_item, parent, false);
        ViewHolder mholder = new ViewHolder(view);
        return mholder;
    }
//负责加载 item 的数据
    @Override
    public void onBindViewHolder(@NonNull MemoAdapter.ViewHolder mholder, int i) {
        final MemoBean memoBean = arr1.get(i);
        mholder.item_title.setText(memoBean.getTitle());
```

```java
            mholder.item_content.setText(memoBean.getContent());
            mholder.item_time.setText(memoBean.getTime());
            Glide.with(mcontext).load(memoBean.getImgpath()).into(mholder.item_img);
        }
//recyView一共有多少个子项
        @Override
        public int getItemCount() {
            return arr1.size();
        }
        public class ViewHolder extends RecyclerView.ViewHolder {
            TextView item_title,item_content,item_time;
            ImageView item_img;
            LinearLayout item_layout;
            public ViewHolder(@NonNull View itemView) {
                super(itemView);
                item_title = itemView.findViewById(R.id.item_title);
                item_content = itemView.findViewById(R.id.item_content);
                item_img = itemView.findViewById(R.id.item_image);
                item_time = itemView.findViewById(R.id.item_time);
                item_layout = itemView.findViewById(R.id.item_layout);
            }
        }
    }
```

上述代码中,最下方首先定义了一个内部类 ViewHolder,ViewHolder 要继承自 RecyclerView.ViewHolder。然后在 ViewHolder 的构造方法中传入一个 itemView 参数,这个参数就是 RecyclerView 子项的最外层布局,通过 itemView.findViewById 方法获取布局中的 TextView 和 ImageView 及 LinearLayout 控件的实例。

由于 MemoAdapter 继承了 RecyclerView.Adapter,那么就必须重写 3 个方法。onCreateViewHolder()方法是用户创建 ViewHolder 实例的,在这个方法中把 recy_item 子布局加载进来。onBindViewHolder()方法用于对 RecyclerView 子项进行赋值,根据子项的位置 i 得到当前项的 MemoBean 实例,然后再将数据显示到对应的控件上。getItem()方法用于告诉 RecyclerView 一共有多少子项,直接返回数据源数组的长度即可。

(4)完善 MainActivity 页面的功能,实现备忘录信息的展示。打开 MainActivity.java 类文件,修改 MainActivity 的代码如下。

```java
public class MainActivity extends AppCompatActivity {
//定义对象
    private Button btn_add;
    private RecyclerView recy_view;
    private MyDbHelper mhelper;
    SQLiteDatabase db;
    @Override
    protected void onCreate(Bundle savedInstanceState) {
        super.onCreate(savedInstanceState);
        setContentView(R.layout.activity_main);
```

```java
        //1.绑定控件
        initView();
        //2.对按钮添加单击事件
        btnonclicknext();
        //3.完善:从数据库获取数据,显示到RecyclerView控件中
        recyDisplay();
    }
    //1.绑定控件
    private void initView() {
        btn_add = findViewById(R.id.button_add);
        recy_view = findViewById(R.id.recy_view);
        mhelper = new MyDbHelper(MainActivity.this);
        db = mhelper.getWritableDatabase();
    }
    //2.对按钮添加单击事件
    private void btnonclicknext() {
        btn_add.setOnClickListener(new View.OnClickListener() {
            @Override
            public void onClick(View view) {
                //单击后跳转到下一页
                Intent intent = new Intent(MainActivity.this,AddInfoActivity.class);
                startActivity(intent);
                finish();
            }
        });
    }
    //3.完善:从数据库获取数据,显示到RecyclerView控件中
    private void recyDisplay() {
        //3.1 准备数据------------标题、内容、图片、时间(类)
        List<MemoBean> arr = new ArrayList();
        //从数据库中取数据
        Cursor cursor = db.rawQuery("select * from tb_memory",null);
        while(cursor.moveToNext()){
            String mytitle = cursor.getString(cursor.getColumnIndex("title"));
            String mycontent = cursor.getString(cursor.getColumnIndex("content"));
            String myimgpath = cursor.getString(cursor.getColumnIndex("imgpath"));
            String mymtime = cursor.getString(cursor.getColumnIndex("mtime"));
            MemoBean memoBean = new MemoBean(mytitle,mycontent,myimgpath,mymtime);
            arr.add(memoBean);
        }
        cursor.close();
        //3.2 子布局 recy_item
        //3.3 数据------桥(适配器 MemoAdapter)-----------------子布局
        MemoAdapter adapter = new MemoAdapter(MainActivity.this,arr);
        //3.4 确定显示的方式
StaggeredGridLayoutManager st = new StaggeredGridLayoutManager(2,StaggeredGridLayoutManager.VERTICAL);
        recy_view.setLayoutManager(st);
        //3.5 让数据显示出来
```

```
            recy_view.setAdapter(adapter);
        }
}
```

(5)备忘录删除功能的实现。单击页面上的某一个备忘录内容,在弹出的对话框中单击"删除"按钮,便可删除备忘录信息,同时刷新 RecyclerView 信息列表以显示最新的备忘录内容。修改每条备忘录的背景颜色,采用随机数字的方式设置背景颜色,同时设置 item 为圆角矩形。双击打开 MemoAdapter 适配器,在 onBindViewHolder()方法中完善代码,具体如下。

```
public class MemoAdapter extends RecyclerView.Adapter<MemoAdapter.ViewHolder> {
    private Context mcontext;
    private List<MemoBean> arr1;
    private MyDbHelper mhelper1;
    private SQLiteDatabase db;

    …//中间的这部分代码保持不变

//负责加载 item 的数据
@RequiresApi(api = Build.VERSION_CODES.JELLY_BEAN)
@Override
public void onBindViewHolder(@NonNull MemoAdapter.ViewHolder mholder, final int i) {
    final MemoBean memoBean = arr1.get(i);
    mholder.item_title.setText(memoBean.getTitle());
    mholder.item_content.setText(memoBean.getContent());
    mholder.item_time.setText(memoBean.getTime());
    Glide.with(mcontext).load(memoBean.getImgpath()).into(mholder.item_img);

    //完善:设置 RecyclerView 中每一个子项的颜色和形状
    Random random = new Random();
    int color = Color.argb(255, random.nextInt(256), random.nextInt(256), random.nextInt(256));
    GradientDrawable gradientDrawable = new GradientDrawable();
    gradientDrawable.setShape(GradientDrawable.RECTANGLE);        //形状
    gradientDrawable.setCornerRadius(10f);                         //设置圆角 Radius
    gradientDrawable.setColor(color);                              //颜色
    mholder.item_layout.setBackground(gradientDrawable);           //设置为 background

    //完善:单击其中的一个子项,弹出删除功能
        mholder.item_layout.setOnClickListener(new View.OnClickListener() {
            @Override
            public void onClick(View view) {
                //弹出对话框,删除
                AlertDialog.Builder dialog = new AlertDialog.Builder(mcontext);
                dialog.setMessage("确定删除吗?");
                dialog.setPositiveButton("确定", new DialogInterface.OnClickListener() {
                    @Override
                    public void onClick(DialogInterface dialogInterface, int abc) {
```

```
                        //从数据库当中删除掉
                        mhelper1 = new MyDbHelper(mcontext);
                        db = mhelper1.getWritableDatabase();
                        db.delete("tb_memory","title = ?",new String[]{arr1.get(i).getTitle
()}); //删除

                        arr1.remove(i); //从 list 动态数组中把该项移除
                        notifyItemRemoved(i); //刷新 RecyclerView 显示的内容
                        dialogInterface.dismiss(); //对话框消失
                    }
                });
                dialog.setNegativeButton("取消",null);
                dialog.setCancelable(false);
                dialog.create();
                dialog.show();
            }
        });
    }
    … //后面的代码保持不变
    …
```

（6）运行程序，单击第一条备忘录，弹出对话框，单击"确定"按钮便将该条备忘录删除，RecyclerView 会及时刷新并显示最新数据，效果如图 5.17 所示。

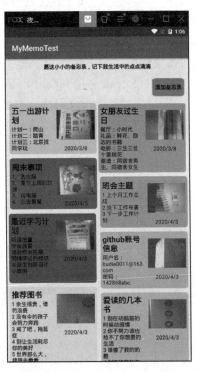

图 5.17 备忘录数据的删除

项目小结

本项目内容紧紧围绕 Android 中的 SharedPreferences 存储、Android 的文件存储、调用摄像头和相册以及运行时权限设置内容来展开,使用户根据存储数据的特点选择不同的存储方式,能够将拍摄的图片根据需要随时存放到内存或外存中。通过学习 Android 运行时权限,保障了用户的信息安全。最后,通过备忘录案例,培养学生具备复杂 App 的设计与开发能力。

习题

1. SharedPreferences 所存储的数据以(　　)的格式保存在 XML 文件中。
   A. "姓-名"　　　　B. "键-值"　　　　C. 文件　　　　D. 字符
2. Android 的几种存储方式中,(　　)是以键值对的方式来存储数据的。
   A. SharedPreferences 存储　　　　B. 文件存储
   C. SQLite 数据库存储　　　　D. SD 卡存储
3. getSharedPreferences("myfile",0)方法中的第二个参数 0 表示以追加模式保存数据。(　　)
   A. 对　　　　B. 错
4. Android 提供了标准的 Java 文件(　　)方式来对文件数据进行读写。
   A. "姓-名"　　　B. "键-值"　　　C. 字符　　　D. 输入输出流
5. Context 类的(　　)方法可以获得文件输入流对象。
   A. openFileOutput　　　B. openFileInput
6. 下面(　　)方法是关闭文件输入/输出流的方法。
   A. close()　　　　B. exit()
7. 打开摄像头和相册的系统页面,属于(　　)跳转。
   A. 显式跳转　　　　B. 隐式跳转
8. 下面(　　)可以打开摄像头。
   A. android.media.action.IMAGE_CAPTURE
   B. android.intent.action.GET_CONTENT
9. 在 Android 项目中,图片加载库有很多选择,常用的是(　　)。
   A. Glide　　　B. Image　　　C. picture　　　D. icon
10. 下面(　　)控件不能直接使用,需要添加(　　)依赖才可使用。
    A. ImageView　　　　B. TextView
    C. EditText　　　　D. RecyclerView

项目 6

多彩水果店

【教学导航】

学习目标	（1）掌握 ListView 控件的使用。 （2）掌握 RecyclerView 控件的使用。 （3）熟悉 Android 系统中的样式和主题。 （4）掌握 Serializable 序列化接口的应用。 （5）培养简单案例的设计开发能力。
教学方法	任务驱动法、理论实践一体化、探究学习法、分组讨论法
课时建议	10 课时

6.1 最常用和最难用的控件——ListView

ListView 可以说是 Android 中最常用的控件之一，应用程序中使用列表的形式来展现一些内容，如查看微信或者 QQ 的聊天记录、查看购物网站的商品列表等。由于手机屏幕空间有限，如果要展示大量数据，可以借助于 ListView 以列表的形式展示，并通过手指上下滑动的方式将屏幕之外的数据滚动到屏幕内显示，同时将屏幕上原有的数据滚动到屏幕外。相比其他的控件而言，ListView 的用法相对复杂很多，因此也可以说是最难用的控件。

6.1.1 ListView 的简单用法

1. 案例分析

（1）界面分析。布局界面中只有一个控件 ListView，是用来显示列表的控件。

（2）设计思路。布局界面中的控件通过添加属性的方式达到用户需求。用一个数组存放要显示的水果的名称，借助于适配器将数据显示在 ListView 上，适配器 ArrayAdapter 中指定 ListView 子项的布局，最后调用 ListView 的 setAdapter()方法，建立 ListView 和数

微课视频

据之间的关联。

2. 实现步骤

(1) 创建一个新的工程,工程名为 ListView1。

(2) 切换工程的 Project 项目结构。选择该模式下方的 app,依次展开,便看到工程的布局界面和工程的类文件,其中,activity_main.xml 是布局界面,MainActivity.java 为类文件。

(3) 修改布局界面。在 app 目录下的结构中,双击 layout 文件夹下的 activity_main.xml 文件,便可打开布局编辑器,切换到 Text 选项卡,输入如下代码。

```xml
<?xml version = "1.0" encoding = "utf-8"?>
<LinearLayout xmlns:android = "http://schemas.android.com/apk/res/android"
    xmlns:app = "http://schemas.android.com/apk/res-auto"
    xmlns:tools = "http://schemas.android.com/tools"
    android:layout_width = "match_parent"
    android:layout_height = "match_parent"
    android:orientation = "vertical"
    tools:context = ".MainActivity">
    <ListView
        android:id = "@+id/list_view"
        android:layout_width = "match_parent"
        android:layout_height = "match_parent"/>
</LinearLayout>
```

上述代码中,根节点为线性布局,方向为 vertical,添加 ListView 控件,设置其高度和宽度为 match_parent,这样 ListView 控件就占满了整个布局,因为在类文件中要调用 ListView,所以给它设置一个 id。

(4) 打开 MainActivity.java 类文件,修改 MainActivity 的代码,具体如下。

```java
public class MainActivity extends AppCompatActivity {
    //1.定义对象
    ListView listView;
    @Override
    protected void onCreate(Bundle savedInstanceState) {
        super.onCreate(savedInstanceState);
        setContentView(R.layout.activity_main);
        //2.绑定控件
        listView = (ListView) findViewById(R.id.list_view);
        //3.准备数据
        String[] data = {"菠萝","芒果","石榴","葡萄","苹果","橙子","西瓜","菠萝","芒果","石榴","葡萄","苹果","橙子","西瓜","菠萝","芒果","石榴","葡萄","苹果","橙子","西瓜"};
        //4.创建适配器
        //参数1:当前的上下文环境
        //参数2:当前列表项所加载的布局文件
        //参数3:数据源
        ArrayAdapter<String> adapter = new ArrayAdapter<>(MainActivity.this,android.R.layout.simple_list_item_1,data);
```

```
            //5.将适配器加载到控件中
            listView.setAdapter(adapter);
        }
}
```

定义一个数组来存放 ListView 中 item 的内容,这些数据可以是从网上下载的,也可以从数据库中读取,这里以一个数组 data 来保存水果的名称,数组中的数据是无法直接传递给 ListView 的,需要借助于适配器。适配器是一个连接数据和 AdapterView(ListView 就是一个典型的 AdapterView)的桥梁,通过它能有效地实现数据与 AdapterView 的分离设置,使 AdapterView 与数据的绑定更加简便,修改更加方便。

常用适配器有 ArrayAdapter、SimpleAdapter 和 SimpleCursorAdapter。ArrayAdapter 最为简单,只能展示一行字;SimpleAdapter 有最好的扩充性,可以自定义各种各样的布局,除了文本外,还可以放 ImageView(图片)、Button(按钮)、CheckBox(复选框)等;SimpleCursorAdapter 可以认为是 SimpleAdapter 对数据库的简单结合,可以方便地把数据库的内容以列表的形式展示出来。但是实际工作中,常用自定义适配器,即继承于 BaseAdapter 的自定义适配器类。

通过实现 ArrayAdapter 的构造方法创建一个 ArrayAdapter 对象,ArrayAdapter 有多个构造方法,最常用三个参数的这种。因为我们准备的数据是字符串类型的,所以将 ArrayAdapter 的泛型指定为 String,在 ArrayAdapter 的构造函数中需要传入三个参数:第一个参数为上下文对象;第二个参数为 ListView 的每一行(也就是 item)的布局资源 id;第三个参数为 ListView 的数据源。我们使用了 android.R.layout.simple_list_item_1 作为 ListView 的每一行(也就是 item)的布局资源 id,这是一个 Android 内置的布局文件,里面只有一个 TextView,可以显示一段文本。最后通过 ListView 的 setAdapter()方法绑定 ArrayAdapter,这样数据和 ListView 的关联就创建好了。

(5)运行程序,效果如图 6.1 所示,可以通过滚动的方式来查看屏幕外的数据。

(6)为列表中选中的项添加单击响应事件,实现单击某个列表项弹出其所对应的水果名称,在上述代码步骤 5 的下方,添加步骤 6,代码如下。

```
        //6.ListView 单击事件
        listView.setOnItemClickListener(new AdapterView.OnItemClickListener() {
            @Override
            public void onItemClick(AdapterView<?> parent, View view, int i, long l) {
                String result = ((TextView)view).getText().toString();
Toast.makeText(MainActivity.this,"您选择的水果是:" + result,Toast.LENGTH_LONG).show();
            }
        });
```

这里使用 setOnItemClickListener()方法为 ListView 注册了一个监听器,当用户单击 ListView 中的任意一个选项时,就会回调 onItemClick()方法,将每一个子项布局 View 强制转为 TextView 类型,通过 getText().toString()方法得到单击项的水果名称,最后通过 Toast 提示显示出来。重新运行程序,并且单击"葡萄",运行效果如图 6.2 所示。

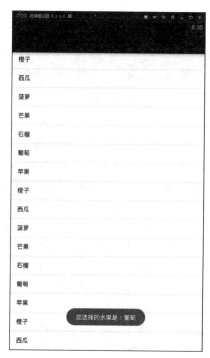

图 6.1　简单 ListView 的运行效果　　　　图 6.2　简单 ListView 的单击事件运行效果

微课视频

6.1.2　定制 ListView 的界面

只能显示一行文本数据的 ListView 太简单了,实用性不强,下面对 ListView 的界面进行定制,满足应用程序图文混排的显示效果。

1. 案例分析

(1) 界面分析。布局界面中只有一个控件 ListView,是用来显示列表的控件。

(2) 设计思路。布局界面中的控件通过添加属性的方式达到用户需求。定义一个实体类 Fruit,作为 ListView 适配器的适配类型;为 ListView 的子项指定一个自定义的布局 fruit_item.xml;创建一个自定义的适配器 FruitAdapter,这个适配器继承自 ArrayAdapter,重写构造方法和 getView()方法;在 MainActivity 中编写代码,初始化水果数据。

2. 实现步骤

(1) 创建一个新的工程,工程名为 ListView2。

(2) 切换工程的 Project 项目结构,选择该模式下方的 app,依次展开,便看到工程的布局界面和工程的类文件,其中,activity_main.xml 是布局界面,MainActivity.java 为类文件。

(3) 修改布局界面。在 app 目录下的结构中,双击 layout 文件夹下的 activity_main.xml 文件,便可打开布局编辑器,切换到 Text 选项卡,输入如下代码。

```xml
<?xml version = "1.0" encoding = "utf-8"?>
<LinearLayout xmlns:android = "http://schemas.android.com/apk/res/android"
    xmlns:app = "http://schemas.android.com/apk/res-auto"
    xmlns:tools = "http://schemas.android.com/tools"
    android:layout_width = "match_parent"
    android:layout_height = "match_parent"
    android:orientation = "vertical"
    tools:context = ".MainActivity">
    <ListView
        android:id = "@+id/list_view"
        android:layout_width = "match_parent"
        android:layout_height = "match_parent"/>
</LinearLayout>
```

上述代码中,根节点为线性布局,方向为 vertical,添加 ListView 控件,设置其高度和宽度为 match_parent,这样 ListView 控件就占满了整个布局,因为在类文件中要调用 ListView,所以给它设置一个 id。

(4) 定义一个实体类,作为 ListView 适配器的适配类型。右击 com.example.listview2 包→New→Java Class,会弹出一个创建类的对话框,将类命名为 Fruit,代码如下。

```java
public class Fruit {
    private int imageID;
    private String name;
    private String price;
    //按 Alt + Insert 组合键,在弹出的快捷菜单中选择 Getter,弹出对话框中的三个属性都选中,
    //单击 OK 按钮,即可出现下面三个方法.
    public int getImageID() {
            return imageID;
        }
    public String getName() {
            return name;
        }
    public String getPrice() {
            return price;
        }
    //按 Alt + Insert 组合键,弹出快捷菜单,选择 Constructor,对话框中的三个属性全选,单击 OK 按
    //钮,即可出现下面的构造方法.
    public Fruit(int imageID, String name, String price) {
        this.imageID = imageID;
        this.name = name;
        this.price = price;
    }
}
```

Fruit 类中有三个字段,imageID 是水果对应图片的资源 id,name 是水果的名称,price 是水果的价格。

(5) 准备好一组水果图片,将其复制粘贴到 res/drawable 文件夹中。

（6）为 ListView 的每一行子项指定要显示的布局样式，右击 Layout→New→Xml→Layout Xml File，会弹出一个 Xml Layout File 对话框，命名为 fruit_item.xml，代码如下。

```xml
<?xml version = "1.0" encoding = "utf-8"?>
<LinearLayout xmlns:android = "http://schemas.android.com/apk/res/android"
    android:layout_width = "match_parent"
    android:orientation = "horizontal"
    android:layout_height = "wrap_content">
    <ImageView
        android:id = "@+id/fruit_image"
        android:src = "@drawable/apple"
        android:layout_width = "100dp"
        android:layout_height = "80dp"/>
    <TextView
        android:id = "@+id/fruit_name"
        android:layout_gravity = "center_vertical"
        android:textSize = "30sp"
        android:textColor = "#000000"
        android:text = "name"
        android:layout_marginLeft = "10dp"
        android:layout_width = "wrap_content"
        android:layout_height = "wrap_content"/>
    <TextView
        android:id = "@+id/fruit_price"
        android:layout_gravity = "center_vertical"
        android:textColor = "#ff0000"
        android:text = "price"
        android:textSize = "30sp"
        android:layout_marginLeft = "10dp"
        android:layout_width = "wrap_content"
        android:layout_height = "wrap_content"/>
</LinearLayout>
```

上述代码中，根节点为线性布局，方向为 horizontal，我们定义了一个 ImageView 用于显示水平的图片，又定义了两个 TextView 用于显示水果的名称和水果的价格，并且垂直居中显示。

（7）因为列表中既有图片又有文字，就不能再用仅仅显示一行字符串的 ArrayAdapter，需要自定义适配器。新建一个类，命名为 FruitAdapter，继承自 ArrayAdapter，并将泛型指定为 Fruit 类，代码如下。

```java
public class FruitAdapter extends ArrayAdapter<Fruit> {
//用于将上下文、ListView 子项布局的 id 和数据都传递过来
    public FruitAdapter(@NonNull Context context, int resource, @NonNull List<Fruit> objects) {
        super(context, resource, objects);
    }
//每个子项被滚动到屏幕内的时候会被调用
```

```java
    @NonNull
    @Override
    public View getView(int position, @Nullable View convertView, @NonNull ViewGroup parent) {
        Fruit fruit = getItem(position);  //得到当前项的Fruit实例
        //为每一个子项加载设定的布局
        View view = LayoutInflater.from(getContext()).inflate(R.layout.fruit_item, parent, false);
        //分别获取ImageView和TextView的实例
        ImageView fruitimage = view.findViewById(R.id.fruit_image);
        TextView fruitname = view.findViewById(R.id.fruit_name);
        TextView fruitprice = view.findViewById(R.id.fruit_price);
        //设置要显示的图片和文字
        fruitimage.setImageResource(fruit.getImageID());
        fruitname.setText(fruit.getName());
        fruitprice.setText(fruit.getPrice());
        return view;
    }
}
```

FruitAdapter重写了父类的一组构造函数,用于将上下文、ListView子项布局的id和数据都传递过来。重写了getView()方法,每个子项被滚动到屏幕内的时候会被调用getView()方法。在getView()方法中,通过getItem(position)方法,根据position单击的位置获得被单击的Fruit实例,然后使用LayoutInflater为ListView子项加载定制好的布局。LayoutInflater的inflate()方法的三个参数分别为:第一个参数是布局;第二个参数是父容器控件;第三个布尔值参数表明是否连接该布局和其父容器控件,在这里设置为false,因为系统已经插入了这个布局到父控件,设置为true将会产生多余的一个View Group。接下来,调用view的findViewById()方法分别获取fruitimage、fruitname及fruitprice,再分别调用setImageResource()和setText()方法将图片和文字显示在对应的控件上,最后将布局返回,这样自定义的适配器就创建好了。

(8) 打开MainActivity.java类文件,修改MainActivity的代码,具体如下。

```java
public class MainActivity extends AppCompatActivity {
    //第一步:定义对象
    ListView listView;
    @Override
    protected void onCreate(Bundle savedInstanceState) {
        super.onCreate(savedInstanceState);
        setContentView(R.layout.activity_main);
        //第二步:绑定控件
        listView = (ListView) findViewById(R.id.list_view);
        //第三步:准备数据
        List<Fruit> fruitlist = new ArrayList<>();
        for (int i = 0; i < 2; i++) {
            Fruit pineapple = new Fruit(R.drawable.pineapple,"菠萝","￥16.9元/ kg ");
            fruitlist.add(pineapple);
```

```
            Fruit mango = new Fruit(R.drawable.mango, "芒果","¥29.9元/kg");
            fruitlist.add(mango);
            Fruit pomegranate = new Fruit(R.drawable.pomegranate, "石榴","¥15元/kg");
            fruitlist.add(pomegranate);
            Fruit grape = new Fruit(R.drawable.grape, "葡萄","¥19.9元/kg");
            fruitlist.add(grape);
            Fruit apple = new Fruit(R.drawable.apple, "苹果","¥20元/kg");
            fruitlist.add(apple);
            Fruit orange = new Fruit(R.drawable.orange, "橙子","¥18.8元/kg");
            fruitlist.add(orange);
            Fruit watermelon = new Fruit(R.drawable.watermelon, "西瓜","¥28.8元/kg");
            fruitlist.add(watermelon);
        }
        //第四步:设计每一个列表项的子布局
        //第五步:定义适配器
         FruitAdapter adapter = new FruitAdapter(MainActivity.this, R.layout.fruit_item, fruitlist);
            listView.setAdapter(adapter);
        }
    }
```

定义一个动态数组 ArrayList 来存放 ListView 中 item 的内容,泛型为新建类 Fruit,在 Fruit 的构造函数中将水果图片的 id、水果名称和水果价格传入,然后将创建好的对象添加到水果列表中。因为数据量较少,不足以填充满整个屏幕,所以使用了 for 循环将数据添加了两遍。接着创建了 FruitAdapter 对象,并将 FruitAdapter 适配器传递给 ListView。

(9) 运行程序,效果如图 6.3 所示,可以通过滚动的方式来查看屏幕外的数据。

目前定制的界面还比较简单,只要修改 frui_item.xml 中的布局,就可以根据需要定制出各种复杂的界面了。

图 6.3　定制 ListView 的界面
　　　 程序运行后效果

微课视频

6.1.3　提升 ListView 的运行效率

目前 ListView 的运行效率是很低的,因为在 FruitAdapter 的 getView()方法中每次都将布局重新加载了一遍,当 ListView 快速滚动的时候这就会成为性能的瓶颈。仔细观察,getView()方法中还有一个 convertView 参数,这个参数用于将之前加载好的布局进行缓存,以便之后可以进行重用。

1. 案例分析

(1) 界面分析。布局界面同定制 ListView 案例的布局。
(2) 设计思路。提升 ListView 的运行效率有两种方法,第一种方法:判断

convertView 是否为空,如果为空则使用 LayoutInflater 去加载布局,如果不为空则直接对 convertView 进行重用。第二种方法:新增一个内部类 ViewHolder,用于对控件的实例进行缓存。当 convertView 为空的时候,创建一个 ViewHolder 对象,并将控件的实例都存放在 ViewHolder 里,然后调用 View 的 setTag()方法,将对象存储在 View 中。如果 convertView 不为空,调用 View 的 getTag()方法,把 ViewHolder 重新取出。

2. 实现步骤

(1) 无须创建新工程,在定制 ListView 的界面的工程 ListView2 中继续完善程序即可。

(2) 修改 FruitAdapter 中的代码,具体如下。

```java
public class FruitAdapter extends ArrayAdapter<Fruit>{
    //用于将上下文、ListView子项布局的 id 和数据都传递过来
    public FruitAdapter(@NonNull Context context, int resource, @NonNull List<Fruit> objects) {
        super(context, resource, objects);
    }
    //每个子项被滚动到屏幕内的时候会被调用
    @NonNull
    @Override
    public View getView(int position, @Nullable View convertView, @NonNull ViewGroup parent) {
        Fruit fruit = getItem(position); //得到当前项的 Fruit 实例
        View view;
        if(convertView == null){
            //为每一个子项加载设定的布局
            view = LayoutInflater.from(getContext()).inflate(R.layout.fruit_item, parent, false);
        }else{
            view = convertView;
        }
        //分别获取 ImageView 和 TextView 的实例
        ImageView fruitimage = view.findViewById(R.id.fruit_image);
        TextView fruitname = view.findViewById(R.id.fruit_name);
        TextView fruitprice = view.findViewById(R.id.fruit_price);
        //设置要显示的图片和文字
        fruitimage.setImageResource(fruit.getImageID());
        fruitname.setText(fruit.getName());
        fruitprice.setText(fruit.getPrice());
        return view;
    }
}
```

这里在 getView()方法中进行了判断,如果 convertView 为空,则使用 LayoutInflater 去加载布局,如果不为空则直接对 convertView 进行重用。这样就大大提高了 ListView 的运行效率,在快速滚动的时候也可以表现出更好的性能。

不过,目前这份代码还是可以继续优化的,虽然现在已经不会再重复去加载布局,但是

每次在 getView()方法中还是会调用 View 的 findViewById()方法来获取一次控件的实例。可以借助一个 ViewHolder 来对这部分性能进行优化,修改 FruitAdapter 中的代码,具体如下。

```java
public class FruitAdapter extends ArrayAdapter<Fruit> {
    //用于将上下文、ListView 子项布局的 id 和数据都传递过来
    public FruitAdapter(@NonNull Context context, int resource, @NonNull List<Fruit> objects) {
        super(context, resource, objects);
    }
    //每个子项被滚动到屏幕内的时候会被调用
    @NonNull
    @Override
    public View getView(int position, @Nullable View convertView, @NonNull ViewGroup parent) {
        Fruit fruit = getItem(position); //得到当前项的 Fruit 实例
        View view;
        //新增一个内部类 ViewHolder,用于对控件的实例进行缓存
        ViewHolder viewHolder;
        if(convertView == null){
            //为每一个子项加载设定的布局
            view = LayoutInflater.from(getContext()).inflate(R.layout.fruit_item, parent, false);
            viewHolder = new ViewHolder();
            //分别获取 ImageView 和 TextView 的实例
            viewHolder.fruitimage = view.findViewById(R.id.fruit_image);
            viewHolder.fruitname = view.findViewById(R.id.fruit_name);
            viewHolder.fruitprice = view.findViewById(R.id.fruit_price);
            view.setTag(viewHolder); //将 viewHolder 存储在 view 中
        }else{
            view = convertView;
            viewHolder = (ViewHolder) view.getTag(); //重新获取 viewHolder
        }
        //设置要显示的图片和文字
        viewHolder.fruitimage.setImageResource(fruit.getImageID());
        viewHolder.fruitname.setText(fruit.getName());
        viewHolder.fruitprice.setText(fruit.getPrice());
        return view;
    }
    private class ViewHolder {
        ImageView fruitimage;
        TextView fruitname;
        TextView fruitprice;
    }
}
```

这里新增了一个内部类 ViewHolder,用于对控件的实例进行缓存。当 convertView 为空的时候,创建一个 ViewHolder 对象,并将控件的实例都存放在 ViewHolder 里,然后调用 View 的 setTag()方法,将 ViewHolder 对象存储在 View 中。当 convertView 不为空的时

候则调用 View 的 getTag()方法,把 ViewHolder 重新取出。这样所有控件的实例都缓存在 ViewHolder 里了,就没有必要每次都通过 findViewById()方法来获取控件实例了。通过这两步的优化之后,ListView 的运行效率就已经非常不错了。

6.1.4 ListView 的单击事件

微课视频

ListView 的滚动毕竟只是满足了人们视觉上的效果,可是如果 ListView 中的子项不能单击的话,这个控件就没有什么实际的用途了。因此,本节就来学习一下 ListView 如何才能响应用户的单击事件。

1. 案例分析

(1)界面分析。布局界面同提升 ListView 的运行效率案例的布局。

(2)设计思路。使用 setOnItemClickListener()方法来为 ListView 注册一个监听器,根据用户单击的 position 参数获取 Fruit 实例,通过 Tosat 提示显示出水果名称。

2. 实现步骤

(1)无须创建新工程,在提升 ListView 的运行效率案例的工程 ListView2 中继续完善程序即可。

(2)修改 MainActivity 中的代码,如下。

```
public class MainActivity extends AppCompatActivity {
    //第一步:定义对象
    ListView listView;
    @Override
    protected void onCreate(Bundle savedInstanceState) {
        super.onCreate(savedInstanceState);
        setContentView(R.layout.activity_main);
        //第二步:绑定控件
        listView = (ListView) findViewById(R.id.list_view);
        //第三步:准备数据
        final List<Fruit> fruitlist = new ArrayList<>();
        for (int i = 0; i < 4 ; i++) {
            Fruit pineapple = new Fruit(R.drawable.pineapple,"菠萝","¥16.9元/ kg ");
            fruitlist.add(pineapple);
            Fruit mango = new Fruit(R.drawable.mango, "芒果","¥29.9元/kg");
            fruitlist.add(mango);
            Fruit pomegranate = new Fruit(R.drawable.pomegranate, "石榴","¥15元/kg");
            fruitlist.add(pomegranate);
            Fruit grape = new Fruit(R.drawable.grape, "葡萄","¥19.9元/kg");
            fruitlist.add(grape);
            Fruit apple = new Fruit(R.drawable.apple, "苹果","¥20元/kg");
            fruitlist.add(apple);
            Fruit orange = new Fruit(R.drawable.orange, "橙子","¥18.8元/kg");
            fruitlist.add(orange);
            Fruit watermelon = new Fruit(R.drawable.watermelon, "西瓜","¥28.8元/kg");
            fruitlist.add(watermelon);
        }
```

```
//第四步:设计每一个列表项的子布局
//第五步:定义适配器
    FruitAdapter adapter = new FruitAdapter(MainActivity.this, R.layout.fruit_item, fruitlist);
    listView.setAdapter(adapter);
//第六步:ListView的单击事件
    listView.setOnItemClickListener(new AdapterView.OnItemClickListener() {
        @Override
        public void onItemClick(AdapterView <?> adapterView, View view, int position, long id) {
            Fruit fruit = fruitlist.get(position);
            Toast.makeText(MainActivity.this,"您选择的水果是:" + fruit.getName(),Toast.LENGTH_LONG).show();
        }
    });
    }
}
```

使用 setOnItemClickListener()方法来为 ListView 注册一个监听器,当用户单击 ListView 的任何一个子项时就会回调 onItemClick()方法,在这个方法中通过 position 参数判断出用户单击的是哪一个子项,然后获取到相应的水果,并通过 Toast 将水果的名字显示出来。重新运行程序,并单击"菠萝",效果如图 6.4 所示。

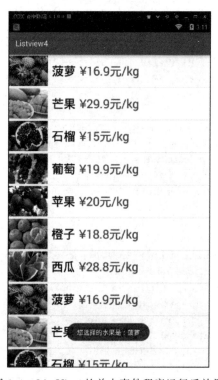

图 6.4 ListView 的单击事件程序运行后效果

6.2 更强大的滚动控件——RecyclerView

RecyclerView 是 Android 5.0 之后 Google 推出的一个用于在有限的窗口中展示大量数据集的控件。它可以称为增强版的 ListView，不仅可以轻松实现和 ListView 一样的效果，同时又优化了 ListView 的不足之处。RecyclerView 比 ListView 更高级且更具灵活性，它是一个用于显示庞大数据集的容器，可通过保持有限数量的视图进行非常有效的滚动操作。如果有数据集合，其中的元素将因用户操作或网络事件而在运行时发生改变，请使用 RecyclerView 控件。

6.2.1 RecyclerView 的简单用法

1. 案例分析

（1）界面分析。布局界面中只有一个控件 RecyclerView，是用来显示列表的控件。

（2）设计思路。布局界面中的控件通过添加属性的方式达到用户需求。为了使 RecyclerView 实现和 ListView 一样的效果，准备一份相同的水果图片、相同的 Fruit 类及相同的定制好的每一行的子布局 fruit_item.xml。自定义适配器 FruitAdapter，将数据源、子布局和 RecyclerView 关联起来，设置 RecyclerView 以线性布局默认垂直方向显示列表内容。

2. 实现步骤

（1）创建一个新的工程，工程名为 RecyclerView 1。

（2）切换工程的 Project 项目结构，选择该模式下方的 app，依次展开，便看到工程的布局界面和工程的类文件，其中，activity_main.xml 是布局界面，MainActivity.java 为类文件。

（3）RecyclerView 属于新增的控件，要想使用首先要在 app/build.gradle 文件中添加依赖库，打开 app 目录下的 build.gradle 文件，在 dependencies 闭包中添加如下内容。

```
dependencies {
    implementation fileTree(dir: 'libs', include: ['*.jar'])
    implementation 'androidx.appcompat:appcompat:1.1.0'
    implementation 'androidx.constraintlayout:constraintlayout:1.1.3'
    testImplementation 'junit:junit:4.12'
    androidTestImplementation 'androidx.test.ext:junit:1.1.1'
    androidTestImplementation 'androidx.test.espresso:espresso-core:3.2.0'
    implementation 'com.android.support:recyclerview-v7:29.0.3'
}
```

添加完之后记得要单击右上角的 Sync Now 来进行同步，此时 recyclerView 控件的依赖才算添加成功。

（4）修改布局界面。在 app 目录下的结构中，双击 layout 文件夹下的 activity_main.xml 文件，便可打开布局编辑器，切换到 Text 选项卡，输入如下代码。

```xml
<?xml version="1.0" encoding="utf-8"?>
<LinearLayout xmlns:android="http://schemas.android.com/apk/res/android"
    xmlns:app="http://schemas.android.com/apk/res-auto"
    xmlns:tools="http://schemas.android.com/tools"
    android:layout_width="match_parent"
    android:layout_height="match_parent"
    android:orientation="vertical"
    tools:context=".MainActivity">
    <androidx.recyclerview.widget.RecyclerView
        android:id="@+id/recyclerview"
        android:layout_width="match_parent"
        android:layout_height="match_parent"/>
</LinearLayout>
```

上述代码中,根节点为线性布局,添加 RecyclerView 控件,设置其高度和宽度为 match_parent,这样 RecyclerView 控件就占满了整个布局,因为在类文件中要调用 RecyclerView,所以给它设置一个 id。需要注意的是,由于 RecyclerView 并不是内置在系统 SDK 当中的,所以需要把完整的包路径写出来。

(5)定义一个实体类,作为 RecyclerView 适配器的适配类型。右击 com.example.recyclerview1 包→New→Java Class,会弹出一个创建类的对话框,将类命名为 Fruit,代码如下。

```java
public class Fruit {
    private int imageID;
    private String name;
    private String price;
    //按 Alt+Insert 组合键,在弹出的快捷菜单中选择 Getter,弹出对话框中的三个属性都选中,
    //单击 OK 按钮,即可出现下面三个方法.
    public int getImageID() {
        return imageID;
    }
    public String getName() {
        return name;
    }
    public String getPrice() {
        return price;
    }
    //按 Alt+Insert 组合键,在弹出的快捷菜单中选择 Constructor,对话框中的三个属性全选,单
    //击 OK 按钮,即可出现下面的构造方法.
    public Fruit(int imageID, String name, String price) {
        this.imageID = imageID;
        this.name = name;
        this.price = price;
    }
}
```

Fruit 类中有三个字段,imageID 是水果对应图片的资源 id,name 是水果的名称,price

是水果的价格。

（6）准备好一组水果图片，将其复制粘贴到 res/drawable 文件夹中。

（7）为 RecyclerView 的每一行子项指定要显示的布局样式，右击 Layout→New→Xml→Layout Xml File，会弹出一个 Xml Layout File 对话框，命名为 fruit_item.xml，代码如下。

```xml
<?xml version = "1.0" encoding = "utf-8"?>
<LinearLayout xmlns:android = "http://schemas.android.com/apk/res/android"
    android:layout_width = "match_parent"
    android:orientation = "horizontal"
    android:layout_height = "wrap_content">
    <ImageView
        android:id = "@+id/fruit_image"
        android:src = "@drawable/apple"
        android:layout_width = "100dp"
        android:layout_height = "80dp"/>
    <TextView
        android:id = "@+id/fruit_name"
        android:layout_gravity = "center_vertical"
        android:textSize = "30sp"
        android:textColor = "#000000"
        android:text = "name"
        android:layout_marginLeft = "10dp"
        android:layout_width = "wrap_content"
        android:layout_height = "wrap_content"/>
    <TextView
        android:id = "@+id/fruit_price"
        android:layout_gravity = "center_vertical"
        android:textColor = "#ff0000"
        android:text = "price"
        android:textSize = "30sp"
        android:layout_marginLeft = "10dp"
        android:layout_width = "wrap_content"
        android:layout_height = "wrap_content"/>
</LinearLayout>
```

上述代码中，根节点为线性布局，方向为 horizontal，定义了一个 ImageView 用于显示水果的图片，又定义了两个 TextView 用于显示水果的名称和水果的价格，并且垂直居中显示。

（8）接下来需要为 RecyclerView 准备一个自定义的适配器，右击 com.example.recyclerview1 包 → New → Java Class，会弹出一个创建类的对话框，将类命名为 FruitAdapter，SuperClass 后面填写继承自父类 RecyclerView.Adapter，单击 OK 按钮，出现如下代码。

```java
public class FruitAdapter extends RecyclerView.Adapter {
}
```

上述代码中，在FruitAdapter类名处有红色的波浪线，用户无须担心，依次完善代码即可。首先将泛型指定为FruitAdapter.ViewHolder，其中，ViewHolder是在FruitAdapter中定义的一个内部类，由于此页面代码较多，可以采用下方步骤提示的快捷键快速生成部分代码，代码如下。

```java
//1.鼠标定位到FruitAdapter类名的末端,添加泛型指定为FruitAdapter.ViewHolder
//2.鼠标定位到FruitAdapter类名代码行的红色波浪线处,按Alt+Enter组合键重写三种方法
//3.鼠标定位到FruitAdapter类名代码行的ViewHolder上,按Alt+Enter组合键创建ViewHolder类
//4.鼠标放到内部类ViewHolder末端,添加RecyclerView.ViewHolder,按Alt+Enter组合键创建构
//造方法
//5.鼠标放在FruitAdapter类中,按Alt+Insert组合键添加构造方法
public class FruitAdapter extends RecyclerView.Adapter<FruitAdapter.ViewHolder> {
    private List<Fruit> mFruitlist;
    //ViewHolder 内部类
    public class ViewHolder extends RecyclerView.ViewHolder {
        ImageView fruitImage;
        TextView fruitName;
        TextView fruitPrice;
        public ViewHolder(@NonNull View view) {
            super(view);
            fruitImage = view.findViewById(R.id.fruit_image);
            fruitName = view.findViewById(R.id.fruit_name);
            fruitPrice = view.findViewById(R.id.fruit_price);
        }
    }
    //适配器的构造方法
    public FruitAdapter(List<Fruit> fruitList) {
        mFruitlist = fruitList;
    }
    //方法1:创建ViewHolder实例,将加载出来的布局传入到构造函数中,最后将ViewHolder实例返回
    @NonNull
    @Override
    public FruitAdapter.ViewHolder onCreateViewHolder(@NonNull ViewGroup parent, int viewType) {
        View view = LayoutInflater.from(parent.getContext()).inflate(R.layout.fruit_item, parent, false);
        ViewHolder holder = new ViewHolder(view);
        return holder;
    }
    //方法2:每个子项滚动到屏幕内执行,通过positon得到当前项的实例,然后显示图片和文字
    @Override
    public void onBindViewHolder(@NonNull FruitAdapter.ViewHolder holder, int position) {
        Fruit fruit = mFruitlist.get(position);
        holder.fruitImage.setImageResource(fruit.getImageID());
        holder.fruitName.setText(fruit.getName());
        holder.fruitPrice.setText(fruit.getPrice());
    }
    //方法3:返回数据源的长度
    @Override
```

```
        public int getItemCount() {
            return mFruitlist.size();
        }
    }
```

首先定义一个内部类 ViewHolder，继承自 RecyclerView.ViewHolder，ViewHolder 的构造函数中要传入一个 View 参数，这个 View 参数就是 RecyclerView 子项最外层的布局。在 ViewHolder 类中包含三个字段：fruitImage 水果图片、fruitName 水果名称和 fruitPrice 水果价格。通过 View 的 findViewById() 方法获取 ImageView 和 TextView 的实例。

FruitAdapter 的构造函数，用于把要展示的数据源传进来，并赋值给一个全局变量 mFruitlist，后续的操作都是在这个数据源上进行的。

由于 FruitAdapter 继承自 RecyclerView.Adapter，所以必须重写 onCreateViewHolder()、onBindViewHolder()、getItemCount() 这三个方法。onCreateViewHolder() 方法用于创建 ViewHolder 实例，在这个方法中将定制好的每一行的布局 fruit_item 加载进来，然后创建一个 ViewHolder 实例，将加载出来的布局传入构造函数中，最后将 ViewHolder 实例返回。onBindViewHolder() 方法用于对 RecyclerView 子项的数据进行赋值，会在每个子项滚动到屏幕内执行，这里通过 positon 参数得到当前项的 Fruit 实例，然后再将水果图片、水果名称和水果价格显示在对应的控件上。getItemCount() 方法是用于告诉 RecyclerView 一共有多少个子项，通过 size() 方法获得 RecyclerView 的数据源长度。

（9）打开 MainActivity.java 类文件，修改 MainActivity 的代码，如下。

```
public class MainActivity extends AppCompatActivity {
    //第一步:定义对象
    RecyclerView recyclerView;
    @Override
    protected void onCreate(Bundle savedInstanceState) {
        super.onCreate(savedInstanceState);
        setContentView(R.layout.activity_main);
        //第二步:绑定控件
        recyclerView = findViewById(R.id.recyclerview);
        //第三步:准备数据
        List<Fruit> fruitlist = new ArrayList<>();
        for (int i = 0; i < 2; i++) {
            Fruit pineapple = new Fruit(R.drawable.pineapple, "菠萝", "￥16.9元/kg");
            fruitlist.add(pineapple);
            Fruit mango = new Fruit(R.drawable.mango, "芒果","￥29.9元/kg");
            fruitlist.add(mango);
            Fruit pomegranate = new Fruit(R.drawable.pomegranate, "石榴","￥15元/kg");
            fruitlist.add(pomegranate);
            Fruit grape = new Fruit(R.drawable.grape, "葡萄","￥19.9元/kg");
            fruitlist.add(grape);
            Fruit apple = new Fruit(R.drawable.apple, "苹果","￥20元/kg");
            fruitlist.add(apple);
```

```
                Fruit orange = new Fruit(R.drawable.orange, "橙子","¥18.8元/kg");
                fruitlist.add(orange);
                Fruit watermelon = new Fruit(R.drawable.watermelon, "西瓜","¥28.8元/kg");
                fruitlist.add(watermelon);
        }
        //第四步:设计每一行的子布局
        //第五步:创建适配器
        FruitAdapter adapter = new FruitAdapter(fruitlist);
        //第六步:将数据以垂直线性布局的方式显示出来
        LinearLayoutManager layoutManager = new LinearLayoutManager(this);
        recyclerView.setLayoutManager(layoutManager);
        recyclerView.setAdapter(adapter);
    }
}
```

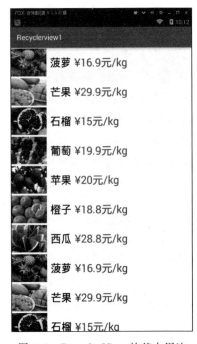

图 6.5　RecyclerView 的基本用法程序运行后效果

首先初始化水果数据,定义一个动态数组 ArrayList 来存放 RecyclerView 中 item 的内容,泛型为新建类 Fruit,在 Fruit 的构造函数中将水果图片的 id、水果名称和水果价格传入,然后将创建好的对象添加到水果列表中。因为数据量较少,不足以填充满整个屏幕,所以使用了 for 循环将数据添加了两遍。接着创建了 FruitAdapter 对象,并将 FruitAdapter 适配器传递给 RecyclerView,实现 RecyclerView 和数据的关联。和 ListView 不同的是,RecyclerView 通过不同的布局管理器来控制 item 的排列顺序,负责 item 元素的布局和复用。RecyclerView 提供了三种布局管理器:LinearLayoutManager 为线性布局,以垂直或水平滚动列表方式显示项目;GridLayoutManager 为网格布局,在网格中显示项目;StaggeredGridLayoutManager 为瀑布流布局,在分散对齐网格中显示项目。这里创建一个 LinearLayoutManager 对象,线性布局默认的是垂直方向,可以实现和 ListView 案例一样的效果,将 LinearLayoutManager 对象通过 setLayoutManager()传递给 RecyclerView,这样 RecyclerView 就以线性布局垂直方向显示列表内容了。

(10) 运行程序,效果如图 6.5 所示,我们使用 RecyclerView 实现了和 ListView 几乎一模一样的效果,虽然代码量没有减少,但是逻辑变得更加清晰了。

6.2.2　实现横向滚动和瀑布流布局

ListView 只能实现纵向滚动的效果,如果想实现横向滚动的话,就只能采用功能更强大的 RecyclerView 控件了。

微课视频

1. 横向滚动案例

1)案例分析

(1)界面分析。布局界面同 RecyclerView 基本用法案例的布局。

(2)设计思路。要想实现横向滚动首先修改 fruit_item.xml 布局里的控件为垂直排列,然后修改布局管理器为线性水平方向。

2)实现步骤

(1)无须创建新工程,在 RecyclerView 基本用法案例的工程中 RecyclerView1 继续完善程序即可。

(2)修改 fruit_item.xml 布局文件,因为目前这个布局里面的三个控件是水平排列的,适用于纵向滚动的场景,如果要实现横向滚动的话,应该把 fruit_item.xml 布局文件中的三个控件改成垂直排列,修改代码如下。

```xml
<?xml version = "1.0" encoding = "utf-8"?>
<LinearLayout xmlns:android = "http://schemas.android.com/apk/res/android"
    android:layout_width = "120dp"
    android:orientation = "vertical"
    android:layout_height = "wrap_content">
    <ImageView
        android:id = "@+id/fruit_image"
        android:src = "@drawable/apple"
        android:layout_gravity = "left"
        android:layout_width = "100dp"
        android:layout_height = "80dp"/>
    <TextView
        android:id = "@+id/fruit_name"
        android:layout_marginLeft = "20dp"
        android:textSize = "30sp"
        android:textColor = "#000000"
        android:text = "name"
        android:layout_marginTop = "10dp"
        android:layout_width = "wrap_content"
        android:layout_height = "wrap_content"/>
    <TextView
        android:id = "@+id/fruit_price"
        android:layout_gravity = "center_horizontal"
        android:textColor = "#ff0000"
        android:text = "price"
        android:textSize = "20sp"
        android:layout_marginTop = "10dp"
        android:layout_width = "wrap_content"
        android:layout_height = "wrap_content"/>
</LinearLayout>
```

可以看到,将线性布局的方向改成垂直方向,并将宽度设成 120dp,这里将宽度设置成固定的值是因为每种水果文字的长度不一样,如果用 wrap_content 的话,RecyclerView 的子项就会有长有短,看起来很不美观;如果用 match_parent 的话,就会导致一个子项占满整个屏幕空间。将 ImageView 控件设置为在布局中左对齐。将显示水果名称的 TextView 控件设置距离左边距 20dp。将显示水果价格的 TextView 控件设置为在布局中水平居中对齐,距离上方水果名称的 TextView 控件 10dp,这样文字间就保持了一定的间距,为了美观将文字大小修改成 20sp。

(3) 修改 MainActivity.java 类文件的代码,修改代码如下。

```java
public class MainActivity extends AppCompatActivity {
    //第一步:定义对象
    RecyclerView recyclerView;
    @Override
    protected void onCreate(Bundle savedInstanceState) {
        super.onCreate(savedInstanceState);
        setContentView(R.layout.activity_main);
        //第二步:绑定控件
        recyclerView = findViewById(R.id.recyclerview);
        //第三步:准备数据
        List<Fruit> fruitlist = new ArrayList<>();
        for (int i = 0; i < 2; i++) {
            Fruit pineapple = new Fruit(R.drawable.pineapple, "菠萝", "￥16.9元/kg");
            fruitlist.add(pineapple);
            Fruit mango = new Fruit(R.drawable.mango, "芒果","￥29.9元/kg");
            fruitlist.add(mango);
            Fruit pomegranate = new Fruit(R.drawable.pomegranate, "石榴","￥15元/kg");
            fruitlist.add(pomegranate);
            Fruit grape = new Fruit(R.drawable.grape, "葡萄","￥19.9元/kg");
            fruitlist.add(grape);
            Fruit apple = new Fruit(R.drawable.apple, "苹果","￥20元/kg");
            fruitlist.add(apple);
            Fruit orange = new Fruit(R.drawable.orange, "橙子","￥18.8元/kg");
            fruitlist.add(orange);
            Fruit watermelon = new Fruit(R.drawable.watermelon, "西瓜","￥28.8元/kg");
            fruitlist.add(watermelon);
        }
        //第四步:设计每一行的子布局
        //第五步:创建适配器
        FruitAdapter adapter = new FruitAdapter(fruitlist);
        //第六步:让数据显示在 RecyclerView 控件上
        LinearLayoutManager layoutManager = new LinearLayoutManager(this);
        //让布局横向排列
        layoutManager.setOrientation(LinearLayoutManager.HORIZONTAL);
        recyclerView.setLayoutManager(layoutManager);
        recyclerView.setAdapter(adapter);
    }
}
```

在 MainActivity.java 中加入一行代码,通过 setOrientation() 方法将 layoutManager 的布局排列方向设置成水平方向,默认的是垂直方向,这样就可以让 RecyclerView 实现横向滚动了。

(4) 重新运行一下程序,效果如图6.6所示。

2. 瀑布流布局案例

1) 案例分析

(1) 界面分析。布局界面中同 RecyclerView 的横向滚动案例的布局。

(2) 设计思路。要想实现瀑布流布局首先修改 fruit_item.xml 布局,然后修改布局管理器为瀑布流布局。

2) 实现步骤

(1) 无须创建新工程,在 RecyclerView 横向滚动的工程中继续完善程序即可。

(2) 修改 fruit_item.xml 布局文件,因为目前这个布局里面的三个控件是水平排列的,适用于纵向滚动的场景,如果要实现横向滚动的话,应该把 fruit_item.xml 布局文件中的三个控件改成垂直排列,修改代码如下。

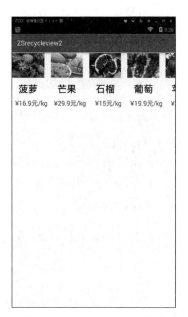

图 6.6 RecyclerView 的横向滚动程序运行后效果

```xml
<?xml version = "1.0" encoding = "utf-8"?>
<LinearLayout xmlns:android = "http://schemas.android.com/apk/res/android"
    android:layout_width = "match_parent"
    android:orientation = "vertical"
    android:layout_margin = "5dp"
    android:layout_height = "wrap_content">
    <ImageView
        android:id = "@+id/fruit_image"
        android:src = "@drawable/apple"
        android:layout_gravity = "left"
        android:layout_width = "100dp"
        android:layout_height = "80dp"/>
    <TextView
        android:id = "@+id/fruit_name"
        android:layout_width = "wrap_content"
        android:layout_height = "wrap_content"
        android:layout_gravity = "left"
        android:layout_marginTop = "10dp"
        android:layout_margin = "15dp"
        android:text = "name"
        android:textColor = "#000000"
        android:textSize = "30sp" />
    <TextView
        android:id = "@+id/fruit_price"
        android:layout_gravity = "left"
        android:textColor = "#ff0000"
        android:text = "price"
        android:textSize = "20sp"
```

```xml
            android:layout_marginTop = "10dp"
            android:layout_width = "wrap_content"
            android:layout_height = "wrap_content"/>
</LinearLayout>
```

首先将 LinearLayout 的宽度由固定值 120dp 修改成 match_parent,因为瀑布流布局的宽度应该根据布局的列数来自动适配,而不是一个固定值。另外设置 layout_margin 属性为 5dp,子项之间保持一定的间距,这样所有的子项就不会紧贴在一起。再将 ImageView 控件和两个 TextView 控件的对齐属性修改成居左对齐。

(3) 修改 MainActivity.java 类文件的代码,修改代码如下。

```java
public class MainActivity extends AppCompatActivity {
    //第一步:定义对象
    RecyclerView recyclerView;
    @Override
    protected void onCreate(Bundle savedInstanceState) {
        super.onCreate(savedInstanceState);
        setContentView(R.layout.activity_main);
        //第二步:绑定控件
        recyclerView = findViewById(R.id.recyclerview);
        //第三步:准备数据
        List< Fruit > fruitlist = new ArrayList<>();
        for (int i = 0; i < 2 ; i++) {
            Fruit pineapple = new Fruit(R.drawable.pineapple, "菠萝", "￥16.9元/kg");
            fruitlist.add(pineapple);
            Fruit mango = new Fruit(R.drawable.mango, "芒果","￥29.9元/kg");
            fruitlist.add(mango);
            Fruit pomegranate = new Fruit(R.drawable.pomegranate, "石榴","￥15元/kg");
            fruitlist.add(pomegranate);
            Fruit grape = new Fruit(R.drawable.grape, "葡萄","￥19.9元/kg");
            fruitlist.add(grape);
            Fruit apple = new Fruit(R.drawable.apple, "苹果","￥20元/kg");
            fruitlist.add(apple);
            Fruit orange = new Fruit(R.drawable.orange, "橙子","￥18.8元/kg");
            fruitlist.add(orange);
            Fruit watermelon = new Fruit(R.drawable.watermelon, "西瓜","￥28.8元/kg");
            fruitlist.add(watermelon);
        }
        //第四步:设计每一行的子布局
        //第五步:创建适配器
        FruitAdapter adapter = new FruitAdapter(fruitlist);
        //第六步:让数据显示在 RecyclerView 控件中
StaggeredGridLayoutManager layoutManager = new
StaggeredGridLayoutManager(3,StaggeredGridLayoutManager.VERTICAL);
recyclerView.setLayoutManager(layoutManager);
recyclerView.setAdapter(adapter);
    }
}
```

上述代码中,在 onCreate()方法中,创建了一个 StaggeredGridLayoutManager 瀑布流布局的实例,StaggeredGridLayoutManager 的构造函数接收两个参数,第一个参数用于指定布局的列数,这里传入 3,表示会把布局分成 3 列;第二个参数用于指定布局的排列方向,这里传入 StaggeredGridLayoutManager.VERTICAL 表示让布局垂直排列,最后再把创建好的实例设置到 RecyclerView 即可。

(4) 重新运行一下程序,效果如图 6.7 所示。

6.2.3 RecyclerView 的单击事件

和 ListView 一样,RecyclerView 也必须要响应单击事件才行,否则就没有实际应用意义了。ListView 依靠子项的 setOnItemClickListener() 方法可以实现子项的单击事件,但是如果想单击子项内部的某个控件,ListView 实现起来就非常复杂了。RecyclerView 直接摒弃了子项单击事件的监听器,所

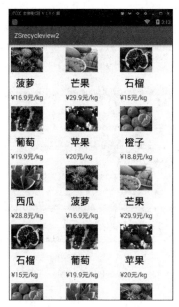

图 6.7 RecyclerView 的瀑布流布局程序运行后效果

微课视频

有的单击事件都由具体的 View 去注册,就可以轻松实现子项内部某个控件的单击事件了。

1. 案例分析

(1) 界面分析。布局界面同 RecyclerView 瀑布流布局案例的布局。

(2) 设计思路。首先在 ViewHolder 中添加了一个保存子项最外层布局的变量 fruitView,然后在 onCreateViewHolder() 方法中添加注册单击事件,分别给子项最外层布局和图片注册单击事件。

2. 实现步骤

(1) 无须创建新工程,在 RecyclerView 瀑布流布局的工程中继续完善程序即可。

(2) 修改 FruitAdapter.java 中的代码,修改如下。

```
public class FruitAdapter extends RecyclerView.Adapter<FruitAdapter.ViewHolder> {
    private List<Fruit> mFruitlist;
public class ViewHolder extends RecyclerView.ViewHolder {
//添加了 fruitView 变量来保存子项最外层布局的实例
    View fruitView;
        ImageView fruitImage;
        TextView fruitName;
        TextView fruitPrice;
        public ViewHolder(@NonNull View view) {
            super(view);
            fruitView = view;
            fruitImage = view.findViewById(R.id.fruit_image);
            fruitName = view.findViewById(R.id.fruit_name);
            fruitPrice = view.findViewById(R.id.fruit_price);
```

```java
        }
    }
    public FruitAdapter(List<Fruit> fruitList) {
        mFruitlist = fruitList;
    }
    //方法1:创建 ViewHolder 实例,将加载出来的布局传入到构造函数中,最后将 ViewHolder 实例返回
    @NonNull
    @Override
    public FruitAdapter.ViewHolder onCreateViewHolder(@NonNull final ViewGroup parent, int viewType) {
        View view = LayoutInflater.from(parent.getContext()).inflate(R.layout.fruit_item, parent, false);
        final ViewHolder holder = new ViewHolder(view);
        //布局的单击事件
        holder.fruitView.setOnClickListener(new View.OnClickListener() {
            @Override
            public void onClick(View view) {
                int position = holder.getAdapterPosition();
                Fruit fruit = mFruitlist.get(position);
                Toast.makeText(view.getContext(),"您单击的布局为:" + fruit.getName(),Toast.LENGTH_LONG).show();
            }
        });
        //图片的单击事件
        holder.fruitImage.setOnClickListener(new View.OnClickListener() {
            @Override
            public void onClick(View view) {
                int position = holder.getAdapterPosition();
                Fruit fruit = mFruitlist.get(position);
                Toast.makeText(view.getContext(),"您单击的图片为:" + fruit.getName(),Toast.LENGTH_SHORT).show();
            }
        });
        return holder;
    }
    //方法2:每个子项滚动到屏幕内执行,通过 positon 得到当前项的实例,然后显示图片和文字
    @Override
    public void onBindViewHolder(@NonNull FruitAdapter.ViewHolder holder, int position) {
        Fruit fruit = mFruitlist.get(position);
        holder.fruitImage.setImageResource(fruit.getImageID());
        holder.fruitName.setText(fruit.getName());
        holder.fruitPrice.setText(fruit.getPrice());
    }
    //方法3:返回数据源的长度
    @Override
    public int getItemCount() {
        return mFruitlist.size();
    }
}
```

首先修改 ViewHolder 类文件，在 ViewHolder 中添加一个变量 fruitView，用来保存子项最外层布局的实例。然后在 onCreateViewHolder()方法中添加注册单击事件，这里分别给子项最外层布局和图片都注册了单击事件，在两个单击事件中，通过 getAdapterPosition()方法获得单击的位置 position，然后通过 position 获得单击的 Fruit 实例，最后通过 Toast 弹出单击的是布局或者是图片两种不同的内容。

（3）重新运行程序，并单击芒果的图片，效果如图 6.8 所示。再单击苹果的文字部分，因为 TextView 并没有注册单击事件，因此单击文字这个事件就会被子项最外面的布局捕获到，效果如图 6.9 所示。

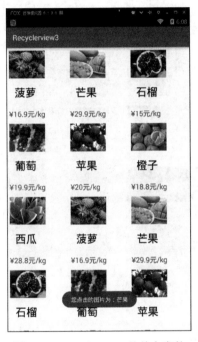
图 6.8　RecyclerView 的单击事件
　　　——单击芒果的图片效果

图 6.9　RecyclerView 的单击事件
　　　——单击苹果的文字效果

6.3　实战案例——多彩水果店 App

前面学习了 UI 开发的基础知识，下面综合运用前面所学的大量知识来完成一个多彩水果店 App 的开发实战，包括用户登录界面，通过输入用户名和密码登录跳转到水果列表界面，在水果列表界面可以滑动手指上下翻看所有水果的图片、水果名称、水果价格及水果介绍，单击某一个选中的水果可以跳转到水果详情页面，详细了解每个品种的水果。

6.3.1　界面分析

本案例中包含三个界面：登录界面、主界面和水果详情界面，效果如图 6.10～图 6.12 所示。

图 6.10　登录界面效果图

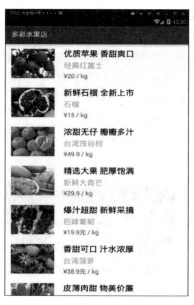
图 6.11　主界面效果图

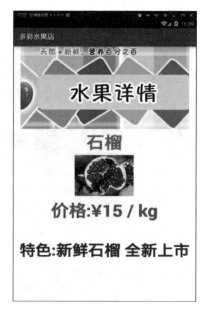
图 6.12　水果详情界面效果图

6.3.2　实现思路

微课视频

第一个界面是登录界面,用来保护用户信息的安全,采用 SharedPreference 键值对存储技术来保存用户名和密码,复选框 CheckBox 供用户选择是否需要记住密码,单击"登录"按钮保存信息并跳转到主界面。第二个界面是主界面,以纵向滚动列表方式显示所有水果信息,采用 RecyclerView 滚动控件实现。第三个界面是水果详情界面,用来查看水果列表中用户单击水果的详细信息。

6.3.3 任务实施

【任务 6-1】 登录界面的设计与功能

(1) 创建一个新的工程,工程名为 Fruitshop。

(2) 切换工程的 Project 项目结构,选择该模式下方的 app,依次展开,便看到工程的布局界面和工程的类文件,其中,activity_main.xml 是布局界面,MainActivity.java 为类文件。

(3) 在工程中添加一个新的页面。右击 com.example.fruitshop 包→New→Activity→Empty Activity,会弹出一个创建活动的对话框,将活动命名为 LoginActivity,默认勾选 Generate Layout File 关联布局界面,布局界面名称为 activity_login,但不要勾选 Launcher Activity。单击 Finish 按钮,便可在工程中完成第二个页面的添加。

(4) 选择该模式下方的 app→src→main→AndroidManifest.xml 依次展开,双击打开 AndroidManifest.xml 配置文件,在 AndroidManifest 文件中含有如下过滤器的 Activity 组件为默认启动类,当程序启动时系统自动调用它,这个功能常用来做启动界面,代码如下。

```xml
<intent-filter>
    <action android:name="android.intent.action.MAIN" />
    <category android:name="android.intent.category.LAUNCHER" />
</intent-filter>
```

<intent-filter>中第一个 action 动作中的.MAIN 为设置程序的主入口,category 语句表示在程序的启动列表中出现该图标的 activity,缺少此 category 在程序启动列表中就找不到该应用的图标,整个 intent-filter 的作用是用于声明 Activity 的 Intent 过滤规则。

要想实现程序从 LoginActivity 启动,修改代码如下。

```xml
<?xml version="1.0" encoding="utf-8"?>
<manifest xmlns:android="http://schemas.android.com/apk/res/android"
    package="com.example.fruitshop">
    <application
        android:allowBackup="true"
        android:icon="@mipmap/ic_launcher"
        android:label="@string/app_name"
        android:roundIcon="@mipmap/ic_launcher_round"
        android:supportsRtl="true"
        android:theme="@style/AppTheme">
        <activity android:name=".LoginActivity">
            <intent-filter>
                <action android:name="android.intent.action.MAIN" />
                <category android:name="android.intent.category.LAUNCHER" />
            </intent-filter>
        </activity>
        <activity android:name=".MainActivity">
        </activity>
    </application>
</manifest>
```

代码中<activity android:name=".LoginActivity">节点中包含<intent-filter>过滤规则，说明当前程序启动时使用 LoginActivity 作为程序主入口。

（5）准备一张图片，图片名为 fruitshop.jpg，将其粘贴到 app 目录结构中 res 下方的 drawable 文件夹下，作为登录页面的背景图片。

（6）修改布局界面。在 app 目录下的结构中，双击 layout 文件夹下的 activity_login.xml 文件，便可打开布局编辑器，切换到 Text 选项卡，输入代码如下。

```xml
<?xml version = "1.0" encoding = "utf-8"?>
<LinearLayout xmlns:android = "http://schemas.android.com/apk/res/android"
    xmlns:app = "http://schemas.android.com/apk/res-auto"
    xmlns:tools = "http://schemas.android.com/tools"
    android:layout_width = "match_parent"
    android:layout_height = "match_parent"
    android:background = "@drawable/fruitshop"
    android:orientation = "vertical" >
    <LinearLayout
        android:layout_width = "match_parent"
        android:layout_height = "wrap_content"
        android:orientation = "horizontal"
        android:layout_marginLeft = "10dp"
        android:layout_marginTop = "280dp">
        <TextView
            android:layout_width = "wrap_content"
            android:layout_height = "wrap_content"
            android:text = "用户名:"
            android:textColor = "#000000"
            android:textSize = "30sp"           />
        <EditText
            android:id = "@+id/editText_inputname"
            android:layout_width = "match_parent"
            android:layout_height = "wrap_content"
            android:textColor = "#000000"
            android:textSize = "30sp"           />
    </LinearLayout>
    <LinearLayout
        android:layout_width = "match_parent"
        android:layout_height = "wrap_content"
        android:orientation = "horizontal"
        android:layout_margin = "10dp">
        <TextView
            android:layout_width = "wrap_content"
            android:layout_height = "wrap_content"
            android:text = "密    码:"
            android:textColor = "#000000"
            android:textSize = "30sp"           />
        <EditText
            android:id = "@+id/editText_inputpwd"
            android:layout_width = "match_parent"
```

```xml
            android:layout_height = "wrap_content"
            android:textColor = "#000000"
            android:inputType = "textPassword"
            android:textSize = "30sp"            />
    </LinearLayout>
    <CheckBox
        android:id = "@+id/checkBox_reme"
        android:layout_width = "wrap_content"
        android:layout_height = "wrap_content"
        android:text = "记住密码"
        android:layout_gravity = "right"
        android:layout_margin = "10dp"/>
    <LinearLayout
        android:layout_width = "match_parent"
        android:layout_height = "wrap_content"
        android:orientation = "horizontal"
        android:layout_margin = "10dp">
        <Button
            android:id = "@+id/button_login"
            android:layout_width = "wrap_content"
            android:layout_height = "wrap_content"
            android:text = "登录"
            android:textColor = "#000000"
            android:textSize = "30sp"
            android:layout_weight = "1"/>
        <Button
            android:layout_width = "wrap_content"
            android:layout_height = "wrap_content"
            android:text = "取消"
            android:textColor = "#000000"
            android:textSize = "30sp"
            android:layout_weight = "1"/>
    </LinearLayout>
</LinearLayout>
```

上述代码中,根节点为线性布局,方向为 vertical,背景图片设置为 fruitshop.jpg,用户名的 TextView 控件和输入用户名的 EditText 控件为嵌套的线性布局,方向为水平方向。同样,密码的 TextView 控件和输入密码的 EditText 控件也是嵌套的水平方向的线性布局。复选框 checkBox 控件,方便用户选择是否记住密码。最下方嵌套的水平方向的线性布局,包含两个 Button 控件,用来实现数据存储和跳转功能。

(7) 打开 LoginActivity.java 类文件,修改 LoginActivity 的代码,具体如下。

```java
public class LoginActivity extends AppCompatActivity {
    //定义对象
    EditText edit_inputname,edit_inputpwd;
    CheckBox check_reme;
    Button btn_login;
    @Override
```

```java
protected void onCreate(Bundle savedInstanceState) {
    super.onCreate(savedInstanceState);
    setContentView(R.layout.activity_login);
    //1.绑定控件
    initView();
    //2.单击"登录"按钮,将用户名和密码保存起来
    btnloginonClick();
    //3.下次启动,直接显示用户名和密码
    displayinfo();
}
//1.绑定控件
private void initView() {
    edit_inputname = findViewById(R.id.editText_inputname);
    edit_inputpwd = findViewById(R.id.editText_inputpwd);
    check_reme = findViewById(R.id.checkBox_reme);
    btn_login = findViewById(R.id.button_login);
}
//2.单击"登录"按钮,将用户名和密码保存起来
private void btnloginonClick() {
    btn_login.setOnClickListener(new View.OnClickListener() {
        @Override
        public void onClick(View view) {
            //保存用户名和密码
            SharedPreferences.Editor editor = getSharedPreferences("myinfo",0).edit();
            editor.putString("name",edit_inputname.getText().toString());
            editor.putString("pwd",edit_inputpwd.getText().toString());
            editor.putBoolean("st",check_reme.isChecked());
            editor.commit();
            //跳转到主页面
            Intent intent = new Intent(LoginActivity.this,MainActivity.class);
            startActivity(intent);
        }
    });
}
//3.下次启动,直接显示用户名和密码
private void displayinfo() {
    String strname = getSharedPreferences("myinfo",0).getString("name","");
    String strpwd = getSharedPreferences("myinfo",0).getString("pwd","");
    Boolean status = getSharedPreferences("myinfo",0).getBoolean("st",false);
    if(status == true){
        edit_inputname.setText(strname);
        edit_inputpwd.setText(strpwd);
        check_reme.setChecked(true);
    }else{
        edit_inputname.setText("");
        edit_inputpwd.setText("");
        check_reme.setChecked(false);
    }
}
}
```

上述代码中,在按钮单击事件内部,首先使用 SharedPreferences 将输入的用户名、密码、复选框的状态存储起来,然后实现页面的跳转,即从当前页面跳转到水果列表主页面。下次启动时,从 SharedPreferences 中获取用户名、密码和复选框的状态等数据信息,根据获取的复选框的状态进行判断,如果为真,说明用户选中了"记住密码",此时将获取的用户名和密码显示到 EditText 控件上,并将复选框设置为选中。如果为假,说明用户没有选中"记住密码",EditText 控件内容设置为空,并将复选框设置为未选中。

（8）设置生成的 App 的名字为"多彩水果店",打开 App→src→main→res→values→strings.xml,修改代码如下。

```
< resources >
    < string name = "app_name">多彩水果店</string >
</resources >
```

（9）运行程序,效果如图 6.13 所示,在界面中输入用户名和密码,选中"记住密码"复选框,单击"登录"按钮,将用户名和密码保存并跳转到主页面,再次运行程序,效果如图 6.14 所示,用户名和密码已经保存在登录界面。

图 6.13　登录界面运行效果图　　　　图 6.14　登录界面记住密码运行效果图

【任务 6-2】 水果列表界面的设计与功能

（1）修改布局界面。在 app 目录下的结构中,双击 layout 文件夹下的 activity_main.xml 文件,便可打开布局编辑器,切换到 Text 选项卡,输入如下代码。

（2）RecyclerView 属于新增的控件,要想使用首先要在 app/build.gradle 文件中添加依赖库,打开 app 目录下的 build.gradle 文件,在 dependencies 闭包中添加如下内容。

```
dependencies {
    implementation fileTree(dir: 'libs', include: ['*.jar'])
    implementation 'androidx.appcompat:appcompat:1.1.0'
    implementation 'androidx.constraintlayout:constraintlayout:1.1.3'
    testImplementation 'junit:junit:4.12'
    androidTestImplementation 'androidx.test.ext:junit:1.1.1'
    androidTestImplementation 'androidx.test.espresso:espresso-core:3.2.0'
    implementation 'com.android.support:recyclerview-v7:29.0.3'
}
```

添加完之后记得要单击一下右上角的 Sync Now 来进行同步。

(3) 修改布局界面。在 app 目录下的结构中，双击 layout 文件夹下的 activity_main.xml 文件，便可打开布局编辑器，切换到 Text 选项卡，输入代码如下。

```
<?xml version="1.0" encoding="utf-8"?>
<LinearLayout xmlns:android="http://schemas.android.com/apk/res/android"
    xmlns:app="http://schemas.android.com/apk/res-auto"
    xmlns:tools="http://schemas.android.com/tools"
    android:orientation="vertical"
    android:layout_width="match_parent"
    android:layout_height="match_parent"
    tools:context=".MainActivity">
    <androidx.recyclerview.widget.RecyclerView
        android:id="@+id/recyclerview"
        android:layout_width="match_parent"
        android:layout_height="match_parent"/>
</LinearLayout>
```

上述代码中，根节点为线性布局，添加 RecyclerView 控件，设置其高度和宽度为 match_parent，这样 RecyclerView 控件就占满了整个布局，因为在类文件中要调用 RecyclerView，所以给它设置一个 id。需要注意的是，由于 RecyclerView 并不是内置在系统 SDK 当中的，所以需要把完整的包路径写出来。

(4) 定义一个实体类，作为 RecyclerView 适配器的适配类型。右击 com.example.fruitshop 包→New→Java Class，会弹出一个创建类的对话框，将类命名为 Fruit，代码如下。

```
public class Fruit {
    private String name;
    private String Intro;
    private int imageId;
    private String Price;
    //下方代码可采用按 Alt+Insert 快捷键快速完成
    public Fruit(String name, int imageId, String Intro, String Price) {
        this.name = name;
        this.imageId = imageId;
        this.Intro = Intro;
        this.Price = Price;
    }
```

```java
    public String getName() {
        return name;
    }
    public int getImageId() {
        return imageId;
    }
    public String getIntro(){
        return Intro;
    }
    public String getPrice(){
        return Price;
    }
}
```

Fruit 类中有 4 个字段,name 是水果的名称,Intro 是水果介绍,imageID 是水果对应图片的资源 id,price 是水果的价格。

(5) 准备好一组水果图片,将其复制粘贴在 res/drawable 文件夹中。

(6) 为 RecyclerView 的每一行子项指定要显示的布局样式,右击 Layout→New→Xml→Layout Xml File,会弹出一个 Xml Layout File 对话框,命名为 fruit_item.xml,代码如下。

```xml
<?xml version = "1.0" encoding = "utf-8"?>
<RelativeLayout xmlns:android = "http://schemas.android.com/apk/res/android"
    android:layout_width = "match_parent"
    android:layout_height = "wrap_content">
    <ImageView
        android:id = "@ + id/fruitImg"
        android:layout_width = "120dp"
        android:layout_height = "120dp"
        android:layout_alignParentLeft = "true"
        android:layout_alignParentStart = "true"
        android:layout_centerVertical = "true"
        android:layout_marginLeft = "16dp"
        android:layout_marginStart = "16dp"/>
    <TextView
        android:id = "@ + id/fruitIntroduce"
        android:layout_toRightOf = "@ + id/fruitImg"
        android:layout_marginTop = "12dp"
        android:layout_marginLeft = "25dp"
        android:text = "TextView"
        android:layout_width = "match_parent"
        android:layout_height = "wrap_content"
        android:textSize = "24sp"
        android:textColor = "#000000" />
    <TextView
        android:id = "@ + id/fruitName"
        android:text = "TextView"
        android:layout_below = "@ + id/fruitIntroduce"
```

```xml
            android:layout_marginTop = "5dp"
            android:layout_alignLeft = "@+id/fruitIntroduce"
            android:layout_width = "match_parent"
            android:layout_height = "wrap_content"
            android:textSize = "22sp"
            android:textColor = "#FFBFBFBF" />
    <TextView
            android:id = "@+id/fruitPrice"
            android:text = "TextView"
            android:layout_below = "@+id/fruitName"
            android:layout_marginTop = "6dp"
            android:layout_alignLeft = "@+id/fruitIntroduce"
            android:layout_width = "match_parent"
            android:layout_height = "wrap_content"
            android:textSize = "18sp"
            android:textColor = "#FFE16531" />
</RelativeLayout>
```

上述代码中,根节点为相对布局,宽度为 match_parent,高度为 wrap_content,如果将高度设置为 match_parent,就会导致一个子项占满整个屏幕。我们定义了一个 ImageView 用于显示水果的图片,又定义了三个 TextView 用于显示水果名称和水果价格及水果介绍。

(7) 接下来需要为 RecyclerView 准备一个自定义的适配器,右击 com.example. fruitshop 包→New→Java Class,会弹出一个创建类的对话框,将类命名为 FruitAdapter,继承自 RecyclerView.Adapter,并将泛型指定为 FruitAdapter.ViewHolder。其中,ViewHolder 是我们在 FruitAdapter 中定义的一个内部类,代码如下。

```java
//1.鼠标定位到 FruitAdapter 类名的末端,添加泛型指定为 FruitAdapter.ViewHolder
//2.鼠标定位到 FruitAdapter 类名代码行的红色波浪线处,按 Alt+Enter 组合键重写三种方法
//3.鼠标定位到 FruitAdapter 类名代码行的 ViewHolder 上,按 Alt+Enter 组合键创建
//ViewHolder 类
//4.鼠标放到内部类 ViewHolder 末端,添加 RecyclerView.ViewHolder,按 Alt+Enter 组合键
//创建构造方法
//5.鼠标放在 FruitAdapter 类中,按 Alt+Insert 组合键添加构造方法
public class FruitAdapter extends RecyclerView.Adapter<FruitAdapter.ViewHolder> {
    private List<Fruit> mFruitList;
//ViewHolder 内部类
    static class ViewHolder extends RecyclerView.ViewHolder {
        ImageView fruitImage;
        TextView fruitName;
        TextView fruitIn;
        TextView fruitPrice;
        public ViewHolder(View view) {
            super(view);
            fruitImage = (ImageView) view.findViewById(R.id.fruitImg);
            fruitName = (TextView) view.findViewById(R.id.fruitName);
            fruitIn = (TextView) view.findViewById(R.id.fruitIntroduce);
            fruitPrice = (TextView) view.findViewById(R.id.fruitPrice);
```

```java
        }
    }
    //适配器的构造方法
    public FruitAdapter(List<Fruit> fruitList) {
        mFruitList = fruitList;
    }
    //方法1:用于创建ViewHolder实例
    @Override
    public ViewHolder onCreateViewHolder(ViewGroup parent, int viewType) {
        final View view = LayoutInflater.from(parent.getContext()).inflate(R.layout.fruit_item, parent, false);
        final ViewHolder holder = new ViewHolder(view);
            return holder;
    }
    //方法2:用于对RecyclerView中子项的数据进行赋值
    @Override
    public void onBindViewHolder(ViewHolder holder, int position) {
        Fruit fruit = mFruitList.get(position);
        holder.fruitImage.setImageResource(fruit.getImageId());
        holder.fruitName.setText(fruit.getName());
        holder.fruitIn.setText(fruit.getIntro());
        holder.fruitPrice.setText(fruit.getPrice());
    }
    //方法3:返回RecyclerView中一共有多少行数据
    @Override
    public int getItemCount() {
        return mFruitList.size();
    }
}
```

首先定义一个内部类 ViewHolder,继承自 RecyclerView.ViewHolder,ViewHolder 的构造函数中要传入一个 View 参数,这个 View 参数就是 RecyclerView 子项最外层的布局,在 ViewHolder 类中包含 4 个字段：fruitImage 为水果图片、fruitName 为水果名称、fruitIn 为水果介绍、fruitPrice 为水果价格。通过 View 的 findViewById()方法获取 ImageView 和 TextView 的实例。

FruitAdapter 的构造函数,用于把要展示的数据源传进来,并赋值给一个全局变量 mFruitlist,后续的操作都是在这个数据源上进行的。

由于 FruitAdapter 继承自 RecyclerView.Adapter,所以必须重写 onCreateViewHolder()、onBindViewHolder()、getItemCount()这三个方法。onCreateViewHolder()方法用于创建 ViewHolder 实例,在这个方法中将定制好的每一行的布局 fruit_item 加载进来,然后创建一个 ViewHolder 实例,将加载出来的布局传入到构造函数中,最后将 ViewHolder 实例返回。onBindViewHolder()方法是用于对 RecyclerView 子项的数据进行赋值,会在每个子项滚动到屏幕内执行,这里通过 positon 参数得到当前项的 Fruit 实例,然后再将水果图片、水果名称、水果介绍和水果价格显示在对应的控件上。getItemCount()方法是用于告诉 RecyclerView 一共有多少个子项,通过 size()方法获得 RecyclerView 的数据源长度。

（8）打开 MainActivity.java 类文件，修改 MainActivity 的代码，具体如下。

```java
public class MainActivity extends AppCompatActivity {
    //1.定义对象
    RecyclerView recyclerView;
    private List<Fruit> fruitList = new ArrayList<Fruit>();
    @Override
    protected void onCreate(Bundle savedInstanceState) {
        super.onCreate(savedInstanceState);
        setContentView(R.layout.activity_main);
        //2.绑定控件
        initView();
        //3.准备数据
        initFruits();
        //4.创建适配器
        FruitAdapter adapter = new FruitAdapter(fruitList);
        //5.让数据显示在控件上
        LinearLayoutManager layoutManager = new LinearLayoutManager(this);
        recyclerView.setLayoutManager(layoutManager);
        recyclerView.setAdapter(adapter);
    }
    //绑定控件
    private void initView() {
        recyclerView = (RecyclerView) findViewById(R.id.recyclerview);
    }
    //准备数据
    private void initFruits() {
        for (int i = 0; i < 5; i++) {
            Fruit apple = new Fruit("陕西红富士", R.drawable.apple, "优质苹果 香甜爽口", "￥20元/kg");
            fruitList.add(apple);
            Fruit shiliu = new Fruit("大理石榴", R.drawable.pomegranate, "新鲜石榴 全新上市", "￥15元/kg");
            fruitList.add(shiliu);
            Fruit chengzi = new Fruit("台湾茂谷柑", R.drawable.orange, "浓甜无渣 瓣瓣多汁", "￥49.9元/kg");
            fruitList.add(chengzi);
            Fruit mangguo = new Fruit("新鲜大青芒", R.drawable.mangguo, "精选大果 肥厚饱满", "￥29.9元/kg");
            fruitList.add(mangguo);
            Fruit putao = new Fruit("巨峰葡萄", R.drawable.putao, "爆汁超甜 新鲜采摘", "￥19.9元/kg");
            fruitList.add(putao);
            Fruit boluo = new Fruit("台湾菠萝", R.drawable.boluo, "香甜可口 汁水浓厚", "￥38.9元/kg");
            fruitList.add(boluo);
            Fruit xigua = new Fruit("早春红玉西瓜", R.drawable.xigua, "皮薄肉甜 物美价廉", "￥28.8元/kg");
            fruitList.add(xigua);
        }
    }
}
```

首先定义对象，定义一个动态数组 ArrayList 来存放 RecyclerView 中 item 的内容，泛型为新建类 Fruit，在 Fruit 的构造函数中将水果名称、水果图片的 id、水果介绍和水果价格传入，然后将创建好的对象添加到水果列表中。因为数据量较少，不足以填充满整个屏幕，所以使用了 for 循环将数据添加了 5 遍。接着创建了 FruitAdapter 对象，并将 FruitAdapter 适配器传递给 RecyclerView，实现 RecyclerView 和数据的关联。这里创建一个 LinearLayoutManager 对象，线性布局默认是垂直方向，将 LinearLayoutManager 对象通过 setLayoutManager() 传递给 RecyclerView，这样 RecyclerView 就以纵向滚动方式显示列表内容。

（9）运行程序，效果如图 6.15 所示。

【任务 6-3】 水果详情界面的设计与功能

（1）在工程中添加一个新的页面。右击 com. example.fruitshop 包→New→Activity→Empty

图 6.15 水果列表界面运行效果图

Activity，会弹出一个创建活动的对话框，将活动命名为 DetailActivity，默认勾选 Generate Layout File 关联布局界面，布局界面名称为 activity_detail，但不要勾选 Launcher Activity。单击 Finish 按钮，便可在工程中完成第 3 个页面的添加。

（2）准备一幅图片 detail.jpg，将其复制粘贴到 res/drawable 文件夹中。

（3）修改布局界面。在 app 目录下的结构中，双击 layout 文件夹下的 activity_detail. xml 文件，便可打开布局编辑器，切换到 Text 选项卡，输入代码如下。

```xml
<?xml version = "1.0" encoding = "utf-8"?>
<LinearLayout xmlns:android = "http://schemas.android.com/apk/res/android"
    xmlns:app = "http://schemas.android.com/apk/res-auto"
    xmlns:tools = "http://schemas.android.com/tools"
    android:orientation = "vertical"
    android:background = "@drawable/detail"
    android:layout_width = "match_parent"
    android:layout_height = "match_parent"
    tools:context = ".DetailActivity">
    <TextView
        android:id = "@+id/name"
        android:layout_marginTop = "280dp"
        android:layout_width = "match_parent"
        android:layout_height = "79dp"
        android:text = "TextView"
        android:gravity = "center"
        android:textSize = "50sp"
        android:textStyle = "bold"
        android:textColor = "#339900"/>
```

```xml
<ImageView
    android:id = "@+id/image"
    android:layout_width = "match_parent"
    android:layout_height = "124dp"
    android:layout_gravity = "right"/>
<TextView
    android:id = "@+id/price"
    android:layout_width = "match_parent"
    android:layout_height = "86dp"
    android:layout_gravity = "left"
    android:gravity = "center"
    android:text = "TextView"
    android:textColor = "#E62413"
    android:textSize = "50sp"
    android:textStyle = "bold" />
<TextView
    android:id = "@+id/intro"
    android:layout_width = "match_parent"
    android:layout_height = "156dp"
    android:gravity = "center"
    android:text = "TextView"
    android:textColor = "#000099"
    android:textSize = "45sp"
    android:textStyle = "bold" />
</LinearLayout>
```

上述代码中，根节点为线性布局，方向为垂直方向，设置布局的背景为图片 detail.jpg，ImageView 控件显示水果图片，3 个 TextView 控件分别显示水果名称、水果价格及水果介绍。

（4）修改 Fruit.java 类文件，代码如下。

```java
public class Fruit implements Serializable {
    private String name;
    private String Intro;
    private int imageId;
    private String Price;
    public Fruit(String name, int imageId,String Intro,String Price) {
        this.name = name;
        this.imageId = imageId;
        this.Intro = Intro;
        this.Price = Price;
    }
    public String getName() {
        return name;
    }
    public int getImageId() {
        return imageId;
    }
```

```
        public String getIntro(){
            return Intro;
        }
        public String getPrice(){
            return Price;
        }
}
```

让 Fruit 类实现 Serializable 序列化接口,在程序中为了能直接以 Java 对象的形式进行保存,然后再重新得到该 Java 对象,这就需要序列化能力。序列化可以看作一种机制,按照一定的格式将 Java 对象的某状态转成介质可接收的形式,以方便存储或传输。序列化时将 Java 对象相关的类信息、属性及属性值等保存起来,反序列化时再根据这些信息构建出 Java 对象。Serializable 接口的作用是:提供一种简单又可扩展的对象保存恢复机制。对于远程调用,能方便对对象进行编码和解码,就像实现对象直接传输一样。可以将对象持久化到介质中,就像实现对象直接存储一样。允许对象自定义外部存储的格式。Serializable 接口的目的是:将对象存储在存储介质中,以便在下次使用的时候,可以很快捷地重建一个副本,便于数据传输,尤其是在远程调用的时候。

(5) 修改 FruitAdapter.java 中的代码,修改如下。

```
public class FruitAdapter extends RecyclerView.Adapter<FruitAdapter.ViewHolder> {
    private List<Fruit> mFruitList;
    static class ViewHolder extends RecyclerView.ViewHolder {
        View fruitView;
        ImageView fruitImage;
        TextView fruitName;
        TextView fruitIn;
        TextView fruitPrice;
        public ViewHolder(View view) {
            super(view);
            fruitView = view;
            fruitImage = (ImageView) view.findViewById(R.id.fruitImg);
            fruitName = (TextView) view.findViewById(R.id.fruitName);
            fruitIn = (TextView)view.findViewById(R.id.fruitIntroduce);
            fruitPrice = (TextView)view.findViewById(R.id.fruitPrice);
        }
    }
    public FruitAdapter(List<Fruit> fruitList) {
        mFruitList = fruitList;
    }
    //方法 1:用于创建 ViewHolder 实例
    @Override
    public ViewHolder onCreateViewHolder(ViewGroup parent, int viewType) {
        final View view = LayoutInflater.from(parent.getContext()).inflate(R.layout.fruit_item, parent, false);
        final ViewHolder holder = new ViewHolder(view);
        //单击事件-跳转到水果详情界面
        holder.fruitView.setOnClickListener(new View.OnClickListener() {
```

```java
        @Override
        public void onClick(View view) {
            int position = holder.getAdapterPosition();
            Intent intent = new Intent(view.getContext(),DetailActivity.class);
            intent.putExtra("myposition",(Serializable)mFruitList.get(position));
            view.getContext().startActivity(intent);
        }
    });
    return holder;
}
//方法 2:用于对 RecyclerView 中子项的数据进行赋值
@Override
public void onBindViewHolder(ViewHolder holder, int position) {
    Fruit fruit = mFruitList.get(position);
    holder.fruitImage.setImageResource(fruit.getImageId());
    holder.fruitName.setText(fruit.getName());
    holder.fruitIn.setText(fruit.getIntro());
    holder.fruitPrice.setText(fruit.getPrice());
}
//方法 3:返回 RecyclerView 中一共有多少行数据
@Override
public int getItemCount() {
    return mFruitList.size();
}
}
```

首先修改 ViewHolder 类文件,在 ViewHolder 中添加一个变量 fruitView,用来保存子项最外层布局的实例。然后在 onCreateViewHolder()方法中添加注册单击事件,这里给子项最外层布局注册了单击事件,通过 getAdapterPosition()方法获得单击的位置 position,通过 Intent 意图类实现页面的跳转,因为是在 FruitAdapter. java 中,所以使用 view.getContext()获取当前上下文环境,DetailActivity. class 为要跳转的页面,跳转的过程中需要传递当前单击的水果实例给水果详情页面,使用 putExtra 在跳转的过程中传递数据,然后通过 position 获得单击的 Fruit 实例,Fruit 类实现 Serializable 序列化接口,最后通过 startActivity()开启跳转。

(6) 双击打开 Detail. java 类文件,代码如下。

```java
public class DetailActivity extends AppCompatActivity {
    //1.定义对象
    private TextView textView;
    private ImageView image;
    private TextView name;
    private TextView price;
    private TextView intro;
    @Override
    protected void onCreate(Bundle savedInstanceState) {
        super.onCreate(savedInstanceState);
        setContentView(R.layout.activity_detail);
```

```
        //2.绑定控件
        initview();
        //3.接收Intent跳转传递过来的数据
        Intent intent = getIntent();
        Fruit fruit = (Fruit) getIntent().getSerializableExtra("myposition");
        //4.将序列化保存的水果信息显示在控件中
        image.setImageResource(fruit.getImageId());
        name.setText("水果:" + fruit.getName());
        price.setText("价格:" + fruit.getPrice());
        intro.setText("特色:" + fruit.getIntro());
    }
    //绑定控件
    private void initview() {
        image = (ImageView) findViewById(R.id.image);
        name  = (TextView)  findViewById(R.id.name);
        price = (TextView)  findViewById(R.id.price);
        intro = (TextView)  findViewById(R.id.intro);
    }
}
```

水果详情页面中首先接收由水果列表界面传递过来的 Fruit 实例,因为传递的 Fruit 实例实现了 Serializable 序列化接口,所以采用 getSerializableExtra() 获取传递过来的数据,最后将序列化保存的水果图片、水果名称、水果价格及水果介绍通过 Fruit 类的 get() 方法获取对应的内容显示在对应的控件中。

(7) 运行程序,在水果列表界面中不论是单击石榴的图片还是石榴的文字,效果如图 6.16 所示。单击图片或者文字这个事件会被子项最外面的布局捕获到,触发布局的单击事件,跳转到单击水果的详情页面。

图 6.16 水果详情界面运行效果图

项目小结

本项目在案例的实施过程中学习了 ListView 的简单用法,定制 ListView 的界面,以及如何提升 ListView 的运行效率。通过 RecyclerView 控件的基本用法实现了和 ListView 一样的效果。RecyclerView 独有的布局管理器可轻松实现 ListView 较难实现的横向滚动和瀑布流布局。RecyclerView 控件使用 View 注册子项单击事件的监听器解决了单击子项里具体元素的单击事件。最后通过多彩水果店项目实战,复习了 SharedPreferences 存储的知识,通过 RecyclerView 控件实现了水果列表的显示,应用 RecyclerView 的单击事件和 Serializable 序列化接口实现了水果详情界面的查看功能。

习题

1. ListView 和 RecyclerView 的功能完全一样,这句话是(　　)。
   A. 正确的　　　　B. 错误的
2. 自定义适配器中重写了(　　)方法,用于在每一个子项被滚动到屏幕内的时候被调用。
   A. ImageView()　　B. getView()　　C. EditText()　　D. listView()
3. 调用 ListView 的(　　)方法,将构建好的适配器对象传递进去,这样 ListView 和数据之间的关联就建立完成了。
   A. setAdapter()　　B. adapter()　　C. setText()　　D. getText()
4. 如果需要定义每一个子项的布局,需要将新写的布局文件放在(　　)工程目录中。
   A. res\drawable　　B. res\string　　C. res\picture　　D. res\layout
5. 为提升 ListView 的运行效率,可以在 getView 方法中进行判断,如果(　　)为空,则使用 LayoutInflater 去加载布局。
   A. view　　　　B. parent　　　　C. convertView　　D. id
6. 可以使用(　　)方法为 ListView 注册一个监听器,实现 ListView 的单击事件。
   A. onItemClick()　　　　　　B. setOnItemClickListener()
   C. onClick()　　　　　　　　D. onClickListener()
7. RecyclerView 控件的 LinearLayoutManager 线性布局排列方式,默认的是(　　)。
   A. 横向排列　　B. 纵向排列　　C. 网格布局　　D. 瀑布流布局
8. RecyclerView 控件 LayoutManager 布局管理器除了线性布局,还有(　　)布局。
   A. 表格　　　　B. 相对　　　　C. 瀑布流　　　　D. 垂直
9. LinearLayoutManager 的(　　)方法可以来设置布局的排列方向。
   A. setOrientation()　　　　B. setAdapter()
   C. setLayoutManager()　　　D. setText()
10. 下面(　　)控件不能直接使用,需要添加依赖才可使用。
    A. ImageView　　B. TextView　　C. EditText　　D. RecyclerView

项目 7

唱歌跳舞小管家——媒体播放器

【教学导航】

学习目标	(1) 掌握 Timer 及 TimerTask 的使用。 (2) 强化 Handler 的工作机制。 (3) 强化 RecyclerView 控件的使用。 (4) 掌握 MediaPlayer 类的使用。 (5) 掌握 SeekBar 控件的使用。 (6) 掌握 VideoView 控件的使用。 (7) 掌握媒体控制柄 mediaController 的使用。 (8) 培养简单案例的设计开发能力。
教学方法	任务驱动法、理论实践一体化、探究学习法、分组讨论法
课时建议	10 课时

7.1 实战案例——简单音乐播放器

综合运用 UI 开发的基础知识,来完成一个简易音乐播放器 App 的开发实战。第一个页面是欢迎界面,第一次打开 App 时从 5 倒计时到 0 跳转到主界面,再次打开 App 时便直接跳转到主界面。第二个页面是主界面,单击进入音乐列表按钮便可跳转到音乐列表界面。第三个页面是音乐列表界面,以纵向滚动列表方式显示所有音乐信息。第四个页面是音乐播放界面,用来播放在音乐列表中单击的歌曲,进度条可以调整播放的进度,并显示音乐总时长和当前播放的时长。

7.1.1 界面分析

本案例中包含 4 个界面:欢迎界面、主界面、音乐列表界面及音乐播放界面,效果如图 7.1~图 7.4 所示。

图 7.1　欢迎界面效果图

图 7.2　主界面效果图

图 7.3　音乐列表界面效果图

图 7.4　音乐播放界面效果图

7.1.2 实现思路

第一个界面是欢迎界面,第一次打开 App 时从 5 倒计时到 0 跳转到主界面,通过 Handler 消息处理器、Timer 及 TimerTask 实现倒计时功能,再次打开 App 时便直接跳转到主界面,通过 SharedPreferences 存储登录的状态,若是第一次登录则跳转到欢迎界面,否则直接跳转到主界面。第二个界面是主界面,通过 Intent 类实现单击进入音乐列表按钮便可跳转到音乐列表界面。第三个界面是音乐列表界面,采用 RecyclerView 滚动控件实现以纵向滚动列表方式显示所有音乐信息。第四个界面是音乐播放界面,用来播放在音乐列表中单击的歌曲,通过 MediaPlayer 类播放音乐列表中选中的音乐,通过 SeekBar 进度条可以调整播放的进度,并显示音乐总时长和当前播放的时长。

7.1.3 任务实施

微课视频

【任务 7.1.1】 欢迎界面的设计与功能

(1) 创建一个新的工程,工程名为 Musicplayer。

(2) 切换工程的 Project 项目结构,选择该模式下方的 app,依次展开,便看到工程的布局界面和工程的类文件,其中,activity_main.xml 是布局界面,MainActivity.java 为类文件。

(3) 在工程中添加一个新的页面。右击 com.example.musicplayer 包 → New → Activity → Empty Activity,会弹出一个创建活动的对话框,将活动命名为 WelcomeActivity,默认勾选 Generate Layout File 关联布局界面,布局界面名称为 activity_welcome,但不要勾选 Launcher Activity。单击 Finish 按钮,便可在工程中完成第二个页面的添加。

(4) 选择该模式下方的 app → src → main → AndroidManifest.xml 依次展开,双击打开 AndroidManifest.xml 配置文件,修改代码如下。

```
<?xml version = "1.0" encoding = "utf - 8"?>
< manifest xmlns:android = "http://schemas.android.com/apk/res/android"
    xmlns:tools = "http://schemas.android.com/tools"
    package = "com.example.musicplayer">
< uses - permission android:name = "android.permission.WRITE_EXTERNAL_STORAGE"/>
< uses - permission android:name = "android.permission.MOUNT_UNMOUNT_FILESYSTEMS"
    tools:ignore = "ProtectedPermissions" />
< application
    android:allowBackup = "true"
    android:icon = "@mipmap/ic_launcher"
    android:label = "@string/app_name"
    android:roundIcon = "@mipmap/ic_launcher_round"
    android:supportsRtl = "true"
    android:theme = "@style/AppTheme">
    < activity android:name = ".WelcomeActivity" >
        < intent - filter >
            < action android:name = "android.intent.action.MAIN" />
```

```xml
                    <category android:name="android.intent.category.LAUNCHER" />
                </intent-filter>
            </activity>
            <activity android:name=".MainActivity">
            </activity>
        </application>
</manifest>
```

代码中<activity android:name=".WelcomeActivity">节点中包含<intent-filter>过滤规则,说明当前程序启动时使用 WelcomeActivity 作为程序主入口。

(5)准备一张图片,图片名为 welcome_picture.jpg,将其粘贴到 app 目录结构中 res 下方的 drawable 文件夹下,作为欢迎页面的背景图片。

(6)修改布局界面。在 app 目录下的结构中,双击 layout 文件夹下的 activity_welcome.xml 文件,便可打开布局编辑器,切换到 Text 选项卡,输入代码如下。

```xml
<?xml version="1.0" encoding="utf-8"?>
<LinearLayout xmlns:android="http://schemas.android.com/apk/res/android"
    xmlns:app="http://schemas.android.com/apk/res-auto"
    xmlns:tools="http://schemas.android.com/tools"
    android:orientation="vertical"
    android:background="@drawable/welcome_picture"
    android:layout_width="match_parent"
    android:layout_height="match_parent"
    tools:context=".MainActivity">
    <TextView
        android:text="欢 迎"
        android:textSize="90dp"
        android:textStyle="bold"
        android:textColor="#ad0000"
        android:layout_marginTop="150dp"
        android:layout_gravity="center"
        android:layout_width="wrap_content"
        android:layout_height="wrap_content"/>
    <TextView
        android:id="@+id/text1"
        android:layout_width="wrap_content"
        android:layout_height="wrap_content"
        android:text="H"
        android:layout_gravity="center"
        android:layout_marginTop="50dp"
        android:textSize="100sp"
        android:textColor="#000000"/>
</LinearLayout>
```

上述代码中,根节点为线性布局,方向为 vertical,背景图片设置为 welcome_picture.jpg,第一个 TextView 控件显示标题,第二个 TextView 控件显示倒计时的数字。

(7)打开 WelcomeActivity.java 类文件,修改 WelcomeActivity 的代码,具体如下。

```java
public class WelcomeActivity extends AppCompatActivity {
    private static final String TAG = "MainActicity";
    private TextView textView1;
    private Timer timer; //创建定时器
    private TimerTask timerTask; //创建定时器任务
    private int count = 5;
    private Handler handler; //消息处理器,专门发送和接收消息
    @Override
    protected void onCreate(Bundle savedInstanceState) {
        super.onCreate(savedInstanceState);
        setContentView(R.layout.activity_welcome);
        initView();     //控件初始化
        initDate();     //数据初始化
        initStatus();   //页面状态初始化
    }
    private void initView() {
        textView1 = findViewById(R.id.text1);
    }
    private void initDate() {
        //1.创建定时器
        timer = new Timer();
        //2.明确定时器要执行的任务
        timerTask = new TimerTask() {
            @Override
            //run()中的代码是定时器要完成的任务,都是耗时的操作,在Android中,耗时的操作都放在子线程
            //中进行
            public void run() {
                //让子线程给主线程发送消息信号,主线程接收到消息信号后就可以更新主界面中的数字显
                //示信息
                if(count!= 0){
                    //给主线程发送消息信息1
                    Message msg = new Message();
                    msg.what = 1; //1表示消息信号
                    handler.sendMessage(msg);
                }else {
                    //给主线程发送消息信息0
                    Message msg = new Message();
                    msg.what = 0; //0表示消息信号
                    handler.sendMessage(msg);
                }
            }
        };
        //3.开启定时器。参数1:定时器任务;参数2:延迟;参数3:变化的周期
        timer.schedule(timerTask,0,1000);
        //4.主线程接收到消息信号对主界面数字显示进行更新
        handler = new Handler(){
            @Override
            public void handleMessage(@NonNull Message msg) {
                super.handleMessage(msg);
```

```java
                //主线程根据接收到的消息进行判断
                switch (msg.what){
                    case 1:
                        count--; //让数字递减
                        textView1.setText(count + "");    //让变化的数字显示在主界面上
                        break;
                    case 0:
                Intent intent = new Intent(WelcomeActivity.this,MainActivity.class);
                                                    //倒计时结束,跳转到主界面

                        startActivity(intent);
                        finish();
                        timer.cancel(); //关闭定时器
                        timerTask.cancel(); //关闭定时器任务
                        break;
                    default:
                        break;
                }
            }
        };
    }
    //5.第一次登录App时显示欢迎界面的倒计时,第二次登录App时直接进入主界面
    private void initStatus() {
        //读取保存的登录状态值
Boolean status = getSharedPreferences("mystatus",MODE_PRIVATE).getBoolean("firstlogin_status",false);
//将第一次登录的状态True保存起来,下次登录后判断状态,如果为True直接跳转到主页面,否则
//从欢迎界面倒计时登录
        SharedPreferences.Editor editor = getSharedPreferences("mystatus",MODE_PRIVATE).edit();
        editor.putBoolean("firstlogin_status",true);
        editor.commit();
    if(status){ //如果status为True
        Intent intent2 = new Intent(WelcomeActivity.this,MainActivity.class);
        startActivity(intent2);
        finish();
        timer.cancel();
        timerTask.cancel();
        }
    }
}
```

在开发中有时会有这样的需求,即在固定的每隔一段时间执行某一个任务。例如,UI上的控件需要随着时间改变,可以使用Java为我们提供的计时器的工具类,即Timer和TimerTask。Timer是一个普通的类,其中有几个重要的方法;而TimerTask则是一个抽象类,其中有一个抽象方法run(),类似线程中的run()方法,使用Timer创建一个它的对象,然后使用这个对象的schedule()方法来完成这种间隔的操作。Timer就是一个线程,使用schedule()方法完成对TimerTask的调度,多个TimerTask可以共用一个Timer,也就

是说，Timer 对象调用一次 schedule()方法就是创建了一个线程，并且调用一次 schedule()后 TimerTask 会无限制地循环下去，使用 Timer 的 cancel()停止操作。timer.schedule()方法有三个参数，第一个参数就是 TimerTask 类型的对象，实现 TimerTask 的 run()方法就是要周期执行的一个任务；第二个参数有两种类型，第一种是 long 类型，表示多长时间后开始执行，另一种是 Date 类型，表示从那个时间后开始执行；第三个参数就是执行的周期，为 long 类型。代码 timer.schedule(timerTask,0,1000)即表示没有延迟，变化周期为 1000ms 即 1s。

Android 中规定只允许 UI 线程修改 Activity 的 UI 组件，当程序第一次启动时，Android 会同时启动一个主线程，用于负责处理与 UI 相关的事件，并把事件分发到对应的组件进行处理后再绘制界面，此时主线程又称为 UI 线程。当在新启动的线程中更新 UI 组件时，需要借助 Handler 的消息传递机制来实现。在 Android 系统中，Handler 是一种用于在线程之间传递消息的机制，使用 Handler 可以在一个线程中发出消息，在另一个线程中接收消息并进行处理。Handler 类中包含发送、接收和处理消息的方法，其中用于接收消息的方法如表 7.1 所示。

表 7.1　Handler 的常用方法

方　　法	功 能 描 述
handleMessage(Message msg)	通过重载该方法来处理消息
hasMessage(int what)	检查消息队列中是否包含 what 所指定的消息
hasMessage(int what,Object object)	检查队列中是否有指定的 what 和指定对象的消息
obtainMessage()	用于获取消息，具有多个重载的方法
sendEmptyMessage(int what)	用于发送空消息
sendEmptyMessageDelayed(int what,long delayMillis)	用于在指定的时间之后发送空消息
sendMessage(Message msg)	立即发送消息
sendMessageDelayed(Message msg,long delayMillis)	用于在指定的时间之后发送空消息

上面代码中当倒计时为 0 时子线程给主线程发送消息信号 1，否则给主线程发送消息信号 0。主线程根据接收到的消息信息做不同的处理，如果接收到的消息信号为 1，则执行 count--，并把变化的数字显示在界面上；如果接收到的消息信号为 0，则通过 Intent 意图类跳转到主界面。

通过 SharedPreferences 将登录状态保存起来，再次登录的时候首先通过 getBoolean()方法获取保存在 SharedPreferences 中的状态值，如果为真，则表明已经登录过，则通过 Intent 意图类跳转到主界面。

(8) 运行程序，效果如图 7.5 所示。

【任务 7.1.2】　主界面的设计与功能

(1) 切换工程的 Project 项目结构。选择该模式下方的 app，依次展开，便看到工程的布局界面和工程的类文件，其中，activity_main.xml 是布局界面，MainActivity.java 为类文件。

图 7.5　欢迎界面运行效果图

微课视频

（2）修改布局界面。在 app 目录下的结构中，双击 layout 文件夹下的 activity_main.xml 文件，便可打开布局编辑器，切换到 Text 选项卡，输入代码如下。

```xml
<?xml version="1.0" encoding="utf-8"?>
<LinearLayout xmlns:android="http://schemas.android.com/apk/res/android"
    xmlns:app="http://schemas.android.com/apk/res-auto"
    xmlns:tools="http://schemas.android.com/tools"
    android:orientation="vertical"
    android:background="@drawable/main_picture"
    android:layout_width="match_parent"
    android:layout_height="match_parent"
    tools:context=".MainActivity">
    <TextView
        android:text="音乐播放器"
        android:textSize="60dp"
        android:textStyle="bold"
        android:textColor="#0000ff"
        android:layout_marginTop="150dp"
        android:layout_gravity="center"
        android:layout_width="wrap_content"
        android:layout_height="wrap_content"/>
    <Button
        android:id="@+id/button1"
        android:layout_width="wrap_content"
        android:layout_height="wrap_content"
        android:text="进入音乐列表"
        android:background="@drawable/button_style"
        android:layout_gravity="center"
        android:layout_marginTop="150dp"
        android:textSize="30sp"
        android:textStyle="bold"
        android:textColor="#000000" />
</LinearLayout>
```

上述代码中，根节点为线性布局，方向为 vertical，TextView 控件显示标题内容，Button 控件用来实现单击跳转到音乐列表界面，Button 控件的 background 属性设置为 @drawable/button_style，表示引用了自定义的 button_style 样式。

（3）准备两张图片，图片名为 grey_picture.jpg、blue_picture.jpg，将其粘贴到 app 目录结构中 res 下方的 drawable 文件夹下，作为 Button 的两幅背景图片，grey_picture.jpg 图片是按钮被单击之后的背景图片，blue_picture.jpg 图片是按钮没有被单击时的背景图片。

（4）自定义 button_style 样式。鼠标右击 drawable → New → Drawable resource file，在新建资源文件对话框中输入文件名"button_style"，节点类型为 selector。代码如下。

```xml
<?xml version="1.0" encoding="utf-8"?>
<selector xmlns:android="http://schemas.android.com/apk/res/android">
    <item android:drawable="@drawable/blue_picture" android:state_pressed="false"/>
    <item android:drawable="@drawable/grey_picture" android:state_pressed="true"/>
</selector>
```

android:state_pressed 设置是否按压,设置为 true 时表示已按压状态,按钮的图片为 grey_picture.jpg;android:state_pressed 设置为 false 时表示没有被按压,按钮的图片为 blue_picture.jpg。

(5)打开 MainActivity.java 类文件,修改 MainActivity 的代码如下。

```java
public class MainActivity extends AppCompatActivity {
    //定义对象
    Button button1;
    @Override
    protected void onCreate(Bundle savedInstanceState) {
        super.onCreate(savedInstanceState);
        setContentView(R.layout.activity_main);
        initView();
        btnclick();
    }
    private void initView() {
        button1 = findViewById(R.id.button1);
    }
    private void btnclick() {
        button1.setOnClickListener(new View.OnClickListener() {
            @Override
            public void onClick(View view) {
                Intent intent = new Intent(MainActivity.this,MusicActivity.class);
                startActivity(intent);
                finish();
            }
        });
    }
}
```

上面代码中给 Button 注册了单击事件,通过 Intent 意图类跳转到音乐列表界面。音乐列表界面类文件名称为 MusicActivity,同学们可以自行添加该页面,也可以参考任务 7.1.3 的步骤 1 创建该页面。

(6)运行程序,效果如图 7.6 所示。

【任务 7.1.3】 音乐列表界面的设计与功能

(1)在工程中添加一个新的页面。右击 com.example. musicplayer 包 → New → Activity → Empty Activity,会弹出一个创建活动的对话框,将活动命名为 MusicActivity,默认勾选 Generate Layout File 关联布局界面,布局界面名称为 activity_music,但不要勾选 Launcher Activity。单击 Finish 按钮,便可在工程中完成第三个页面的添加。

(2)RecyclerView 属于新增的控件,要想使用它首先要在 app/build.gradle 文件中添加依赖库,打开 app 目录

图 7.6 主界面运行效果图

微课视频

下的 build.gradle 文件,在 dependencies 闭包中添加如下内容。

```
dependencies {
    implementation fileTree(dir: 'libs', include: ['*.jar'])
    implementation 'androidx.appcompat:appcompat:1.1.0'
    implementation 'androidx.constraintlayout:constraintlayout:1.1.3'
    testImplementation 'junit:junit:4.12'
    androidTestImplementation 'androidx.test.ext:junit:1.1.1'
    androidTestImplementation 'androidx.test.espresso:espresso-core:3.2.0'
    implementation 'com.android.support:recyclerview-v7:29.0.3'
}
```

添加完之后记得要单击右上角的 Sync Now 来进行同步。

(3) 修改布局界面。在 app 目录下的结构中,双击 layout 文件夹下的 activity_music.xml 文件,便可打开布局编辑器,切换到 Text 选项卡,输入代码如下。

```xml
<?xml version="1.0" encoding="utf-8"?>
<LinearLayout xmlns:android="http://schemas.android.com/apk/res/android"
    xmlns:app="http://schemas.android.com/apk/res-auto"
    xmlns:tools="http://schemas.android.com/tools"
    android:orientation="vertical"
    android:layout_width="match_parent"
    android:layout_height="match_parent"
    tools:context=".MusicActivity">
    <TextView
        android:text="音乐列表"
        android:textSize="50sp"
        android:layout_gravity="center"
        android:textStyle="bold"
        android:textColor="#000078"
        android:layout_width="wrap_content"
        android:layout_height="wrap_content"/>
    <androidx.recyclerview.widget.RecyclerView
        android:id="@+id/recyclerview"
        android:layout_width="match_parent"
        android:layout_height="match_parent"/>
</LinearLayout>
```

上述代码中,根节点为线性布局,方向为 vertical,TextView 控件显示标题内容,添加 RecyclerView 控件,设置其高度和宽度为 match_parent,这样 RecyclerView 控件就占满了整个布局。因为在类文件中要调用 RecyclerView,所以给它设置一个 id。需要注意的是,由于 RecyclerView 并不是内置在系统 SDK 当中的,所以需要把完整的包路径写出来。

(4) 准备一张图片,图片名为 music_icon.jpg,将其粘贴到 app 目录结构中 res 下方的 drawable 文件夹下,作为音乐图标。

(5) 定义一个实体类,作为 RecyclerView 适配器的适配类型。右击 com.example.musicplayer 包→New→Java Class,会弹出一个创建类的对话框,将类命名为 Music,代码如下。

```java
public class Music {
    private String musicTitle;
    private String singerName;
    public String getMusicTitle() {
        return musicTitle;
    }
    public String getSingerName() {
        return singerName;
    }
    public Music(String musicTitle, String singerName) {
        this.musicTitle = musicTitle;
        this.singerName = singerName;
    }
}
```

Music类中有两个字段：musicTitle 是音乐的名字，singerName 是歌手的名字。

（6）为 RecyclerView 的每一行子项指定要显示的布局样式，右击 Layout→New→Xml→Layout Xml File，会弹出一个 Xml Layout File 对话框，命名为 music_item.xml，代码如下。

```xml
<?xml version="1.0" encoding="utf-8"?>
<LinearLayout xmlns:android="http://schemas.android.com/apk/res/android"
    android:layout_width="match_parent"
    android:orientation="horizontal"
    android:layout_height="wrap_content">
    <ImageView
        android:src="@drawable/music_icon"
        android:layout_width="70dp"
        android:layout_height="70dp"/>
    <LinearLayout
        android:orientation="vertical"
        android:layout_width="wrap_content"
        android:layout_height="wrap_content">
        <TextView
            android:id="@+id/music_title"
            android:textStyle="bold"
            android:text="音乐的名字"
            android:textSize="30sp"
            android:textColor="#000000"
            android:layout_width="wrap_content"
            android:layout_height="wrap_content"/>
        <TextView
            android:id="@+id/singer_name"
            android:textStyle="bold"
            android:text="歌手的名字"
            android:textSize="20sp"
            android:textColor="#0000ff"
            android:layout_width="wrap_content"
```

```xml
            android:layout_height = "wrap_content"/>
        </LinearLayout>
</LinearLayout>
```

上述代码中,根节点为线性布局,方向为 horizontal,宽度设置为 match_parent,高度设置为 wrap_content,高度如果设置为 match_parent,就会出现一个子项占满整个屏幕。ImageView 控件显示音乐的图标,ImageView 控件右侧嵌套的垂直方向的线性布局,包含两个 TextView 控件,分别为音乐的名字和歌手的名字。

(7) 接下来需要为 RecyclerView 准备一个自定义的适配器。右击 com.example.musicplayer 包→New→Java Class,会弹出一个创建类的对话框,将类命名为 MusicAdapter,继承自 RecyclerView.Adapter,并将泛型指定为 MusicAdapter.ViewHolder,其中,ViewHolder 是在 MusicAdapter 中定义的一个内部类,代码如下。

```java
public class MusicAdapter extends RecyclerView.Adapter<MusicAdapter.ViewHolder> {
    Context mcontext;
    List<Music> myMusicList;
    public class ViewHolder extends RecyclerView.ViewHolder{
        View musicview;
        TextView music_title;
        TextView singer_name;
        public ViewHolder(@NonNull View view) {
            super(view);
            musicview = view;
            music_title = view.findViewById(R.id.music_title);
            singer_name = view.findViewById(R.id.singer_name);
        }
    }
    public MusicAdapter(List<Music> musicList) {
        myMusicList = musicList;
    }
//方法1:用于创建 ViewHolder 实例
    @NonNull
    @Override
    public MusicAdapter.ViewHolder onCreateViewHolder(@NonNull ViewGroup parent, int viewType) {
        View view = LayoutInflater.from(parent.getContext()).inflate(R.layout.music_item, parent,false);
        final ViewHolder holder = new ViewHolder(view);
        return holder;
    }
//方法2:用于对 RecyclerView 中子项的数据进行赋值
    @Override
    public void onBindViewHolder(@NonNull MusicAdapter.ViewHolder holder, int position) {
        Music musicinfo = myMusicList.get(position);
        holder.music_title.setText(musicinfo.getMusicTitle());
        holder.singer_name.setText(musicinfo.getSingerName());
    }
```

```
//方法3:返回数据源长度
    @Override
    public int getItemCount() {
        return myMusicList.size();
    }
}
```

首先定义了一个内部类 ViewHolder,继承自 RecyclerView.ViewHolder。ViewHolder 的构造函数中要传入一个 View 参数,这个 View 参数就是 RecyclerView 子项最外层的布局,在 ViewHolder 类中包含 3 个字段:musicview 为子项最外层布局,music_title 为音乐的名字,singer_name 为歌手的名字。通过 view 的 findViewById()方法获取 TextView 的实例。

MusicAdapter 的构造函数,用于把要展示的数据源传进来,并赋值给一个全局变量 myMusicList,后续的操作都是在这个数据源上进行的。

由于 MusicAdapter 继承自 RecyclerView.Adapter,所以必须重写 onCreateViewHolder()、onBindViewHolder()、getItemCount()这三个方法。onCreateViewHolder()方法是用于创建 ViewHolder 实例的,在这个方法中将定制好的每一行的布局 music_item 加载进来,然后创建一个 ViewHolder 实例,将加载出来的布局传入到构造函数中,最后将 ViewHolder 实例返回。onBindViewHolder()方法是用于对 RecyclerView 子项的数据进行赋值的,会在每个子项滚动到屏幕内时执行,这里通过 positon 参数得到当前项的 Music 实例,然后再将音乐的名字和歌手的名字显示在对应的控件上。getItemCount()方法是用于告诉 RecyclerView 一共有多少个子项,通过 size()方法获得 RecyclerView 的数据源长度。

(8)选择该模式下方的 app→src→main→AndroidManifest.xml 依次展开,双击打开 AndroidManifest.xml 配置文件,修改代码如下。

```xml
<?xml version = "1.0" encoding = "utf-8"?>
<manifest xmlns:android = "http://schemas.android.com/apk/res/android"
    xmlns:tools = "http://schemas.android.com/tools"
    package = "com.example.musicplayer">
    <uses-permission android:name = "android.permission.WRITE_EXTERNAL_STORAGE"/>
    <uses-permission android:name = "android.permission.MOUNT_UNMOUNT_FILESYSTEMS"
        tools:ignore = "ProtectedPermissions" />
    <application
        android:allowBackup = "true"
        android:icon = "@mipmap/ic_launcher"
        android:label = "@string/app_name"
        android:roundIcon = "@mipmap/ic_launcher_round"
        android:supportsRtl = "true"
        android:theme = "@style/AppTheme">
        <activity android:name = ".MusicActivity"></activity>
        <activity android:name = ".WelcomeActivity" >
            <intent-filter>
                <action android:name = "android.intent.action.MAIN" />
```

```xml
                <category android:name = "android.intent.category.LAUNCHER" />
            </intent-filter>
        </activity>
        <activity android:name=".MainActivity">
        </activity>
    </application>
```

因为要读写 SD 卡上的歌曲，所以需要在配置文件中添加读写 SD 卡的权限。

（9）准备好一些 MP3 音乐，将其拖动到夜神模拟器中。如果是高版本的夜神模拟器则默认保存在 /storage/emulated/0/Pictures 文件夹中，如果是低版本的夜神模拟器需要存放在 sdcard 目录下的 music 文件夹中。

（10）打开 MusicActivity.java 类文件，修改 MusicActivity 的代码如下。

```java
public class MusicActivity extends AppCompatActivity {
    //定义对象
    RecyclerView recyclerView;
    private static final String TAG = "MusicActivity";
    @Override
    protected void onCreate(Bundle savedInstanceState) {
        super.onCreate(savedInstanceState);
        setContentView(R.layout.activity_music);
        initView();
        initData();
    }
    private void initView() {
        recyclerView = findViewById(R.id.recyclerview);
    }
    private void initData() {
        List<Music> musicList = new ArrayList<>();
        Cursor cursor = getContentResolver().query(MediaStore.Audio.Media.EXTERNAL_CONTENT_URI, null, null, null, MediaStore.Audio.Media.DEFAULT_SORT_ORDER);
        Log.d(TAG, "initData:我查询获取到的歌曲共:" + cursor.getCount() + "首");
        while (cursor.moveToNext()){
            String mymusictitle = cursor.getString(cursor.getColumnIndex(MediaStore.Audio.Media.TITLE));
            String mysingername = cursor.getString(cursor.getColumnIndex(MediaStore.Audio.Media.ARTIST));
            Music music = new Music(mymusictitle,mysingername);
            musicList.add(music);
        }
        cursor.close();
        MusicAdapter adapter = new MusicAdapter(musicList);
        StaggeredGridLayoutManager layoutManager = new StaggeredGridLayoutManager(1,StaggeredGridLayoutManager.VERTICAL);
        recyclerView.setLayoutManager(layoutManager);
        recyclerView.setAdapter(adapter);
    }
}
```

首先定义对象,定义一个动态数组 ArrayList 来存放 RecyclerView 中 item 的内容,泛型为新建类 Music。ContentResolver 是内容解析器,Android 中程序间数据的共享是通过 Provider/Resolver 进行的。提供数据(内容)的就叫 Provider,Resovler 提供接口对这个内容进行解读。在这里,系统提供了音乐的 Provider,那么就需要构建一个 Resolver 来读取音乐的内容。query()方法有 5 个参数:第 1 个参数 table,是根据 Uri 确定的数据库表;第 2 个参数 projection 是要查询的列;第 3 个参数 selection 是查询条件;第 4 个参数 selectionArgs,填充 where 查找条件中的占位符"?";第 5 个参数 order,是指定的排序方式,这里设定为升序排序。首先通过 log 日志类打印出来查询到的歌曲数目,获取到音乐的名字和歌手的名字,传入到 Music 的构造函数中,然后将创建好的对象添加到音乐列表中,通过游标指针 cursor 依次向下移动,把模拟器中所有的歌曲都添加到音乐列表中。接着创建了 MusicAdapter 对象,并将 MusicAdapter 适配器传递给 RecyclerView,实现 RecyclerView 和数据的关联。这里创建一个 StaggeredGridLayoutManager 对象,设置列数为 1 列,方向为垂直,将 StaggeredGridLayoutManager 对象通过 setLayoutManager()传递给 RecyclerView,这样 RecyclerView 就以瀑布流布局显示音乐列表内容。

图 7.7 音乐列表界面运行效果图

(11) 运行程序,效果如图 7.7 所示。

【任务 7.1.4】 音乐播放界面的设计与功能

(1) 在工程中添加一个新的页面。右击 com.example.musicplayer 包 → New → Activity → Empty Activity,会弹出一个创建活动的对话框,将活动命名为 PlayerActivity,默认勾选 Generate Layout File 关联布局界面,布局界面名称为 activity_player,但不要勾选 Launcher Activity。单击 Finish 按钮,便可在工程中完成第四个页面的添加。

微课视频

(2) 修改布局界面。在 app 目录下的结构中,双击 layout 文件夹下的 activity_player.xml 文件,便可打开布局编辑器,切换到 Text 选项卡,输入代码如下。

```xml
<?xml version = "1.0" encoding = "utf - 8"?>
<LinearLayout xmlns:android = "http://schemas.android.com/apk/res/android"
    xmlns:app = "http://schemas.android.com/apk/res - auto"
    xmlns:tools = "http://schemas.android.com/tools"
    android:layout_width = "match_parent"
    android:layout_height = "match_parent"
    android:orientation = "vertical"
    tools:context = ".PlayerActivity">
    <TextView
        android:id = "@ + id/title"
        android:text = "获取到的音乐标题"
```

```xml
            android:textColor = "#000000"
            android:textSize = "35sp"
            android:gravity = "center"
            android:layout_margin = "5dp"
            android:layout_width = "match_parent"
            android:layout_height = "wrap_content"/>
        <TextView
            android:id = "@+id/name"
            android:text = "获取到的歌手名字"
            android:textColor = "#000000"
            android:textSize = "35sp"
            android:gravity = "center"
            android:layout_margin = "5dp"
            android:layout_width = "match_parent"
            android:layout_height = "wrap_content"/>
        <ImageView
            android:src = "@drawable/music_icon"
            android:layout_width = "wrap_content"
            android:layout_height = "wrap_content"/>
        <LinearLayout
            android:orientation = "horizontal"
            android:layout_marginLeft = "5dp"
            android:layout_marginRight = "5dp"
            android:weightSum = "1"
            android:layout_width = "match_parent"
            android:layout_height = "wrap_content">
            <TextView
                android:id = "@+id/current_time"
                android:text = "0:00"
                android:textSize = "25sp"
                android:layout_width = "wrap_content"
                android:layout_height = "wrap_content"/>
            <SeekBar
                android:id = "@+id/seek_bar"
                android:layout_weight = "1"
                android:layout_gravity = "center_vertical"
                android:layout_width = "wrap_content"
                android:layout_height = "wrap_content"/>
            <TextView
                android:id = "@+id/total_time"
                android:text = "8:23"
                android:textSize = "25sp"
                android:layout_width = "wrap_content"
                android:layout_height = "wrap_content"/>
        </LinearLayout>
</LinearLayout>
```

上述代码中,根节点为线性布局,方向为 vertical,两个 TextView 控件分别显示音乐的名字和歌手的名字,ImageView 控件显示音乐的图标,图片内容是固定不变的。最下方嵌套水平方向的线性布局,包括两个 TextView 控件,分别显示当前的播放时长和歌曲的总时长。SeekBar 控件可以手动改变相应的值。音频播放和音乐播放时,大多数时候都要用到 SeekBar。SeekBar 其实是 ProgressBar 的子类,但是 ProgressBar 是不跟用户交互的,只是提供显示任务进度的作用,而 SeekBar 则是对 ProgressBar 的功能进行扩充,使其可以和用户进行交互,即响应用户的单击和拖动事件。SeekBar 的几个重要属性如表 7.2 所示。

表 7.2 SeekBar 属性

属性	描述
android:max	设置值的大小
android:thumb="@drawable/"	显示的可拖动图标
android:thumbOffset	拖动图标的偏量值
android:progress	设置 SeekBar 当前的默认值,范围为 0~max
android:secondaryProgress	设置默认显示的值为多少,范围为 0~max
android:progressDrawable	调用自己定义的 SeekBar 图标

(3) 打开 MusicAdapter.java 类文件,修改 MusicAdapter 的代码如下。

```java
public class MusicAdapter extends RecyclerView.Adapter<MusicAdapter.ViewHolder>{
    Context mcontext;
    List<Music> myMusicList;
public class ViewHolder extends RecyclerView.ViewHolder{
        View musicview;
        TextView music_title;
        TextView singer_name;
        public ViewHolder(@NonNull View view) {
            super(view);
            musicview = view;
            music_title = view.findViewById(R.id.music_title);
            singer_name = view.findViewById(R.id.singer_name);
        }
    }
    public MusicAdapter(List<Music> musicList) {
        myMusicList = musicList;
    }
    //方法1:用于创建 ViewHolder 实例
    @NonNull
    @Override
    public MusicAdapter.ViewHolder onCreateViewHolder(@NonNull ViewGroup parent, int viewType) {
        View view = LayoutInflater.from(parent.getContext()).inflate(R.layout.music_item, parent,false);
        final ViewHolder holder = new ViewHolder(view);
        //单击任意歌曲跳转到播放界面
        holder.musicview.setOnClickListener(new View.OnClickListener() {
            @Override
            public void onClick(View view) {
                int position = holder.getAdapterPosition(); //返回数据在适配器中的位置
```

```java
                    Intent intent = new Intent(view.getContext(),PlayerActivity.class);
                    intent.putExtra("my",position);
                    view.getContext().startActivity(intent);
                }
            });
            return holder;
        }
        //方法2:用于对RecyclerView中子项的数据进行赋值
        @Override
        public void onBindViewHolder(@NonNull MusicAdapter.ViewHolder holder, int position) {
            Music musicinfo = myMusicList.get(position);
            holder.music_title.setText(musicinfo.getMusicTitle());
            holder.singer_name.setText(musicinfo.getSingerName());
        }
        //方法3:返回数据源长度
        @Override
        public int getItemCount() {
            return myMusicList.size();
        }
    }
```

onCreateViewHolder()方法中添加注册单击事件，这里给子项最外层布局注册了单击事件，通过 getAdapterPosition()方法获得单击的位置 position，通过 Intent 意图类实现页面的跳转，因为是在 MusicAdapter.java 中，所以使用 view.getContext()获取当前上下文环境，PlayerActivity.class 为要跳转的页面，跳转的过程中需要传递当前单击的 Music 实例给音乐播放页面，使用 putExtra()在跳转的过程中传递数据，然后通过 position 获取单击的 Music 实例，最后通过 startActivity()开启跳转。

（4）打开 PlayerActivity.java 类文件，PlayerActivity 的代码如下。

```java
public class PlayerActivity extends AppCompatActivity {
    //定义对象
    private static final String TAG = "PlayerActivity";
    private TextView title;
    private TextView name;
    private Cursor cursor;
    private MediaPlayer mediaPlayer;
    private SeekBar seekBar;
    private TextView current_time;
    private TextView total_time;
    private Handler mhandler;
    @Override
    protected void onCreate(Bundle savedInstanceState) {
        super.onCreate(savedInstanceState);
        setContentView(R.layout.activity_player);
        initView();        //绑定控件
        initData();        //获取并显示歌曲标题和歌手名字
        initPlay();        //播放歌曲
```

```java
        initSeek();        //初始化进度条
        moveSeek();        //拖动滑动条
        initUpdate();      //实时更新滑动条的当前时间
    }
    private void initView() {
        title = findViewById(R.id.title);
        name = findViewById(R.id.name);
        seekBar = findViewById(R.id.seek_bar);
        current_time = findViewById(R.id.current_time);
        total_time = findViewById(R.id.total_time);
    }
    private void initData() {
        int position = getIntent().getIntExtra("my",0);
        Log.d(TAG, "initData: " + position); cursor = getContentResolver().query(MediaStore.Audio.Media.EXTERNAL_CONTENT_URI, null, null, null, MediaStore.Audio.Media.DEFAULT_SORT_ORDER);
        cursor.moveToPosition(position);
        String mytitle = cursor.getString(cursor.getColumnIndex(MediaStore.Audio.Media.TITLE));
        String myname = cursor.getString(cursor.getColumnIndex(MediaStore.Audio.Media.ARTIST));
        title.setText(mytitle);
        name.setText(myname);
    }
    private void initPlay() {
        mediaPlayer = new MediaPlayer();
        String path = cursor.getString(cursor.getColumnIndex(MediaStore.Audio.Media.DATA));
        mediaPlayer.reset();   //清空里面的其他歌曲
        try {
            mediaPlayer.setDataSource(path);
            mediaPlayer.prepare();   //准备就绪
            mediaPlayer.start();     //开始唱歌
        } catch (IOException e) {
            e.printStackTrace();
        }
    }
    private void initSeek() {
        seekBar.setMax(mediaPlayer.getDuration());  //获取音频文件总时长
        seekBar.setProgress(mediaPlayer.getCurrentPosition());  //获取当前播放的进度值
        total_time.setText(toTime(mediaPlayer.getDuration()));
        current_time.setText(toTime(mediaPlayer.getCurrentPosition()));

    }
    private String toTime(int getDutation) {
        int time = getDutation/1000;   //毫秒转换为秒
        int minute = time/60;   //取整:求出分钟
        int second = time % 60;  //取余:求出秒
        String mm = String.format("%01d:%02d",minute,second);  //指定显示的格式
        return mm;
```

```java
    }
    private void moveSeek() {
        seekBar.setOnSeekBarChangeListener(new SeekBar.OnSeekBarChangeListener() {
            @Override
            public void onProgressChanged(SeekBar seekBar, int position, boolean b) {
                if(b){
                    mediaPlayer.seekTo(position); //音频从拖到的位置处开始播放
                    initSeek();
                }
            }
            @Override
            public void onStartTrackingTouch(SeekBar seekBar) {

            }
            @Override
            public void onStopTrackingTouch(SeekBar seekBar) {

            }
        });
    }
    private void initUpdate() {
        Timer timer = new Timer();
        TimerTask timerTask = new TimerTask() {
            @Override
            public void run() {
                Message msg = new Message();
                msg.what = 11;
                mhandler.sendMessage(msg);
            }
        };
        timer.schedule(timerTask,0,1000);
        mhandler = new Handler(){
            @Override
            public void handleMessage(@NonNull Message msg) {
                super.handleMessage(msg);
                switch (msg.what){
                    case 11:
                        initSeek(); //重新执行进度条的初始化代码
                        break;
                    default:
                        break;
                }
            }
        };
    }
    //按返回键停止播放
    @Override
    protected void onDestroy() {
        super.onDestroy();
```

```
        mediaPlayer.stop();
    }
}
```

在 initData()方法中,首先接收由音乐列表界面传递过来的 Music 实例,采用 getIntExtra()获取传递过来的数据,通过游标指针 cursor 的 moveToPosition()方法定位到音乐列表界面单击的音乐的名字和歌手的名字,通过 cursor 的 getString()方法查询到音乐的名字和歌手的名字,显示到对应的 TextView 控件上。

Android 提供了常见的音频、视频的编码、解码机制,借助于多媒体类 MediaPlayer 的支持,开发人员可以很方便地在应用中播放音频、视频。initSing()方法中通过 MediaPlayer 实现音乐播放功能。首先创建 MediaPlayer 对象,获取歌曲的路径存放在 Sing 类型的变量 path 中;然后调用 setDataSource()方法来设置音频文件的路径;再调用 prepare()方法使 MediaPlayer 进入到准备状态;最后调用 start()方法就可以播放音频了。如果播放器中有其他的歌曲,需要使用 reset()方法清空里面的歌曲。

在 initSeek()方法中,通过 SeekBar 的 setMax()方法获取音频文件的总时长,通过 SeekBar 的 setProgress()方法获取当前播放的进度值,再将总时长和当前播放的进度值显示在对应的控件上。因为当前的总时长和当前播放的进度值的单位为毫秒,故通过 toTime()方法将其转换为分钟和秒的时间显示方式。首先将毫秒转换为秒,再通过取整运算求出分钟,最后通过取余运算求出秒,并指定时间的显示格式。

在 moveSeek()方法中,setOnSeekBarChangeListener()是当进度改变后用于通知客户端的回调函数。这包括用户通过手势、方向键或轨迹球触发的改变,以及编程触发的改变。onProgressChanged()方法中有三个参数:第一个参数 seekBar 为当前被修改进度的 SeekBar;第二个参数 progress 为当前的进度值,此值的取值范围为 0~max;第三个参数是布尔值,如果是用户触发的改变则返回 True。在 onProgressChanged()方法中判断如果拖动了进度条,即 b 的值为 true,则音频从拖到的位置处开始播放,这样就实现了拖动滑动条改变播放音乐播放的进度。

在 initUpdate()方法中通过 Timer 和 TimerTask 实时更新滑动条的当前播放时长,TimerTask 每秒完成一次更新,子线程给主线程发送消息信号 11,主线程根据接收到的消息信号执行不同的处理。如果接收到消息信号为 11,则重新执行 initSeek()方法,将获取到的当前播放时长显示在对应的控件上。

在音乐播放过程中如果想按下返回键让音乐停止播放,可以重写 onDestroy()方法,通过 stop()方法停止音乐的播放。

(5)运行程序,单击音乐列表界面的"梁祝化蝶"子项的图片或者文字,运行效果如图 7-8 所示。

图 7.8 音乐列表界面运行效果图

7.2 实战案例——简单视频播放器

综合运用 UI 开发的基础知识,来完成一个简易视频播放器 App 的开发实战。第一个界面是主界面,单击"进入视频列表"跳转到视频列表界面;第二个界面是视频列表界面,以纵向滚动列表方式显示所有视频信息;第三个界面是视频播放界面,用来播放在视频列表中单击的视频,进度条可以调整视频播放的进度,并具有暂停和播放功能。

7.2.1 界面分析

本案例中包含三个界面:主界面、视频列表界面、视频播放界面,效果如图 7.9～图 7.11 所示。

图 7.9　主界面效果图　　　图 7.10　视频列表界面效果图　　　图 7.11　视频播放界面效果图

7.2.2 实现思路

第一个界面是主界面,通过 Intent 类实现单击"进入视频列表"按钮便可跳转到视频列表界面;第二个界面是视频列表界面,采用 RecyclerView 滚动控件实现以纵向滚动列表方式显示所有视频信息;第三个界面是视频播放界面,用来播放在视频列表中单击的视频,通过 VideoView 类播放音乐列表中选中的视频,通过媒体控制柄 mediaController 调整播放的进度,并显示视频总时长和当前播放的时长。

7.2.3 任务实施

【任务 7.2.1】 主界面的设计与功能
(1) 创建一个新的工程,工程名为 Videoplayer。

微课视频

(2) 切换工程的 Project 项目结构,选择该模式下方的 app,依次展开,便看到工程的布局界面和工程的类文件,其中,activity_main.xml 是布局界面,MainActivity.java 为类文件。

(3) 准备好图片 bg_picture.jpg 和 video_icon.jpg,将其复制粘贴到 res/drawable 文件夹中。

(4) 修改布局界面。在 app 目录下的结构中,双击 layout 文件夹下的 activity_main.xml 文件,便可打开布局编辑器,切换到 Text 选项卡,输入代码如下。

```xml
<?xml version = "1.0" encoding = "utf-8"?>
<LinearLayout xmlns:android = "http://schemas.android.com/apk/res/android"
    xmlns:app = "http://schemas.android.com/apk/res-auto"
    xmlns:tools = "http://schemas.android.com/tools"
    android:orientation = "vertical"
    android:background = "@drawable/bg_picture"
    android:layout_width = "match_parent"
    android:layout_height = "match_parent"
    tools:context = ".MainActivity">
    <Button
        android:id = "@+id/button"
        android:text = "进入视频列表"
        android:textColor = "#000000"
        android:textSize = "30sp"
        android:textStyle = "bold"
        android:layout_gravity = "center"
        android:layout_marginTop = "60dp"
        android:layout_width = "wrap_content"
        android:layout_height = "wrap_content"/>
</LinearLayout>
```

上述代码中,根节点为线性布局,方向为 vertical,布局的背景设置为 bg_picture.jpg 图片,单击 Button 跳转到视频列表界面。

(5) 打开 MainActivity.java 类文件,修改 MainActivity 的代码如下。

```java
public class MainActivity extends AppCompatActivity {
    //定义对象
    Button button;
    @Override
    protected void onCreate(Bundle savedInstanceState) {
        super.onCreate(savedInstanceState);
        setContentView(R.layout.activity_main);
        //绑定控件
        button = findViewById(R.id.button);
        //单击跳转事件
        button.setOnClickListener(new View.OnClickListener() {
            @Override
            public void onClick(View view) {
```

```
            Intent intent = new Intent(MainActivity.this,VideoActivity.class);
            startActivity(intent);
        }
    });
    }
}
```

给 Button 控件注册监听事件,通过 Intent 意图类实现从当前页面跳转到 VideoActivity.class 视频列表页面。视频列表页面请同学们自行添加,也可参考任务 7.2.2 步骤 1 完成页面的添加。

(6) 运行程序,效果如图 7.12 所示。

微课视频

【任务 7.2.2】 视频列表界面的设计与功能

(1) 在工程中添加一个新的页面。右击 com.example.videoplayer 包 → New → Activity → Empty Activity,会弹出一个创建活动的对话框,将活动命名为 VideoActivity,默认勾选 Generate Layout File 关联布局页面,布局界面名称为 activity_video,但不要勾选 Launcher Activity。单击 Finish 按钮,便可在工程中完成第二个页面的添加。

(2) RecyclerView 属于新增的控件,要想使用它首先要在 app/build.gradle 文件中添加依赖库,打开 app 目录下的 build.gradle 文件,在 dependencies 闭包中添加如下内容。

图 7.12 主界面运行效果图

```
dependencies {
    implementation fileTree(dir: 'libs', include: ['*.jar'])
    implementation 'androidx.appcompat:appcompat:1.1.0'
    implementation 'androidx.constraintlayout:constraintlayout:1.1.3'
    testImplementation 'junit:junit:4.12'
    androidTestImplementation 'androidx.test.ext:junit:1.1.1'
    androidTestImplementation 'androidx.test.espresso:espresso-core:3.2.0'
    implementation 'com.android.support:recyclerview-v7:29.0.3'
}
```

添加完之后记得要单击右上角的 Sync Now 来进行同步。

(3) 修改布局界面。在 app 目录下的结构中,双击 layout 文件夹下的 activity_video.xml 文件,便可打开布局编辑器,切换到 Text 选项卡,输入代码如下。

```
<?xml version = "1.0" encoding = "utf-8"?>
<LinearLayout xmlns:android = "http://schemas.android.com/apk/res/android"
    xmlns:app = "http://schemas.android.com/apk/res-auto"
    xmlns:tools = "http://schemas.android.com/tools"
```

```xml
    android:orientation = "vertical"
    android:layout_width = "match_parent"
    android:layout_height = "match_parent"
    tools:context = ".VideoActivity">
    <TextView
        android:text = "视频列表"
        android:textColor = "#00007f"
        android:textSize = "40sp"
        android:textStyle = "bold"
        android:layout_gravity = "center"
        android:layout_width = "wrap_content"
        android:layout_height = "wrap_content"/>
    <androidx.recyclerview.widget.RecyclerView
        android:id = "@+id/recyclerview"
        android:layout_width = "match_parent"
        android:layout_height = "wrap_content"/>
</LinearLayout>
```

上述代码中，根节点为线性布局，方向为 vertical，TextView 控件显示标题内容，添加 RecyclerView 控件，设置其宽度为 match_parent，高度为 wrap_content。因为在类文件中要调用 RecyclerView，所以给它设置一个 id。需要注意的是，由于 RecyclerView 并不是内置在系统 SDK 当中的，所以需要把完整的包路径写出来。

（4）定义一个实体类，作为 RecyclerView 适配器的适配类型。右击 com.example.videoplayer 包→New→Java Class，会弹出一个创建类的对话框，将类命名为 Video，代码如下。

```java
public class Video {
    private String videoName;
    private String videoUrl;
    //下方的代码可通过快捷键 Alt+Insert 快速生成，分别是 Getter 方法和 Constructor 方法
    public String getVideoName() {
        return videoName;
    }
    public String getVideoUrl() {
        return videoUrl;
    }
    public Video(String videoName, String videoUrl) {
        this.videoName = videoName;
        this.videoUrl = videoUrl;
    }
}
```

Video 类中有两个字段：videoName 是视频的名字，videoUrl 是视频的地址。

（5）为 RecyclerView 的每一行子项指定要显示的布局样式，右击 Layout→New→Xml→Layout Xml File，会弹出一个创建的 Xml Layout File 对话框，命名为 video_item.xml，代码如下。

```xml
<?xml version="1.0" encoding="utf-8"?>
<LinearLayout xmlns:android="http://schemas.android.com/apk/res/android"
    android:layout_width="match_parent"
    android:orientation="horizontal"
    android:layout_height="wrap_content">
    <ImageView
        android:src="@drawable/video_icon"
        android:layout_width="60dp"
        android:layout_height="60dp"
        android:layout_marginLeft="5dp"
        android:layout_marginTop="5dp"/>
    <TextView
        android:id="@+id/video_name"
        android:text="视频的名字"
        android:textColor="#000000"
        android:textSize="30sp"
        android:textStyle="bold"
        android:layout_gravity="center"
        android:layout_marginLeft="5dp"
        android:layout_width="wrap_content"
        android:layout_height="wrap_content"/>
</LinearLayout>
```

上述代码中,根节点为线性布局,方向为 horizontal,宽度设置为 match_parent,高度设置为 wrap_content,高度如果设置为 match_parent,就会出现一个子项占满整个屏幕。ImageView 控件显示视频的图标,ImageView 控件右侧的 TextView 控件用来显示视频的名称。

(6) 接下来需要为 RecyclerView 准备一个自定义的适配器,右击 com.example.videoplayer 包→New→Java Class,会弹出一个创建类的对话框,将类命名为 VideoAdapter,继承自 RecyclerView.Adapter,并将泛型指定为 VideoAdapter.ViewHolder,其中,ViewHolder 是在 VideoAdapter 中定义的一个内部类,代码如下。

```java
public class VideoAdapter extends RecyclerView.Adapter<VideoAdapter.ViewHolder>{
    List<Video> myvideolist;
    public VideoAdapter(List<Video> myvideolist) {
        this.myvideolist = myvideolist;
    }
    //方法1:用于创建 ViewHolder 实例
    @NonNull
    @Override
    public VideoAdapter.ViewHolder onCreateViewHolder(@NonNull ViewGroup parent, int viewType) {
        View view = LayoutInflater.from(parent.getContext()).inflate(R.layout.video_item, parent, false);
        ViewHolder holder = new ViewHolder(view);
        return holder;
    }
```

```java
//方法2:用于对RecyclerView中子项的数据进行赋值
@Override
public void onBindViewHolder(@NonNull VideoAdapter.ViewHolder holder, int position) {
    Video video = myvideolist.get(position);
    holder.video_name.setText(video.getVideoName());
}
//方法3:返回RecyclerView中数据源长度
@Override
public int getItemCount() {
    return myvideolist.size();
}
//内部类:ViewHolder类
public class ViewHolder extends RecyclerView.ViewHolder {
    TextView video_name;
    View videoview;
    public ViewHolder(@NonNull View view) {
        super(view);
        video_name = itemView.findViewById(R.id.video_name);
        videoview = view;
    }
}
}
```

首先定义了一个内部类 ViewHolder,继承自 RecyclerView。ViewHolder。ViewHolder 的构造函数中要传入一个 View 参数,这个 View 参数就是 RecyclerView 子项最外层的布局,在 ViewHolder 类中包含两个字段:video_name 为视频的名字,videoview 为视频。通过 view 的 findViewById()方法获取 TextView 的实例。

VideoAdapter 的构造函数,用于把要展示的数据源传进来,并赋值给一个全局变量 myVideoList,后续的操作都是在这个数据源上进行的。

由于 VideoAdapter 继承自 RecyclerView. Adapter,所以必须重写 onCreateViewHolder()、onBindViewHolder()、getItemCount()这三个方法。onCreateViewHolder()方法是用于创建 ViewHolder 实例的,在这个方法中将定制好的每一行的布局 video_item 加载进来,然后创建一个 ViewHolder 实例,将加载出来的布局传入到构造函数中,最后将 ViewHolder 实例返回。onBindViewHolder()方法是用于对 RecyclerView 子项的数据进行赋值的,会在每个子项滚动到屏幕内时执行,这里通过 positon 参数得到当前项的 Video 实例,然后再将视频的名称显示在对应的控件上。getItemCount()方法是用于告诉 RecyclerView 一共有多少个子项,通过 size()方法获得 RecyclerView 的数据源长度。

(7) 选择该模式下方的 app→src→main→AndroidManifest. xml 依次展开,双击打开 AndroidManifest. xml 配置文件,修改代码如下。

```xml
<?xml version = "1.0" encoding = "utf-8"?>
< manifest xmlns:android = "http://schemas.android.com/apk/res/android"
    xmlns:tools = "http://schemas.android.com/tools"
    package = "com.example.videoplayer">
    <!-- 允许联网 -->
```

```xml
<uses-permission android:name="android.permission.INTERNET" />
<!-- 获取 SD 卡写的权限,用于文件上传和下载 -->
<uses-permission android:name="android.permission.WRITE_EXTERNAL_STORAGE" />
<uses-permission android:name="android.permission.MOUNT_UNMOUNT_FILESYSTEMS"
    tools:ignore="ProtectedPermissions" />
<application
    android:allowBackup="true"
    android:icon="@mipmap/ic_launcher"
    android:label="@string/app_name"
    android:roundIcon="@mipmap/ic_launcher_round"
    android:supportsRtl="true"
    android:theme="@style/AppTheme"
    tools:ignore="GoogleAppIndexingWarning">
    <activity android:name=".PlayActivity"></activity>
    <activity android:name=".VideoActivity" />
    <activity android:name=".MainActivity">
        <intent-filter>
            <action android:name="android.intent.action.MAIN" />
            <category android:name="android.intent.category.LAUNCHER" />
        </intent-filter>
    </activity>
</application>
</manifest>
```

因为要读写 SD 卡上的歌曲,所以需要在配置文件中添加权限,如下。

```xml
<uses-permission android:name="android.permission.WRITE_EXTERNAL_STORAGE"/>
<uses-permission android:name="android.permission.MOUNT_UNMOUNT_FILESYSTEMS"
    tools:ignore="ProtectedPermissions" />
```

因为要访问网络上的视频资源,所以需要在配置文件中,开通网络权限,代码如下。

```xml
<uses-permission android:name="android.permission.INTERNET"/>
```

(8) 下面添加三种类型的视频:第一种是网络视频资源,第二种是本工程中的视频,第三种是模拟器中的视频。网络资源视频,准备好视频的 URL 地址即可。本工程中的视频,需要提前将视频文件存放在本工程中,右击 res→New→Directory,会弹出一个创建文件夹的对话框,将文件夹命名为 raw,将准备好的视频文件"chinastory.mp4"和"chinesesing.mp4"复制粘贴到该文件夹下。第三种为模拟器中的视频,将准备好的"爱读书 爱生活.mp4"和"阅读陪伴 共抗疫情.mp4"两个视频选中,拖动到夜神模拟器中。如果是高版本的夜神模拟器则默认保存在/storage/emulated/0/Pictures 文件夹中,如果是低版本的夜神模拟器需要存放在 sdcard 目录下的 Movies 文件夹中。

(9) 打开 VideoActivity.java 类文件,修改 VideoActivity 的代码如下。

```java
public class VideoActivity extends AppCompatActivity {
    //定义对象
    RecyclerView recyclerView;
    List<Video> videolist;
```

```java
    VideoAdapter adapter;
    private static final String TAG = "VideoActivity";
    @Override
    protected void onCreate(Bundle savedInstanceState) {
        super.onCreate(savedInstanceState);
        setContentView(R.layout.activity_video);
        initview(); //控件初始化
        initData();//数据初始化
    }
    private void initview() {
        recyclerView = findViewById(R.id.recyclerview);
    }
    private void initData() {
        videolist = new ArrayList<>();
        //1.获取网络视频
Video firstvideo = new Video("玩具总动员", "http://vfx.mtime.cn/Video/2019/03/21/mp4/190321153853126488.mp4");
        videolist.add(firstvideo);
Video secondvideo = new Video("紧急救援", "http://vfx.mtime.cn/Video/2019/03/19/mp4/190319222227698228.mp4");
        videolist.add(secondvideo);
        //2.获取本工程中的视频
Video thirdvideo = new Video("大国工匠","android.resource://" + getPackageName() + "/" + R.raw.chinastory);
        videolist.add(thirdvideo);
Video fourthvideo = new Video("义勇军进行曲","android.resource://" + getPackageName() + "/" + R.raw.chinesesing);
        videolist.add(fourthvideo);
        //3.获取模拟器中的视频
Video fifthvideo = new Video("爱读书 爱生活 ", "file:///storage/emulated/0/Pictures/爱读书 爱生活.mp4"); videolist.add(fifthvideo);
Video sixthvideo = new Video("阅读陪伴 共抗疫情", "file:///storage/emulated/0/Pictures/阅读陪伴 共抗疫情.mp4");
        videolist.add(sixthvideo);
        //让数据显示到RecyclerView控件上
        adapter = new VideoAdapter(videolist);
StaggeredGridLayoutManager layoutManager = new StaggeredGridLayoutManager(1,StaggeredGridLayoutManager.VERTICAL);
        recyclerView.setLayoutManager(layoutManager);
        recyclerView.setAdapter(adapter);
    }
}
```

首先定义对象,定义一个动态数组 ArrayList 来存放 RecyclerView 中 item 的内容,泛型为新建类 Video。initData()数据初始化的方法中,在 Video 的构造函数中将视频名称、视频的 URL 传入,然后将创建好的视频实例添加到视频列表中。本工程中视频的 URL 为"android.resource://"＋getPackageName()＋"/"＋R.raw.视频名,意为安卓资源＋包名＋ res→raw 文件夹中的视频文件,模拟器中视频的 URL 为 file:///storage/emulated/0/Pictures/视频名。

mp4。一部分手机将 eMMC 存储挂载到 /mnt/external_sd、/mnt/sdcard2 等节点,而将外置的 SD 卡挂载到 Environment.getExternalStorageDirectory()这个节点,此时,调用 Environment.getExternalStorageDirectory(),则返回外置的 SD 的路径。URL 地址为:Environment.getExternalStorageDirectory().getPath() + "/文件所在子目录/视频文件名.MP4"。

接着创建了 VideoAdapter 对象,并将 VideoAdapter 适配器传递给 RecyclerView,实现 RecyclerView 和数据的关联。这里创建一个 StaggeredGridLayoutManager 对象,设置列数为一列,方向为垂直,将 StaggeredGridLayoutManager 对象通过 setLayoutManager() 传递给 RecyclerView,这样 RecyclerView 就以瀑布流布局显示视频列表内容。

(10) 运行程序,效果如图 7.13 所示。

图 7.13 视频列表界面运行效果图

【任务 7.2.3】 视频播放界面的设计与功能

(1) 在工程中添加一个新的页面。右击 com.example.videoplayer 包 → New → Activity → Empty Activity,会弹出一个创建活动的对话框,将活动命名为 PlayActivity,默认勾选 Generate Layout File 关联布局界面,布局界面名称为 activity_play,但不要勾选 Launcher Activity。单击 Finish 按钮,便可在工程中完成第三个页面的添加。

(2) 修改布局界面。在 app 目录下的结构中,双击 layout 文件夹下的 activity_play.xml 文件,便可打开布局编辑器,切换到 Text 选项卡,输入代码如下。

```xml
<?xml version = "1.0" encoding = "utf-8"?>
<LinearLayout xmlns:android = "http://schemas.android.com/apk/res/android"
    xmlns:app = "http://schemas.android.com/apk/res-auto"
    xmlns:tools = "http://schemas.android.com/tools"
    android:orientation = "vertical"
    android:layout_width = "match_parent"
    android:layout_height = "match_parent"
    tools:context = ".PlayActivity">
    <TextView
        android:id = "@+id/my_videoname"
        android:layout_width = "wrap_content"
        android:layout_height = "wrap_content"
        android:text = "视频名称"
        android:textStyle = "bold"
        android:textSize = "30sp"
        android:textColor = "#00008F"
        android:layout_gravity = "center"
        android:layout_marginTop = "30dp"/>
    <VideoView
```

```xml
        android:id = "@ + id/my_videoview"
        android:layout_width = "match_parent"
        android:layout_height = "400dp"/>
</LinearLayout>
```

上述代码中,根节点为线性布局,方向为 vertical,TextView 控件显示视频的名称,另一个控件是视频播放控件 VideoView,可以直接在手机上面开辟一个视频播放的 UI,播放视频,程序中利用 VideoView 控件进行播放比较简单。VideoView 几个重要的方法如下。

```java
//指定需要播放的视频的地址?
videoView.setVideoURI(Uri.parse(" android.resource://poet.android.factory/" + R.raw.demo));
//设置视频路径
videoView.setVideoPath();
//设置播放器的控制条?
videoView.setMediaController(new MediaController(this));
//开始播放视频?
videoView.start();
```

(3) 打开 VideoAdapter.java 类文件,修改 VideoAdapter 的代码,如下。

```java
public class VideoAdapter extends RecyclerView.Adapter<VideoAdapter.ViewHolder>{
    List<Video> myvideolist;
    public VideoAdapter(List<Video> myvideolist) {
        this.myvideolist = myvideolist;
    }
    //方法 1:用于创建 ViewHolder 实例
    @NonNull
    @Override
    public VideoAdapter.ViewHolder onCreateViewHolder(@NonNull ViewGroup parent, int viewType) {
        View view = LayoutInflater.from(parent.getContext()).inflate(R.layout.video_item,parent,false);
        final ViewHolder holder = new ViewHolder(view);
        //单击任意视频跳转到播放界面
        holder.videoview.setOnClickListener(new View.OnClickListener() {
            @Override
            public void onClick(View view) {
                int position = holder.getAdapterPosition(); //返回数据在适配器中的位置
                Video video = myvideolist.get(position);
                String myvideoname = video.getVideoName();
                String myvideourl = video.getVideoUrl();
                Intent intent = new Intent(view.getContext(),PlayActivity.class);
                intent.putExtra("videoname",myvideoname);
                intent.putExtra("videourl",myvideourl);
                view.getContext().startActivity(intent);
            }
        });
```

```java
        return holder;
    }
    //方法2:用于对RecyclerView中子项的数据进行赋值
    @Override
    public void onBindViewHolder(@NonNull VideoAdapter.ViewHolder holder, int position) {
        Video video = myvideolist.get(position);
        holder.video_name.setText(video.getVideoName());
    }
    //方法3:返回RecyclerView中数据源长度
    @Override
    public int getItemCount() {
        return myvideolist.size();
    }
    public class ViewHolder extends RecyclerView.ViewHolder {
        TextView video_name;
        View videoview;
        public ViewHolder(@NonNull View view) {
            super(view);
            video_name = itemView.findViewById(R.id.video_name);
            videoview = view;
        }
    }
}
```

微课视频

首先修改 ViewHolder 类文件,在 ViewHolder 中添加一个变量 videoView,用来保存子项最外层布局的实例。onCreateViewHolder()方法中添加注册单击事件,这里给子项最外层布局注册了单击事件,通过 getAdapterPosition()方法获得单击的位置 position,通过 Intent 意图类实现页面的跳转,因为是在 VideoAdapter.java 中,所以使用 view.getContext()方法获取当前上下文环境,PlayActivity.class 为要跳转的页面,跳转的过程中需要传递当前单击的 Video 实例给视频播放页面,使用 putExtra()方法在跳转的过程中传递数据,然后通过 position 参数获得单击的 Video 实例,最后通过 startActivity()方法开启跳转。

(4) 打开 PlayActivity.java 类文件,PlayActivity 的代码如下。

```java
public class PlayActivity extends AppCompatActivity {
    //定义对象
    private TextView my_videoname;
    private VideoView my_videoview; //视频播放器
    @Override
    protected void onCreate(Bundle savedInstanceState) {
        super.onCreate(savedInstanceState);
        setContentView(R.layout.activity_play);
        initView(); //控件初始化
        initData(); //数据初始化
    }
    private void initView() {
        my_videoname = findViewById(R.id.my_videoname);
        my_videoview = findViewById(R.id.my_videoview);
```

```java
    }
    private void initData() {
        //1.获取从音乐列表传过来的视频名称和视频地址
        String videoname = getIntent().getStringExtra("videoname");
        String videourl = getIntent().getStringExtra("videourl");
        //2.将视频名称显示在文本框中,将视频地址关联到播放器中
        my_videoname.setText(videoname);
        my_videoview.setVideoPath(videourl);
        //3.启动视频播放器播放视频
        my_videoview.start();
    }
}
```

在 initData()方法中,首先接收由音乐列表界面传递过来的 Video 实例,采用 getStringExtra()方法获取传递过来的视频名称和视频 URL,通过 setText()方法将视频名称显示在对应的文本框控件中,通过 setVideoPath()方法将视频地址关联到播放器中。最后通过 start()方法启动视频播放器播放视频。

(5) 为了显示视频播放进度,添加媒体控制柄 mediaController,并通过 setMediaController()方法将视频播放器和媒体控制柄关联起来,通过 setMediaPlayer()方法将媒体控制柄和视频播放器关联起来。代码如下。

```java
public class PlayActivity extends AppCompatActivity {
//定义对象
    private TextView my_videoname;
    private VideoView my_videoview; //视频播放器
    private MediaController mediaController; //媒体控制柄
    @Override
    protected void onCreate(Bundle savedInstanceState) {
        super.onCreate(savedInstanceState);
        setContentView(R.layout.activity_play);
        initView(); //控件初始化
        initData(); //数据初始化
    }
    private void initView() {
        my_videoname = findViewById(R.id.my_videoname);
        my_videoview = findViewById(R.id.my_videoview);
        mediaController = new MediaController(this);
    }
    private void initData() {
        //1.获取从音乐列表传过来的视频名称和视频地址
        String videoname = getIntent().getStringExtra("videoname");
        String videourl = getIntent().getStringExtra("videourl");
        //2.将视频名称显示在文本框中,将视频地址关联到播放器中
        my_videoname.setText(videoname);
        my_videoview.setVideoPath(videourl);
        //视频播放器和媒体控制柄关联起来
        my_videoview.setMediaController(mediaController);
        //媒体控制柄和视频播放器关联起来
```

```
        mediaController.setMediaPlayer(my_videoview);
    //3.启动视频播放器播放视频
        my_videoview.start();
    }
}
```

（6）运行程序,单击视频列表界面中的"爱读书 爱生活",运行效果如图 7.14 所示。

图 7.14　视频播放界面运行效果图

项目小结

本项目在简单音乐播放器的设计与实现的案例实施过程中学习了 Timer 及 TimerTask 的用法,强化了 Handler 的工作机制。通过 RecyclerView 控件实现了音乐列表的显示,通过 MediaPlayer 类播放音乐列表中选中的音乐,通过 SeekBar 进度条可以调整音乐播放的进度,并显示音乐总时长和当前播放的时长。最后通过简单视频播放器的设计与实现的项目实战,复习了 RecyclerView 控件,实现了视频列表的显示,通过 VideoView 控件实现视频的播放功能,通过媒体控制柄 mediaController 调整视频播放的进度,并显示视频总时长和当前播放的时长。

习题

1. 视频播放器中存放在本工程中的视频存放在 res 目录下的(　　)文件夹中。
　　A. drawble　　　　B. raw　　　　　C. values　　　　D. layout
2. 视频播放控件 VideoView 可以使用(　　)方法设置播放器的控制条。

A. setMediaController() B. setVideoURI()
C. setVideoPath() D. start()

3. Message 是在（ ）之间传递的消息，它可以在内部携带少量的信息，用于交换数据。

A. 线程 B. 程序 C. 主线程 D. 子线程

4. Handler 主要用于发送和接收消息，发送消息一般是使用 Handler 的（ ）方法。

A. handlerMessage() B. sendMessage()
C. send() D. get()

5. Timer 就是一个线程，使用（ ）方法完成对 TimerTask 的调度。

A. start() B. open() C. cancle() D. schedule()

6. 音乐播放器的播放进度可以通过（ ）控件来实现。

A. TextView B. SeekBar C. ImageView D. RecycleView

7. 在 Android 的界面控件中有一个视频播放控件（ ），可以直接在手机上面开辟一个视频播放的 UI，播放视频。

A. Button B. MediaPlay C. TextView D. VideoView

8. 默认的 MediaController 有后退、（ ）、播放和快进按钮，还有一个清除和进度条组合控件，可以用来定位到视频中的任何一个位置。

A. 暂停 B. 定位 C. 返回 D. 确定

9. 视频播放器中的视频来源可以是网络视频、本工程中的视频、模拟器中保存的视频以及存放在 Bmob 后端云中的视频。这句话是（ ）。

A. 正确的 B. 错误的

10. 视频播放器的播放进度可以通过（ ）类来实现。

A. SeekBar B. MediaController
C. MediaPlayer D. RecycleView

项目 8

我的第一桶金——理财通App的设计与实现

【教学导航】

学习目标	(1) 了解 Android 系统的数据存储方式。 (2) 理解使用 SQLite 数据库存储数据。 (3) 熟练掌握数据库的增、删、改、查操作。 (4) 培养复杂案例的设计开发能力。
教学方法	任务驱动法、理论实践一体化、探究学习法、分组讨论法
课时建议	16 课时

微课视频

8.1 使用数据库存储数据

作为一个完整的应用程序,数据存储操作是必不可少的。学习 Android 相关知识,数据存储是其中的重点之一。

8.1.1 Android 系统的数据存储方式

Android 系统提供了四种数据存储方式,分别是:SharedPreferences、File 存储、SQLite 存储、ContentProvider。

(1) SQLite:SQLite 是一个轻量级的数据库,支持基本 SQL 语法,是常被采用的一种数据存储方式。Android 为此数据库提供了一个名为 SQLiteDatabase 的类,封装了一些操作数据库的 API。

(2) SharedPreferences:除了 SQLite 数据库外,SharedPreferences 是另一种常用的数据存储方式,其本质就是一个 XML 文件,常用于存储较简单的参数设置。

(3) File 存储:即常说的文件(I/O)存储方法,常用于存储大量的数据。

（4）ContentProvider：Android 系统中能实现所有应用程序共享的一种数据存储方式，由于数据在各应用间通常是私密的，所以此存储方式较少使用，但是又是必不可少的一种存储方式，例如音频、视频、图片、通讯录等，一般都可以采用此种方式进行存储。每个 ContentProvider 都会对外提供一个公共的 URI(包装成 Uri 对象)，如果应用程序有数据需要共享时，就需要使用 ContentProvider 为这些数据定义一个 URI，然后其他的应用程序就通过 ContentProvider 传入这个 URI 来对数据进行操作。URI 由"content：//"、数据的路径、标识 ID(可选)3 个部分组成。

8.1.2　使用 SQLite 数据库存储数据

Android 的数据库有两种，一种是 Android 内置的数据库 SQLite，另一种是一款开源的 Android 数据库框架 LitePal。SQLite 是 Android 系统内置的数据库，用户无须安装任何插件便可直接使用。它占用资源很少，通常只要几百 KB 就够了，运算速度非常快，所以它是一款轻量级的关系型数据库，特别适合在移动设备上使用。SQLite 不仅支持标准的 SQL 语法，还遵循了数据库的 ACID 事务，所以只要以前使用过其他的关系型数据库，就可以很快地上手 SQLite，而 SQLite 又比一般的数据库要简单得多，甚至不用设置用户名和密码就可以使用。

随着 Android 的发展，Github 上出现了成百上千个 Android 开源项目，LitePal 就是其中的一个非常优秀的开源项目。因为是开源的，所有的用户使用时都可以随时下载，原来需要写很久才能实现的功能，使用开源库可能短短几分钟就能实现了，提高了工作效率。代码运行时，稳定性一直是每个开发者的追求，开源项目的代码都是经过时间验证的，通常比自己的代码要稳定很多，因此，很多公司为了追求开发效率以及项目的稳定性，都会选择使用开源库。LitePal 是一款开源的数据库框架，它采用了对象关系映射的模式，并将平时开发中最常用的一些数据库功能进行了封装，使得不用编写一行 SQL 语句就可以完成各种建表和增删改查的操作，感兴趣的同学可以通过百度查找资料自行学习 LitePal 开源库的用法。

8.1.3　创建和升级数据库

微课视频

SQLite 数据库文件存放位置为/data/data/< package name >/databases/，其中，package name 是创建工程的包名。查看数据库的方法：在手机或者模拟器中安装一个 RE 文件管理器应用，双击打开模拟器，将模拟器挂载为可读写模式，按照文件存放的路径/data/data/< package name >/databases/一直单击，便可打开工程的数据库。SQLite 内部只支持 null(空)、integer(整型)、real(浮点数)、text(文本)/String(文本)/ varchar(文本)和 blob(二进制)这五种数据类型。

1. 创建数据库和升级数据库常用的方法

Android 为了让用户更加方便地管理数据库，专门提供了一个 SQLiteOpenHelper 帮助类，借助于这个类就可以非常简单地对数据库进行创建和升级，其基本用法如表 8.1 所示。

表 8.1 创建数据库和升级数据库的常用方法

方法	功能描述	实例
SQLiteOpenHelper	抽象类，用来管理数据库，使用时必须自己创建一个类去继承它	public class MyDBOpenHelper extends SQLiteOpenHelper {…} 其中，MyDBOpenHelper 便是自己创建的数据库帮助类

SQLiteOpenHelper (Context context, String name, SQLiteDatabase. CursorFactory factory, int version)	SQLiteOpenHelper 提供的构造器，用于传递当前上下文对象以及 SQLite 数据库版本信息	//定义数据库名和版本号
		private static final String DBNAME="student.db";
		private static final int VERSION=1;
		public MyDBOpenHelper(Context context) {
		    super(context, DBNAME, null, VERSION); }

public void onCreate (SQLiteDatabase db)	创建数据库	@Override
		 public void onCreate(SQLiteDatabase db) {
		  }

db.execSQL()	在数据库中创建数据表	db.execSQL("create table stu_info(id INTEGER primary key autoincrement, sno varchar(10), name varchar(10), sex varchar(4), professional varchar(10), deparment varchar(20))");
public void onUpgrade (SQLiteDatabase db, int oldVersion, int newVersion)	升级数据库	@Override public void onUpgrade(SQLiteDatabase db, int oldVersion, int newVersion) { }

getWritableDatabase()	获取一个读写操作的数据库对象	private MyDBOpenHelper mhelper;//定义数据库帮助类对象
		private SQLiteDatabase db;//定义一个可以操作的数据库对象
		mhelper=new MyDBOpenHelper(MainActivity.this);
		//实例化数据库帮助类
		db=mhelper.getWritableDatabase();//创建数据库，获取数据库的读写权限

getReadableDatabase()	获取一个读操作的数据库对象	…前几行同上，此处省略 db=mhelper.getReadableDatabase();;//创建数据库，获取数据库的读权限

上述方法中，SQLiteOpenHelper 是 Android 中专门提供的一个帮助类，借助这个类就可以非常简单地对数据库进行创建和升级。但是 SQLiteOpenHelper 是一个抽象类，如果用户想使用它的话，就需要创建一个自己的帮助类去继承它。

SQLiteOpenHelper 提供了两个构造器，一般使用参数少一点儿的那个构造方法即可，用于传递当前上下文对象以及 SQLite 数据库版本信息，在 SQLiteOpenHelper 的继承类的构造函数中会调用它，该构造函数也用于在其他页面中对数据库帮助类进行实例化。

SQLiteOpenHelper 中有两个抽象方法，分别是 onCreate()和 onUpgrade()，用户必须在自己的帮助类里面重写这两个方法，然后分别在这两个方法中去实现创建、升级数据库的逻辑。

SQLiteOpenHelper 中还有两个非常重要的实例方法：getWritableDatabase()和

getReadableDatabase()。这两个方法都可以创建或者打开一个现有的数据库(如果数据库已存在则直接打开,否则创建一个新的数据库),并返回一个可对数据库进行读写操作的对象,如表 8.1 中的 db,利用这个对象便可以对数据库进行增删改查操作了。

2. 创建数据库案例

1) 案例分析

(1) 界面分析。程序创建好的默认界面,没有做任何改动。

(2) 设计思路。由于创建数据库是通过 SQLiteOpenHelper 帮助类实现的,因此用户首先需要创建一个自己的帮助类去继承它,其次添加自己创建的帮助类的构造方法,用于传递当前上下文对象以及 SQLite 数据库版本信息。最后添加该类的两个抽象方法 onCreate()和 onUpgrade(),在这两个方法中去实现创建、升级数据库的逻辑。

2) 实现步骤

(1) 创建一个新的工程,工程名为 StudentInfoTest,即学生信息管理系统。

(2) 切换工程的 Project 项目结构,选择该模式下方的 app,依次展开,便看到工程的布局界面和工程的类文件,其中,activity_main.xml 是布局界面,MainActivity.java 为类文件。

(3) 创建数据库帮助类文件。选中 MainActivity.java 类文件上方的包名 com.example.studentinfotest,右键单击→New→Java Class,在 Name 文本框中输入自己创建的数据库的类文件名"MyDBOpenHelper",在 Superclass 文本框中输入继承的父类名"SQLiteOpenHelper",单击 OK 按钮,如图 8.1 所示。

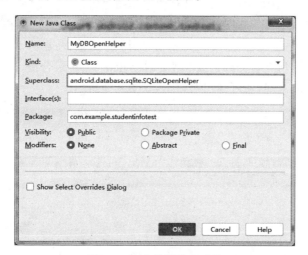

图 8.1 创建数据库帮助类

(4) 查看创建好的数据库帮助类文件。在工程 MainActivity 类文件的下方,便出现了自己创建好的数据库帮助类,代码如下:

```
public class MyDBOpenHelper extends SQLiteOpenHelper {

}
```

上述代码中，MyDBOpenHelper 是自己创建的数据库帮助类，为了达到见名知意的目的，习惯上用户都会在类名中含有 DBHelper 等典型字母，为了让自己创建的类具有创建数据库和升级数据库的特性，让其继承 SQLiteOpenHelper 即可。

（5）添加 SQLiteOpenHelper 类的两个抽象方法。由于 SQLiteOpenHelper 是一个抽象类，需要添加该类内部的两个抽象方法。方法为：鼠标定位到 MyDBOpenHelper 类名所在行的红色波浪线处，按 Alt＋Enter 组合键，选中 Implement methods，按 Enter 键，便可弹出对话框，同时选中 onCreate()和 onUpgrade()两种方法，单击 OK 按钮即可，代码如下。

```
public class MyDBOpenHelpert extends SQLiteOpenHelper {
    //创建数据库
    @Override
    public void onCreate(SQLiteDatabase db) {

    }
    //升级数据库
    @Override
    public void onUpgrade(SQLiteDatabase db, int oldVersion, int newVersion) {

    }
}
```

上述代码中，onCreate()用来实现创建数据库，onUpgrade()用来实现升级数据库，用户需要分别在这两个方法中去实现创建、升级数据库的逻辑。

（6）添加数据库帮助类的构造方法。每一个类都有构造方法，该构造函数也用于在其他页面中对数据库帮助类进行实例化。添加构造函数的方法为：鼠标再次定位到 MyDBOpenHelper 类名所在行的红色波浪线处，按 Alt＋Enter 组合键，选中 Create constructor matching super，按 Enter 键，便可弹出对话框，如图 8.2 所示。

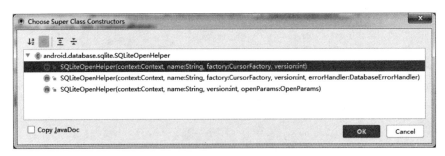

图 8.2　数据库类文件的构造方法

在图 8.2 中，选择参数较少的第一种构造方法，单击 OK 按钮，类文件中自动出现构造方法代码，具体如下。

```
public class MyDBOpenHelper extends SQLiteOpenHelper {
    public MyDBOpenHelper(Context context, String name, SQLiteDatabase.CursorFactory factory, int version) {
        super(context, name, factory, version);
```

```
        }
    //创建数据库
    …(此处代码略)
    //升级数据库
    …(此处代码略)
}
```

上述代码中,构造方法中接收 4 个参数:参数 1 为 Context;参数 2 为数据库名;参数 3 为在查询数据的时候返回一个自定义的 Cursor,一般都是传入 null;参数 4 为数据库的版本号,可用于对数据库进行升级操作。在实际使用过程中,为了方便数据库帮助类的实例化,会适当修改其构造方法代码,具体如下。

```
public class MyDBOpenHelper extends SQLiteOpenHelper {
    //定义数据库名和版本号
        private static final String DBNAME = "student.db";
        private static final int VERSION = 1;
        public MyDBOpenHelper(Context context) {
            super(context, DBNAME, null, VERSION);
        }
    //创建数据库
    …(此处代码略)
    //升级数据库
    …(此处代码略)
}
```

上述代码中,定义了一个字符串常量 DBNAME 和整型常量 VERSION,其中,student.db 是数据库名,1 是数据库的版本号。

(7) 创建数据表。在 onCreate()方法中,调用 SQLiteDatabase 的 execSQL()方法去执行创建表的语句,添加代码如下。

```
public class MyDBOpenHelper extends SQLiteOpenHelper {
    //定义数据库名和版本号
        private static final String DBNAME = "student.db";
        private static final int VERSION = 1;
        public MyDBOpenHelper(Context context) {
            super(context, DBNAME, null, VERSION);
        }
        //创建数据库
        @Override
        public void onCreate(SQLiteDatabase db) {
    //创建数据表
    db.execSQL("create table stu_info(id INTEGER primary key autoincrement, sno varchar(10), name varchar(10), sex varchar(4), professional varchar(10), deparment varchar(20) )");
        }
    //升级数据库
        @Override
```

```
        public void onUpgrade(SQLiteDatabase db, int oldVersion, int newVersion) {

        }
}
```

只要对 SQL 方面的知识稍微有一些了解,上面的建表语句理解起来就不难。SQLite 数据类型很简单,integer 表示整型,real/double 表示浮点型,text/String/varchar 表示文本字符串类型,blob 表示二进制类型。上述建表语句中使用了 primary key 将 id 列设为主键,并用 autoincrement 关键字表示 id 列是自增长的。

(8) 运行程序,并通过 RE 文件管理器查看数据库。由于工程创建完成后并未在界面中添加任何控件,运行程序之后,界面中会显示默认的控件 helloworld。在模拟器中打开 RE 文件管理器(如果还没有安装,在百度中搜索 RE 文件管理器,下载 apk,将 apk 拖放到模拟器中即可完成安装),挂载为可读写模式,根据数据库的存储路径依次操作:data/data/<package name>/databases/数据库名,单击 data,在打开的模拟器页面中再次单击 data,在列表中找到工程的包名 com.example.studentinfotest,如图 8.3 所示。

图 8.3　程序运行效果及查看数据库

由于数据库类文件在 MainActivity 页面中还未使用,因此目前数据库并未真正的开始创建,所以在 RE 文件管理器中未能看到数据库,在后续内容当中来完善数据库创建及数据表中数据的增删改查。

微课视频

8.1.4　添加数据

当使用 SQLiteOpenHelper 的 getReadableDatabase()或者 getWritableDatabase()方法获取到 SQLiteDatabase 对象,借助这个对象就可以对数据进行 CRUD 操作了。其中,C 代表添加(Create),R 代表查询(Retrieve),U 代表更新(Update),D 代表删除(Delete),添加数据所使用的方法如下。

1. 添加数据常用方法

系统提供了添加数据的常用方法,如表 8.2 所示。

表 8.2 添加数据常用方法

方法	功能描述	实例

insert(String table,String nullColumnHack,ContentValues values)	添加数据	ContentValues values=new ContentValues();//用 value 表示一行
		values.put("sno",edit_onesno.getText().toString());
		values.put("name",edit_onename.getText().toString());
		values.put("sex",edit_onesex.getText().toString());
		values.put("professional",edit_onepro.getText().toString());
		values.put("deparment",edit_onedep.getText().toString());
		db.insert("stu_info",null,values);

表 8.2 中,insert()方法接收 3 个参数:参数 1 为表名;参数 2 为在未指定添加数据的情况下给某些可为空的列自动复制 NULL,一般用不到这个功能,直接传入 null 即可;参数 3 是一个 Content Values 对象,用来表示一行数据,它提供了一系列的 put()方法重载,用于向 Content Values 这行数据中的每一列添加数据,这里只需要将表中的每个列名以及相应的待添加数据传入即可。

2. 添加数据案例

1) 案例分析

(1) 界面分析。该布局界面为信息添加界面,用来完成学生信息的添加。布局界面中有九个控件,TextView 控件用来显示标题,EditText 控件用来输入内容,两个 Button 控件用来添加学生信息和清空学生信息,下一页按钮用来实现页面的切换。

(2) 设计思路。布局界面中的控件通过添加属性的方式达到用户需求,在"添加"按钮单击事件内部,数据库的 insert()方法,将所输入的用户信息保存到数据库的 stu_info 表中;单击"清除"按钮,清空输入的学生信息;单击"下一页"按钮,跳转到下一页,达到用户的需求。

2) 实现步骤

(1) 无须创建新工程,在创建数据库工程 StudentInfoTest 中完善程序,完成学生信息的添加。

(2) 在工程中添加新的页面。右击 com.example.studentinfotest 包→New→Activity→Empty Activity,会弹出一个创建活动的对话框,将活动命名为 SecondActivity,默认勾选 Generate Layout File 关联布局界面,布局界面名称为 activity_second,但不要勾选 Launcher Activity。单击 Finish 按钮,便可在工程中完成第二个页面的添加。同理,添加第三个页面,类文件为 ThirdActivity,布局界面名称为 activity_third。继续添加第四个页面,类文件为 FourActivity,布局界面名称为 activity_four。

(3) 准备 4 张图片,图片名为 addbg.jpg、querybg.jpg、modifybg.jpg、deletebg.jpg,将其粘贴到 app 目录结构中 res 下方的 drawable 文件夹下,图片分别作为学生信息添加页面、学生信息查询页面、学生信息修改页面、学生信息删除页面的背景图片。

(4) 设计布局界面。双击 layout 文件夹下的 activity_main.xml 文件,便可打开布局编辑器,在布局界面上添加 7 个控件,代码如下。

```xml
<?xml version = "1.0" encoding = "utf-8"?>
<LinearLayout xmlns:android = "http://schemas.android.com/apk/res/android"
    xmlns:app = "http://schemas.android.com/apk/res-auto"
    xmlns:tools = "http://schemas.android.com/tools"
    android:layout_width = "match_parent"
    android:layout_height = "match_parent"
    android:orientation = "vertical"
    android:background = "@drawable/addbg"
    tools:context = ".MainActivity">

    <TextView
        android:layout_width = "wrap_content"
        android:layout_height = "wrap_content"
        android:text = "信息添加页面"
        android:textSize = "30sp"
        android:textStyle = "bold"
        android:textColor = "#000000"
        android:layout_gravity = "center"
        android:layout_margin = "80dp"/>
    <EditText
        android:id = "@+id/editText_onesno"
        android:layout_width = "match_parent"
        android:layout_height = "wrap_content"
        android:hint = "学号"
        android:textSize = "25sp"
        android:textColor = "#000000"/>
    <EditText
        android:id = "@+id/editText_onename"
        android:layout_width = "match_parent"
        android:layout_height = "wrap_content"
        android:hint = "姓名"
        android:textSize = "25sp"
        android:textColor = "#000000"/>
    <EditText
        android:id = "@+id/editText_onesex"
        android:layout_width = "match_parent"
        android:layout_height = "wrap_content"
        android:hint = "性别"
        android:textSize = "25sp"
        android:textColor = "#000000"/>
    <EditText
        android:id = "@+id/editText_onepro"
        android:layout_width = "match_parent"
        android:layout_height = "wrap_content"
        android:hint = "专业班级"
        android:textSize = "25sp"
```

```xml
        android:textColor = "#000000"/>
    <EditText
        android:id = "@+id/editText_onedep"
        android:layout_width = "match_parent"
        android:layout_height = "wrap_content"
        android:hint = "所属系部"
        android:textSize = "25sp"
        android:textColor = "#000000"/>
    <LinearLayout
        android:layout_width = "match_parent"
        android:layout_height = "wrap_content"
        android:orientation = "horizontal">
        <Button
            android:id = "@+id/button_oneadd"
            android:layout_width = "wrap_content"
            android:layout_height = "wrap_content"
            android:text = "添加"
            android:textSize = "25sp"
            android:textColor = "#000000"
            android:layout_weight = "1"/>
        <Button
            android:id = "@+id/button_oneclear"
            android:layout_width = "wrap_content"
            android:layout_height = "wrap_content"
            android:text = "清除"
            android:textSize = "25sp"
            android:textColor = "#000000"
            android:layout_weight = "1"/>
    </LinearLayout>
    <Button
        android:id = "@+id/button_onenext"
        android:layout_width = "wrap_content"
        android:layout_height = "wrap_content"
        android:text = "下一页"
        android:textSize = "25sp"
        android:textColor = "#000000"
        android:layout_gravity = "right"
        android:layout_marginTop = "30dp"/>
</LinearLayout>
```

(5) 实现学生信息的添加功能。重点关注定义数据库有关的两个对象、数据库的初始化方法和数据添加的方法,具体如下。

```java
public class MainActivity extends AppCompatActivity {
    //定义对象
    private EditText edit_onesno,edit_onename,edit_onesex,edit_onepro,edit_onedep;
    private Button btn_oneadd,btn_oneclear,btn_onenext;
    private MyDBOpenHelper mhelper;  //定义数据库帮助类对象
```

```java
        private SQLiteDatabase db;  //定义一个可以操作的数据库对象
        @Override
        protected void onCreate(Bundle savedInstanceState) {
            super.onCreate(savedInstanceState);
            setContentView(R.layout.activity_main);
            //1.绑定控件
            initView();
            //2."添加"按钮功能的实现
            btnAdd();
            //3."清除"和"下一页"按钮的功能
            btnClearNext();
        }
        //1.绑定控件
        private void initView() {
            edit_onesno = findViewById(R.id.editText_onesno);
            edit_onename = findViewById(R.id.editText_onename);
            edit_onesex = findViewById(R.id.editText_onesex);
            edit_onepro = findViewById(R.id.editText_onepro);
            edit_onedep = findViewById(R.id.editText_onedep);
            btn_oneadd = findViewById(R.id.button_oneadd);
            btn_oneclear = findViewById(R.id.button_oneclear);
            btn_onenext = findViewById(R.id.button_onenext);
            mhelper = new MyDBOpenHelper(MainActivity.this);  //实例化数据库帮助类
            db = mhelper.getWritableDatabase();  //创建数据库,获取数据库的读写权限
        }
        //2."添加"按钮功能的实现
        private void btnAdd() {
            btn_oneadd.setOnClickListener(new View.OnClickListener() {
                @Override
                public void onClick(View v) {
                    //定义一个对象,构建一行数据
                    ContentValues values = new ContentValues();  //用value表示一行
                    values.put("sno",edit_onesno.getText().toString());  //把输入的学号放到sno列
                    values.put("name",edit_onename.getText().toString());  //把输入的姓名放到name列
                    values.put("sex",edit_onesex.getText().toString());  //把输入的性别放到sex列
                    values.put("professional",edit_onepro.getText().toString());
                                          //把输入的专业放到professional列
                    values.put("deparment",edit_onedep.getText().toString());
                                          //把输入的系部放到department列
        //将这一行数据存放到数据库的数据表中
                    db.insert("stu_info",null,values);
                    Toast.makeText(MainActivity.this,"添加成功",Toast.LENGTH_SHORT).show();
                }
            });
        }
        //3."清除"和"下一页"按钮的功能
        private void btnClearNext() {
            //"清除"按钮的功能
            btn_oneclear.setOnClickListener(new View.OnClickListener() {
```

```java
            @Override
            public void onClick(View v) {
                edit_onesno.setText("");
                edit_onename.setText("");
                edit_onesex.setText("");
                edit_onepro.setText("");
                edit_onedep.setText("");
            }
        });
        //"下一页"按钮的功能
        btn_onenext.setOnClickListener(new View.OnClickListener() {
            @Override
            public void onClick(View v) {
                Intent intent = new Intent(MainActivity.this,SecondActivity.class);
                startActivity(intent);
                finish();
            }
        });
    }
}
```

上述代码中,首先定义了数据库帮助类对象 mhelper 和一个可供操作的数据库对象 db。其次,通过 new MyDBOpenHelper(MainActivity.this)实例化数据库帮助类,并调用数据库帮助类对象的 getWritableDatabase()方法创建数据库,获取数据库的读写权限。此时数据库已经真正的创建成功,在 RE 文件管理器中,可以通过路径 data/data/< package name >/databases/数据库名/数据表名进行查看。

向数据库的数据表 stu_info 中添加数据的过程为:首先使用 ContentValues 来对要添加的一行数据进行组装,它提供了一系列的 put()方法向一行中的每一列字段添加数据。由于 id 列设置为自增长,所以 id 那一列并没有给它赋值。代码 values.put("sno",edit_onesno.getText().toString())中,edit_onesno.getText().toString()方法获取输入的学号,sno 表示数据表中的学号列,所以整行代码的含义是:调用 ContentValues 对象的 put()方法将输入的学号放到数据库中数据表的 sno 列,同理,获取输入的用户名放到数据表中的 name 列;获取输入的性别存放到数据表中的 sex 列;获取输入的专业放到数据表中的 professional 列;获取输入的系部存放到数据表中的 deparment 列。最后调用数据库的 insert 方法,将组装好的这一行数据添加到 stu_info 表中。

(6) 运行程序,效果如图 8.4 所示。

(7) 查看保存到数据表中的学生信息。在模拟器中找到文件管理器,双击打开文件管理器,挂载为可读写模式,根据数据库的存储路径依次操作 data/data/com.example.studentinfotest/databases/student.db/stu_info,其中,databases 是数据库,student.db 是数据库的名称,stu_info 是数据表名称。双击 stu_info 将其打开,便可查看刚刚保存到数据表中的文件内容,如图 8.5 所示。

图 8.4　学生信息添加界面　　　图 8.5　从模拟器中查看数据库中数据表里面的数据

微课视频

8.1.5　查询数据

完成了信息的添加之后,用户可以进行查询操作,从添加的众多信息中查找自己所需要的用户信息。可以使用 Android 封装好的方法 query(),也可以直接传入 SQL 语句来操作,查询语句所使用的方法如下。

1. 查询数据常用方法

系统提供了查询数据的常用方法,如表 8.3 所示。

表 8.3　查询数据常用方法

方　　法	功能描述	实　　例
public Cursor query(String table, String [ ] columns, String selection, String[] selectionArgs, String groupBy, String having, String orderBy)	查询数据(使用 Android 提供的 API 查询)	Cursor cursor =db.query("stu_info",new String [ ] { " sno "," name "," sex ", "professional"," deparment"}," sno=?", new String[ ]{edit_twosno.getText(). toString()},null,null,null);
public Cursor query(String sql, String[] selectionArgs)	查询数据(直接使用 SQL 语句查询)	Cursor cursor = db.rawQuery("select * from stu_info where sno=?",new String[] {edit_twosno.getText().toString()});

表 8.3 中介绍了两种查询数据的方法,query()方法是使用 Android 提供的 API 来查询,该方法需要传入 7 个参数,7 个参数各自的含义不同。其中,参数 1 为表名;参数 2 为查

询的列名；参数 3 为查询的条件；参数 4 为条件的参数（即查询条件的具体的值）；参数 5 表示是否需要分组；参数 6 为对分组后的结果进一步约束（例如，是否需要对分组后的数据分类汇总）；参数 7 表示查询结果的排序方式。虽然 query()方法的参数非常多，但是不要对它产生畏惧，因为我们不必为每条查询语句都指定所有的参数，多数情况下只需要传入少数几个参数就可以完成查询操作了，例如，db. query("stu_info",null,null,null,null,null,null)表示查询 stu_info 这张表当中的所有数据。

Android 考虑到用户的编程习惯，同样也提供了 db. rawQuery()方法，用户可以直接通过 SQL 来查询数据。该方法接收两个参数：参数 1 为 sql 语句；参数 2 为查询条件的具体的值。根据用户的需要，完成 SQL 语句的书写，如果没有查询条件，参数 2 直接写为 null 就可以了。例如，db. rawQuery("select * from stu_info",null)表示查询 stu_info 这张表当中的所有数据。

2. 查询数据案例

1) 案例分析

（1）界面分析。该布局界面为信息查询界面，用来完成学生信息的查询。布局界面中有四个控件：EditText 控件用来输入查询的条件，TextView 控件用来显示查询的结果，两个 Button 控件用来完成查询和跳转到下一页。

（2）设计思路。布局界面中的控件通过添加属性的方式达到用户需求，在"查询"按钮单击事件内部，调用 SQLiteDatabase 的 query()方法或 rawQuery()方法，输入学生的学号，便可查询到该学号的学生信息并显示到 TextView 控件上。单击"下一页"按钮，跳转到下一页，达到用户的需求。

2) 实现步骤

（1）无须创建新工程，在创建数据库工程 StudentInfoTest 中完善程序，完成学生信息的查询。

（2）设计布局界面。双击 layout 文件夹下的 activity_ second. xml 文件，便可打开布局编辑器，在布局界面上添加四个控件，代码如下。

```xml
<?xml version = "1.0" encoding = "utf - 8"?>
< LinearLayout xmlns:android = "http://schemas.android.com/apk/res/android"
    xmlns:app = "http://schemas.android.com/apk/res - auto"
    xmlns:tools = "http://schemas.android.com/tools"
    android:layout_width = "match_parent"
    android:layout_height = "match_parent"
    android:orientation = "vertical"
    android:background = "@drawable/querybg"
    tools:context = ".SecondActivity">

    < TextView
        android:layout_width = "wrap_content"
        android:layout_height = "wrap_content"
        android:text = "信息查询页面"
        android:textSize = "30sp"
        android:textStyle = "bold"
```

```xml
        android:textColor = "#000000"
        android:layout_gravity = "center"
        android:layout_margin = "80dp"/>
    <EditText
        android:id = "@+id/editText_twosno"
        android:layout_width = "match_parent"
        android:layout_height = "wrap_content"
        android:hint = "请输入要查询的学号"
        android:textColor = "#000000"
        android:textSize = "25sp"/>
    <Button
        android:id = "@+id/button_twoquery"
        android:layout_width = "match_parent"
        android:layout_height = "wrap_content"
        android:text = "查询"
        android:textSize = "25sp"/>
    <TextView
        android:id = "@+id/textView_tworesult"
        android:layout_width = "wrap_content"
        android:layout_height = "wrap_content"
        android:text = "显示查询结果"
        android:textSize = "25sp"
        android:textColor = "#000000"/>
    <Button
        android:id = "@+id/button_twonext"
        android:layout_width = "wrap_content"
        android:layout_height = "wrap_content"
        android:text = "下一页"
        android:textSize = "25sp"
        android:layout_gravity = "right"
        android:layout_marginTop = "30dp"/>
</LinearLayout>
```

（3）实现学生信息的查询功能。重点关注定义数据库有关的两个对象、数据库的初始化方法和数据查询的方法，如下。

```java
public class SecondActivity extends AppCompatActivity {
    //定义对象
    private EditText edit_twosno;
    private Button btn_twoquery,btn_twonext;
    private TextView txt_tworesult;
    private MyDBOpenHelper mhelper; //定义一个数据库帮助类对象
    private SQLiteDatabase db; //定义一个操作的数据库的类对象
    @Override
    protected void onCreate(Bundle savedInstanceState) {
        super.onCreate(savedInstanceState);
        setContentView(R.layout.activity_second);
        //1.控件初始化
        initView();
```

```java
        //2."查询"按钮功能的实现
        btnQuery();
        //3."下一页"按钮功能的实现
        btnNext();
    }
    //1.控件初始化
    private void initView() {
        edit_twosno = findViewById(R.id.editText_twosno);
        btn_twoquery = findViewById(R.id.button_twoquery);
        txt_tworesult = findViewById(R.id.textView_tworesult);
        btn_twonext = findViewById(R.id.button_twonext);
        mhelper = new MyDBOpenHelper(SecondActivity.this);  //实例化数据库帮助类对象
        db = mhelper.getWritableDatabase();  //获取数据库的读写权限
    }
    //2."查询"按钮功能的实现
    private void btnQuery() {
        btn_twoquery.setOnClickListener(new View.OnClickListener() {
            @Override
            public void onClick(View v) {
                //开始查询
                Cursor cursor = db.rawQuery("select * from stu_info where sno = ?",new String[]{edit_twosno.getText().toString()});
                if(cursor.getCount()!= 0){//判断结果集中是否有数据。有:查询成功;
                                        //无:查询失败
                    Toast.makeText(SecondActivity.this,"查询成功",Toast.LENGTH_SHORT).show();
                    //循环遍历结果集,取出数据,显示出来
                    while (cursor.moveToNext()){
                        String mysno = cursor.getString(cursor.getColumnIndex("sno"));
                        String myname = cursor.getString(cursor.getColumnIndex("name"));
                        String mysex = cursor.getString(cursor.getColumnIndex("sex"));
                        String mypro = cursor.getString(cursor.getColumnIndex("professional"));
                        String mydep = cursor.getString(cursor.getColumnIndex("deparment"));
                        txt_tworesult.setText(mysno + "\n" + myname + "\n" + mysex + "\n" + mypro + "\n" + mydep);
                    }
                }else{
                    Toast.makeText(SecondActivity.this,"没有查询到该学号的学生",Toast.LENGTH_SHORT).show();
                    txt_tworesult.setText("");
                }
            }
        });
    }
    //3."下一页"按钮功能的实现
    private void btnNext() {
        btn_twonext.setOnClickListener(new View.OnClickListener() {
```

```
        @Override
        public void onClick(View v) {
            Intent intent = new Intent(SecondActivity.this,ThirdActivity.class);
            startActivity(intent);
            finish();
        }
    });
  }
}
```

上述代码中,首先定义了数据库帮助类对象和一个可供操作的数据库对象,对象名为 mhelper 和 db。其次通过 new MyDBOpenHelper(SecondActivity.this)实例化数据库帮助类,并调用数据库帮助类对象的 getWritableDatabase()方法获取数据库的读写权限。

从数据库的数据表 stu_info 中查询数据的过程为:首先在"查询"按钮的单击事件里面调用了 SQLiteDatabase 的 rawQuery()方法查询数据,这里的 rawQuery()方法非常简单,参数 1 是查询的 SQL 语句,参数 2 是条件参数的具体值。

查完之后就得到了一个 Cursor 对象,接着调用它的 getCount()方法判断获取数据的条数,如果查询到的数据条数为 0,说明没有查询到数据,弹出提示告诉用户"没有查询到该学号的学生",同时清空输入框,便于用户重新输入新的学号进行查询。如果查询到的数据条数不为 0,说明查询到了数据,接着调用它的 moveToNext()方法将数据的指针移动到第一行的位置,然后进入一个循环当中,去遍历查询到的每一行数据。在这个循环当中,可以通过 Cursor 的 getColumnIndex()方法根据列名获取列的索引下标,然后调用 Cursor.getXXX()方法获取该列对应的值,例如,cursor.getString(cursor.getColumnIndex("sno"))获取该行学号列对应的值,同理,获取姓名列对应的值,获取性别列对应的值,获取专业班级列对应的值,获取院系列对应的值,最后将获取的值显示到 TextView 控件中。为了清晰地显示数据,利用\n 将该行的用户信息分别换行显示。

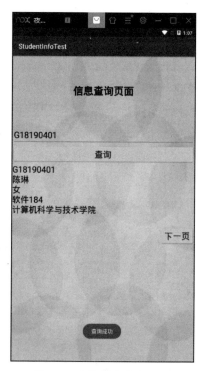

图 8.6 查询学生信息界面

(4) 运行程序,效果如图 8.6 所示。

微课视频

8.1.6 修改数据

学生信息的添加和查询完成之后,本节学习修改表中的已有数据。SQLiteDatabase 中也提供了一个非常好用的 update()方法,用于对数据进行更新。

1. 更新数据常用方法

系统提供了更新数据的常用方法，如表 8.4 所示。

表 8.4 更新数据常用方法

方　法	功能描述	实　例
update（String table，ContentValues values，String whereClause，String[] whereArgs）	更新数据（修改数据）	ContentValues values = new ContentValues（）；values.put（"deparment"，edit_threedep.getText（）.toString（））；db.update（"stu_info"，values，"sno=?"，new String[]｛edit_threesno.getText（）.toString（）｝）；

表 8.4 中，update（）方法接收 4 个参数，其中，参数 1 为表名；参数 2 为 ContentValues 对象，要把更新数据在这里组装进去；参数 3 为更新的条件；参数 4 为条件的参数（更新条件的具体值）。第三、第四个参数用于约束更新某一行或者某几行中的数据，不指定的话默认就是更新所有行。

2. 修改数据案例

1）案例分析

（1）界面分析。该布局界面为信息更新界面，用来完成学生信息的修改。布局界面中有八个控件，TextView 控件用来显示页面的标题，一个 EditText 控件用来输入查询的条件，另外三个 EditText 控件用来显示查询的结果，三个 Button 控件用来完成查询、修改和跳转到下一页。

（2）设计思路。布局界面中的控件通过添加属性的方式达到用户需求，在"查询"按钮单击事件内部，调用 SQLiteDatabase 的数据库的 rawQuery（）方法，输入学生的学号，便可查询到该学号的学生信息并显示到 EditeText 控件上。可直接手动修改 EditText 中显示的学生信息，单击"修改"按钮便可完成该学号学生信息的修改。单击"下一页"按钮，跳转到下一页，达到用户的需求。

2）实现步骤

（1）无须创建新工程，在创建数据库工程 StudentInfoTest 中完善程序，完成学生信息的修改。

（2）设计布局界面。双击 layout 文件夹下的 activity_third.xml 文件，便可打开布局编辑器，在布局界面上添加八个控件，代码如下。

```
<?xml version = "1.0" encoding = "utf-8"?>
<LinearLayout xmlns:android = "http://schemas.android.com/apk/res/android"
    xmlns:app = "http://schemas.android.com/apk/res-auto"
    xmlns:tools = "http://schemas.android.com/tools"
    android:layout_width = "match_parent"
    android:layout_height = "match_parent"
    android:orientation = "vertical"
    android:background = "@drawable/modifybg"
    tools:context = ".ThirdActivity">
    <TextView
        android:layout_width = "wrap_content"
```

```xml
        android:layout_height = "wrap_content"
        android:text = "信息修改页面"
        android:textSize = "30sp"
        android:textStyle = "bold"
        android:textColor = "#000000"
        android:layout_gravity = "center"
        android:layout_margin = "80dp"/>
    <LinearLayout
        android:layout_width = "match_parent"
        android:layout_height = "wrap_content"
        android:orientation = "horizontal"
        android:layout_marginBottom = "30dp">
        <EditText
            android:id = "@+id/editText_threeinputsno"
            android:layout_width = "wrap_content"
            android:layout_height = "wrap_content"
            android:hint = "请输入要查询的学号"
            android:textSize = "25sp"/>
        <Button
            android:id = "@+id/button_threequery"
            android:layout_width = "match_parent"
            android:layout_height = "wrap_content"
            android:text = "查询"
            android:textSize = "25sp"/>
    </LinearLayout>
    <EditText
        android:id = "@+id/editText_threesno"
        android:layout_width = "match_parent"
        android:layout_height = "wrap_content"
        android:hint = "学号"
        android:textSize = "25sp"/>
    <EditText
        android:id = "@+id/editText_threename"
        android:layout_width = "match_parent"
        android:layout_height = "wrap_content"
        android:hint = "姓名"
        android:textSize = "25sp"/>
    <EditText
        android:id = "@+id/editText_threedep"
        android:layout_width = "match_parent"
        android:layout_height = "wrap_content"
        android:hint = "所属系部"
        android:textSize = "25sp"/>
    <Button
        android:id = "@+id/button_threemodify"
        android:layout_width = "wrap_content"
        android:layout_height = "wrap_content"
        android:text = "修改"
        android:textSize = "25sp"
```

```xml
            android:layout_gravity = "right"
            android:layout_marginTop = "30dp"/>
    <Button
        android:id = "@+id/button_threenext"
        android:layout_width = "wrap_content"
        android:layout_height = "wrap_content"
        android:text = "下一页"
        android:textSize = "25sp"
        android:layout_gravity = "right"/>
</LinearLayout>
```

（3）实现学生信息的修改功能。重点关注定义数据库有关的两个对象、数据库的初始化方法和数据先查询后修改的方法，如下。

```java
public class ThirdActivity extends AppCompatActivity {
    //定义对象
    EditText edit_threeinputsno,edit_threesno,edit_threename,edit_threedep;
    Button btn_threequery,btn_threemodify,btn_threenext;
    MyDBOpenHelper mhelper;  //定义一个数据库帮助类对象
    SQLiteDatabase db;  //定义一个操作的数据库的类对象
    @Override
    protected void onCreate(Bundle savedInstanceState) {
        super.onCreate(savedInstanceState);
        setContentView(R.layout.activity_third);
        //1.控件初始化
        initView();
        //2."查询"按钮功能的实现
        btnQuery();
        //3."修改"按钮功能的实现
        btnModify();
        //4."下一步"按钮功能的实现
        btnNext();
    }
    //1.控件初始化
    private void initView() {
        edit_threeinputsno = findViewById(R.id.editText_threeinputsno);
        edit_threesno = findViewById(R.id.editText_threesno);
        edit_threename = findViewById(R.id.editText_threename);
        edit_threedep = findViewById(R.id.editText_threedep);
        btn_threequery = findViewById(R.id.button_threequery);
        btn_threemodify = findViewById(R.id.button_threemodify);
        btn_threenext = findViewById(R.id.button_threenext);
        mhelper = new MyDBOpenHelper(ThirdActivity.this);  //实例化数据库帮助类对象
        db = mhelper.getWritableDatabase();  //获取数据库的读写权限
    }
    //2."查询"按钮功能的实现
    private void btnQuery() {
        btn_threequery.setOnClickListener(new View.OnClickListener() {
            @Override
```

```java
        public void onClick(View v) {
            //先查询显示,再修改
            Cursor cursor = db.rawQuery("select * from stu_info where sno = ?",new String[]{edit_threeinputsno.getText().toString()});
            if(cursor.getCount()!= 0){
                Toast.makeText (ThirdActivity.this,"查询成功",Toast.LENGTH_SHORT).show();
                while(cursor.moveToNext()){
                    String mysno = cursor.getString(cursor.getColumnIndex("sno"));
                    String myname = cursor.getString(cursor.getColumnIndex("name"));
                    String mydep = cursor.getString(cursor.getColumnIndex("deparment"));
                    edit_threesno.setText(mysno);
                    edit_threename.setText(myname);
                    edit_threedep.setText(mydep);
                }
            }else{
                Toast.makeText (ThirdActivity.this,"没有查询到该学号的学生",Toast.LENGTH_SHORT).show();
                edit_threesno.setText("");
                edit_threename.setText("");
                edit_threedep.setText("");
            }
        }
    });
}
//3."修改"按钮功能的实现
private void btnModify() {
    btn_threemodify.setOnClickListener(new View.OnClickListener() {
        @Override
        public void onClick(View v) {
            //修改
            ContentValues values = new ContentValues();
            values.put("sno",edit_threesno.getText().toString());
            values.put("name",edit_threename.getText().toString());
            values.put("deparment",edit_threedep.getText().toString());
            db.update("stu_info",values,"sno = ?",new String[]{edit_threeinputsno.getText().toString()});
            Toast.makeText (ThirdActivity.this,"修改成功",Toast.LENGTH_SHORT).show();
        }
    });
}
//4."下一页"按钮功能的实现
private void btnNext() {
    btn_threenext.setOnClickListener(new View.OnClickListener() {
        @Override
        public void onClick(View v) {
            Intent intent = new Intent(ThirdActivity.this,FourActivity.class);
            startActivity(intent);
            finish();
```

```
                }
            });
        }
    }
```

上述代码中,首先根据输入的学生的学号查询学生的信息,查询的学生信息会显示到对应的 EditText 控件上,用户可以在 EditText 控件上直接修改该生信息,但此时数据库中该生信息并未真的修改,只有单击"修改"按钮之后,才可最终将修改后的信息更新到数据库中。代码 db.update("stu_info",values,"sno=?",new String[]{edit_threesno.getText().toString()})中,update()方法中接收 3 个参数:参数 1 为表名(学生信息表 stu_info);参数 2 为更新后的值(更新后的数据组装到了 ContentValues 对象中);参数 3 为更新的条件(根据学生学号进行更新);参数 4 为条件参数的具体值(学号为"查询"按钮前所输入的学号)。

(4) 运行程序,修改前后的学生信息如图 8.7 和图 8.8 所示。

(5) 查看修改后的数据表中的学生信息。在模拟器中找到文件管理器,双击打开文件管理器,挂载为可读写模式,根据数据库的存储路径依次操作 data/data/com.example.studentinfotest/ databases/ student.db/stu_info,双击 stu_info 将其打开,便可查看刚刚修改的数据表中的文件内容,如图 8.9 所示。

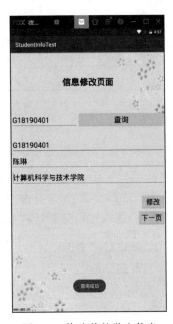

图 8.7 修改前的学生信息

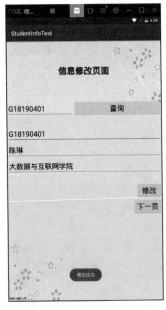

图 8.8 修改后的学生信息

图 8.9 从模拟器中查看数据表中的数据

8.1.7 删除数据

微课视频

删除数据即从表中删除数据,相比较前面学习的添加、查询、修改,删除数据是最简单的一种操作了。SQLiteDatabase 中也提供了一个 delete()方法,专门用于删除数据。

1. 删除数据常用方法

系统提供了删除数据的常用方法,如表 8.5 所示。

表 8.5 删除数据常用方法

方 法	功能描述	实 例
delete(String table, String whereClause, String[] whereArgs)	删除数据	db.delete("stu_info","sno=?",new String[]{edit_foursno.getText().toString()});

表 8.5 中,delete()方法接收 3 个参数,其中,参数 1 为表名;参数 2 为删除的条件;参数 3 为条件的参数(删除条件的取值)。第二、第三个参数又是用于约束某一行或者某几行中的数据,不指定的话默认就是删除所有行。

2. 删除数据案例

1) 案例分析

(1) 界面分析。该布局界面为信息删除界面,用来完成学生信息的删除。布局界面中有三个控件,TextView 控件用来显示页面的标题,EditText 控件用来输入删除的条件,Button 控件用来完成删除。

(2) 设计思路。布局界面中的控件通过添加属性的方式达到用户需求,在"删除"按钮单击事件内部,调用 SQLiteDatabase 的 delete()方法,删除指定学号的学生信息,达到用户的需求。

2) 实现步骤

(1) 无须创建新工程,在创建数据库工程 StudentInfoTest 中完善程序,完成学生信息的删除。

(2) 设计布局界面。双击 layout 文件夹下的 activity_four.xml 文件,便可打开布局编辑器,在布局界面上添加三个控件,代码如下。

```xml
<?xml version = "1.0" encoding = "utf-8"?>
<LinearLayout xmlns:android = "http://schemas.android.com/apk/res/android"
    xmlns:app = "http://schemas.android.com/apk/res-auto"
    xmlns:tools = "http://schemas.android.com/tools"
    android:layout_width = "match_parent"
    android:layout_height = "match_parent"
    android:orientation = "vertical"
    android:background = "@drawable/deletebg"
    tools:context = ".FourActivity">

    <TextView
        android:layout_width = "wrap_content"
```

```xml
            android:layout_height = "wrap_content"
            android:text = "信息删除页面"
            android:textSize = "30sp"
            android:textStyle = "bold"
            android:textColor = "#000000"
            android:layout_gravity = "center"
            android:layout_margin = "80dp"/>
    <EditText
            android:id = "@+id/editText_foursno"
            android:layout_width = "match_parent"
            android:layout_height = "wrap_content"
            android:hint = "请输入要删除的学号"
            android:textSize = "25sp"
            android:textColor = "#000000"/>

    <Button
            android:id = "@+id/button_fourdelete"
            android:layout_width = "wrap_content"
            android:layout_height = "wrap_content"
            android:text = "删除"
            android:textSize = "25sp"
            android:textColor = "#000000"
            android:layout_gravity = "right"/>
</LinearLayout>
```

（3）实现学生信息的删除功能。重点关注定义数据库有关的两个对象、数据库的初始化方法和数据删除的方法，如下。

```java
public class FourActivity extends AppCompatActivity {
//定义对象
    EditText edit_foursno;
    Button btn_fourdelete;
    MyDBOpenHelper mhelper; //定义一个数据库帮助类对象
    SQLiteDatabase db; //定义一个操作的数据库的类对象
    @Override
    protected void onCreate(Bundle savedInstanceState) {
        super.onCreate(savedInstanceState);
        setContentView(R.layout.activity_four);
        //1.控件初始化
        initView();
        //2."删除"按钮功能的实现
        btnDelete();
    }
    //1.控件初始化
    private void initView() {
        edit_foursno = findViewById(R.id.editText_foursno);
        btn_fourdelete = findViewById(R.id.button_fourdelete);
        mhelper = new MyDBOpenHelper(FourActivity.this); //实例化数据库帮助类对象
        db = mhelper.getWritableDatabase(); //获取数据库的读写权限
```

```
        }
    //2."删除"按钮功能的实现
    private void btnDelete() {
        btn_fourdelete.setOnClickListener(new View.OnClickListener() {
            @Override
            public void onClick(View v) {
    //删除
                db.delete("stu_info","sno = ?",new String[]{edit_foursno.getText().toString()});
                Toast.makeText (FourActivity.this,"删除成功",Toast.LENGTH_SHORT ).show();
            }
        });
    }
}
```

可以看到,在删除按钮单击事件的内部,指明删除 stu_info 表中的数据,并且通过第二个和第三个参数来指定仅删除指定学号的学生的信息,单击删除按钮后,这条学生信息会被删除掉。

(4) 运行程序,效果如图 8.10 所示。从模拟器的 RE 文件管理器中查看数据库,发现数据表中的该条学生记录已被删除,如图 8.11 所示。

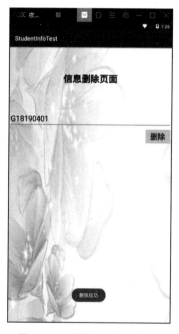

图 8.10　删除学生信息界面

图 8.11　从模拟器中查看数据表中的数据

8.2　实战案例——理财通 App 设计与实现

在经济大爆炸的今天,很多人的存款只是手机银行卡界面中的一串数字,习惯了钱包空空如也,手持一部手机就可消费遍天下。一个月下来,生活中的开销使我们大笔的钞票未经过手的触摸就花完了,面对一连串的账单却再也回忆不起来钱到底花在哪里了?理财记账软件,是一款金融理财应用,为用户提供便捷的日常记账功能,方便用户通过软件了解日常收入和支出情况,不仅可以培养用户的理财头脑,还能够帮助用户养成良好的理财习惯。

8.2.1 作品分析

微课视频

【任务描述】

对工程整体功能模块、界面布局原型、数据库表、收入类型和支出类型、工程的目录结构、工程的界面布局等,以图片或表格的样式展示给用户,让用户对整个工程有一个全局的概念。

【设计思路】

以组织结构图、图片、表格形式展示,形象直观,更易于用户理解和掌握。

【任务实施】

1. 功能模块

本案例中包含:欢迎界面、登录界面、注册界面、主界面、新增收入界面、收入明细界面、新增支出界面、支出明细界面、数据分析界面、系统设置界面、收入管理界面和支出管理界面等 12 个界面。效果如图 8.12 所示。

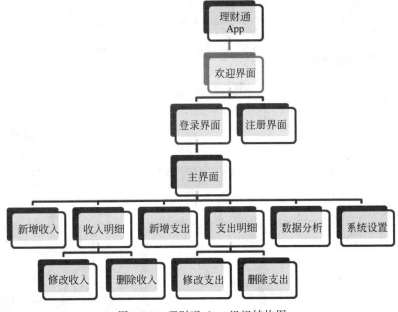

图 8.12 理财通 App 组织结构图

2. 界面原型图

界面原型图如图 8.13 所示。

3. 数据库表

(1) 用户表:主要包括自动 ID 编号、用户名、用户密码、用户邮箱、联系电话等,如表 8.6 所示。

(a) 欢迎界面

(b) 注册界面

(c) 登录界面

(d) 主界面

(e) 新增收入界面

(f) 收入明细界面

(g) 新增支出界面

(h) 支出明细界面

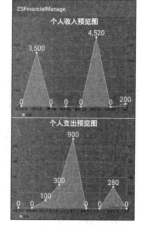
(i) 数据分析界面

图 8.13　界面原型图

(j) 系统设置界面　　　　(k) 收入管理界面　　　　(l) 支出管理界面

图 8.13 （续）

表 8.6　用户表

字段类型	数据类型（长度）	说　　明
id	integer	自动编号（主键）
name	varchar(10)	用户名
pwd	varchar(15)	用户密码
email	varchar(50)	用户邮箱
phone	varchar(11)	联系电话

（2）收入信息表：主要包括自动 ID 编号、收入金额、收入日期、收入类型、付款方、备注等，如表 8.7 所示。

表 8.7　收入信息表

字段类型	数据类型（长度）	说　　明
id	integer	自动编号（主键）
inmoney	double	收入金额
intime	varchar(20)	收入日期
intype	varchar(30)	收入类型
inpayer	varchar(100)	付款方
inremark	varchar(500)	备注

（3）支出信息表：主要包括自动 ID 编号、支出金额、支出日期、支出类型、收款方、备注等，如表 8.8 所示。

表 8.8 支出信息表

字段类型	数据类型(长度)	说　　明
id	integer	自动编号(主键)
outmoney	double	支出金额
outtime	varchar(20)	支出日期
outtype	varchar(30)	支出类型
outpayer	varchar(100)	收款方
outremark	varchar(500)	备注

4. 收入类型和支出类型

收入类型和支出类型如表 8.9 所示。

表 8.9　Spinner 下拉列表中的收入类型和支出类型

类　　型	收入类型子项
收入类型(incometype)	---请选择---
	学习-奖金
	补助-奖金
	比赛-奖励
	业余-兼职
	基本-工资
	福利-分红
	加班-津贴
	其他
类　　型	支出类型子项
支出类型(paytype)	---请选择---
	电影-娱乐
	美食-畅饮
	欢乐-购物
	手机-充值
	交通-出行
	教育-培训
	社交-礼仪
	生活-日用
	其他

5. 工程的目录结构

工程的目录结构如图 8.14 所示。

6. 工程的界面布局

工程的界面布局如图 8.15 所示。

项目8 我的第一桶金——理财通App的设计与实现

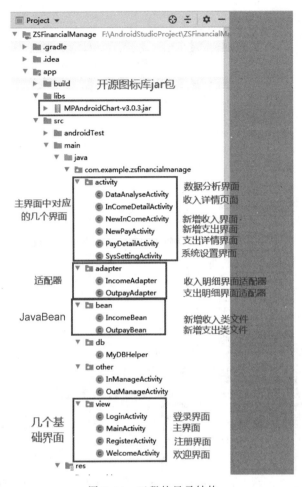

图 8.14 工程的目录结构

8.2.2 制作 App 的欢迎界面

微课视频

【任务描述】

单击程序图标,即可启动程序,首先映入眼帘的是一个动感十足的欢迎界面,欢迎界面采用 Android Material Design 设计理念,手绘原创图标并设计私人定制界面,单击"知道了"按钮便可进入登录界面。

【设计思路】

布局界面中的控件通过添加属性的方式达到用户需求,在功能上,单击"知道了"按钮,跳转到登录界面,达到用户的需求。

【任务实施】

(1) 创建一个新的工程,工程名为 ZSFinacialManage。

(2) 切换工程的 Project 项目结构,选择该模式下方的 app,依次展开,便看到工程的布局界面和工程的类文件,其中,activity_main.xml 是布局界面,MainActivity.java 为类文件。

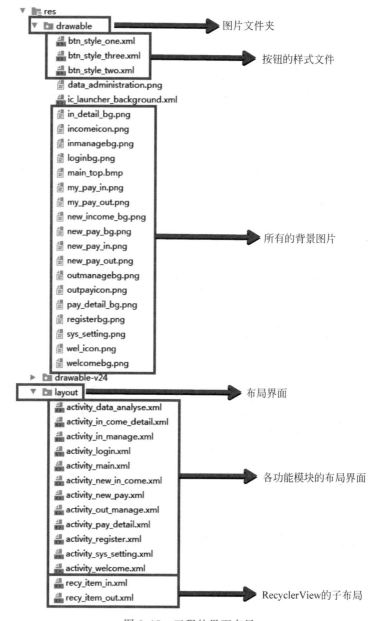

图 8.15 工程的界面布局

（3）在工程中添加新的页面。右击 com. example. zsfinacialmanage 包 → New → Activity → Empty Activity，会弹出一个创建活动的对话框，将活动命名为 WelcomeActivity，默认勾选 Generate Layout File 关联布局界面，布局界面名称为 activity_welcome，但不要勾选 Launcher Activity。单击 Finish 按钮，便可在工程中完成第二个页面的添加。同理，添加第三个页面，类文件名称为 RegisterActivity，其对应的布局界面名称为 activity_register。继续添加第四个页面，类文件名称为 LoginActivity，其对应的布局界面名称为 activity_login。完成效果如图 8.16 所示。

（4）修改工程从欢迎界面启动。打开工程的配置文件 AndroidManifest. xml，修改欢迎界面 WelcomeActivity 的开始节点和结束节点,将控制工程启动的四行代码放置在欢迎界面节点内部,如图 8.17 所示。

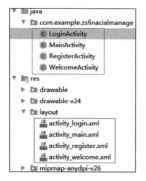

图 8.16　工程中新增的三个页面

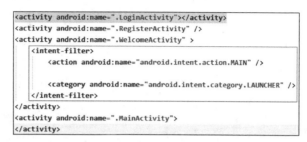

图 8.17　设置工程从欢迎界面启动

(5)准备图片,将其粘贴到 app 目录结构中 res 下的 drawable 文件夹下,如图 8.18 所示。

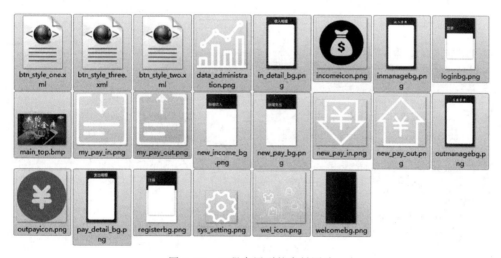

图 8.18　工程中用到的素材图片

其中,btn_style_one、btn_style_two、btn_style_three 是按钮的三种样式。样式文件创建的方法:在 app 目录结构下找到 res/drawable,右击 drawable 文件夹,选择 New→Drawable resource file,输入样式名称"btn_style_one",选择节点类型 shape,单击 OK 按钮。btn_style_one.xml 样式文件代码如下。

```xml
<?xml version = "1.0" encoding = "utf - 8"?>
< shape xmlns:android = "http://schemas.android.com/apk/res/android">
    < stroke android:color = "#ffffff" android:width = "1dp"/>
    < solid android:color = "#575d91"/>
    < corners android:radius = "20dp"/>
</shape>
```

同理,btn_style_two.xml 样式文件代码如下。

```xml
<?xml version = "1.0" encoding = "utf-8"?>
<shape xmlns:android = "http://schemas.android.com/apk/res/android">
    <stroke android:color = "#6A6969" android:width = "1dp"/>
    <solid android:color = "#eeeeee"/>
    <corners android:radius = "20dp"/>
</shape>
```

同理,btn_style_three.xml 样式文件如下。

```xml
<?xml version = "1.0" encoding = "utf-8"?>
<shape xmlns:android = "http://schemas.android.com/apk/res/android">
    <stroke android:color = "#ffffff" android:width = "1dp"/>
    <corners android:radius = "20dp"/>
</shape>
```

(6) 设计 WelcomeActivity 的布局界面。双击 layout 文件夹下的 activity_welcome.xml 文件,便可打开布局编辑器,修改布局类型为 LinearLayout,添加方向属性为 vertical。依次添加控件,代码如下。

```xml
<?xml version = "1.0" encoding = "utf-8"?>
<LinearLayout xmlns:android = "http://schemas.android.com/apk/res/android"
    xmlns:app = "http://schemas.android.com/apk/res-auto"
    xmlns:tools = "http://schemas.android.com/tools"
    android:layout_width = "match_parent"
    android:layout_height = "match_parent"
    android:orientation = "vertical"
    android:background = "@drawable/welcomebg"
    tools:context = ".WelcomeActivity">
    <ImageView
        android:layout_width = "wrap_content"
        android:layout_height = "400dp"
        android:src = "@drawable/wel_icon"/>
    <TextView
        android:layout_width = "wrap_content"
        android:layout_height = "wrap_content"
        android:text = "钱都花哪了?"
        android:textSize = "30sp"
        android:textColor = "#ffffff"
        android:layout_gravity = "center"
        android:layout_marginTop = "30dp"/>
    <TextView
        android:layout_width = "wrap_content"
        android:layout_height = "wrap_content"
        android:text = "记录生活,点亮财富人生!"
        android:textSize = "20sp"
        android:textColor = "#ffffff"
        android:layout_gravity = "center"
        android:layout_marginTop = "30dp"/>
```

项目8 我的第一桶金——理财通App的设计与实现

```
    <Button
        android:id = "@ + id/bt_know_wel"
        android:layout_width = "wrap_content"
        android:layout_height = "wrap_content"
        android:text = "知道了"
        android:textSize = "20sp"
        android:textColor = "#ffffff"
        android:layout_gravity = "center"
        android:layout_marginTop = "30dp"
        android:background = "@drawable/btn_style_three"/>
</LinearLayout>
```

布局代码编写成功后,可以看到界面的预览效果,如图 8.19 所示。

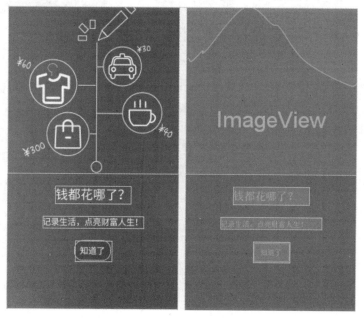

图 8.19 欢迎界面布局设计效果图

(7) 实现 WelcomeActivity 页面的功能。打开 WelcomeActivity.java 类文件,添加如下代码。

```
public class WelcomeActivity extends AppCompatActivity {
    //定义对象
    Button bt_know_wel;
    @Override
    protected void onCreate(Bundle savedInstanceState) {
        super.onCreate(savedInstanceState);
        setContentView(R.layout.activity_welcome);
        //绑定控件
        bt_know_wel = findViewById(R.id.bt_know_wel);
        //按钮单击事件
        bt_know_wel.setOnClickListener(new View.OnClickListener() {
```

```
            @Override
            public void onClick(View v) {
                Intent intent = new Intent(WelcomeActivity.this,LoginActivity.class);
                startActivity(intent);
                finish();
            }
        });
    }
}
```

上述代码中,给按钮添加单击事件,在按钮单击事件的内部,完成从欢迎界面到登录界面的跳转。用户可以运行程序,运行效果如图 8.20 所示。

微课视频

8.2.3 注册界面的设计与功能

【任务描述】

在登录界面中,显示"注册"和"登录"两个按钮,用户第一次运行该程序,需要先注册账号。此时在登录界面中单击"注册"按钮,便可打开注册界面。在注册界面中,输入用户信息,单击"注册"按钮,便可将输入的学生信息保存到数据库的用户表当中。单击"取消"按钮,关闭本页面,返回登录界面。

【设计思路】

布局界面中的控件通过添加属性的方式达到用户需求,在功能上,单击"注册"按钮,将输入的用户信息保存到数据库的用户表中;单击"取消"按钮,跳转到登录界面,达到用户的需求。

图 8.20 欢迎界面运行效果

【任务实施】

(1) 注册界面布局设计。

双击 layout 文件夹下的 activity_register.xml 文件,便可打开布局编辑器,修改布局类型为 LinearLayout,添加方向属性为 vertical。依次添加控件,代码如下。

```xml
<?xml version = "1.0" encoding = "utf-8"?>
<LinearLayout xmlns:android = "http://schemas.android.com/apk/res/android"
    xmlns:app = "http://schemas.android.com/apk/res-auto"
    xmlns:tools = "http://schemas.android.com/tools"
    android:layout_width = "match_parent"
    android:layout_height = "match_parent"
    android:orientation = "vertical"
    android:background = "@drawable/registerbg"
    tools:context = ".RegisterActivity">
<LinearLayout
```

```xml
        android:layout_width = "match_parent"
        android:layout_height = "wrap_content"
        android:orientation = "horizontal"
        android:layout_marginTop = "260dp"
        android:layout_marginLeft = "60dp"
        android:layout_marginRight = "60dp">
        <TextView
            android:layout_width = "wrap_content"
            android:layout_height = "wrap_content"
            android:text = "用户名"
            android:textSize = "15sp"
            android:textColor = "#000000"/>
        <EditText
            android:id = "@+id/et_name_rg"
            android:layout_width = "match_parent"
            android:layout_height = "wrap_content"
            android:background = "@null"
            android:gravity = "center"
            android:textColor = "#000000"/>
</LinearLayout>
<View
    android:layout_width = "match_parent"
    android:layout_height = "2dp"
    android:background = "#3700b3"
    android:layout_marginRight = "60dp"
    android:layout_marginLeft = "60dp"/>
<LinearLayout
    android:layout_width = "match_parent"
    android:layout_height = "wrap_content"
    android:orientation = "horizontal"
    android:layout_marginTop = "20dp"
    android:layout_marginLeft = "60dp"
    android:layout_marginRight = "60dp">
    <TextView
        android:layout_width = "wrap_content"
        android:layout_height = "wrap_content"
        android:text = "密    码"
        android:textSize = "15sp"
        android:textColor = "#000000"/>
    <EditText
        android:id = "@+id/et_pwd_rg"
        android:layout_width = "match_parent"
        android:layout_height = "wrap_content"
        android:background = "@null"
        android:gravity = "center"
        android:textColor = "#000000"
        android:inputType = "textPassword"/>
</LinearLayout>
<View
```

```xml
        android:layout_width = "match_parent"
        android:layout_height = "2dp"
        android:background = "#3700b3"
        android:layout_marginRight = "60dp"
        android:layout_marginLeft = "60dp"/>
<LinearLayout
    android:layout_width = "match_parent"
    android:layout_height = "wrap_content"
    android:orientation = "horizontal"
    android:layout_marginTop = "20dp"
    android:layout_marginLeft = "60dp"
    android:layout_marginRight = "60dp">
    <TextView
        android:layout_width = "wrap_content"
        android:layout_height = "wrap_content"
        android:text = "邮    箱"
        android:textSize = "15sp"
        android:textColor = "#000000"/>
    <EditText
        android:id = "@+id/et_email_rg"
        android:layout_width = "match_parent"
        android:layout_height = "wrap_content"
        android:background = "@null"
        android:gravity = "center"
        android:textColor = "#000000"/>
</LinearLayout>
<View
    android:layout_width = "match_parent"
    android:layout_height = "2dp"
    android:background = "#3700b3"
    android:layout_marginRight = "60dp"
    android:layout_marginLeft = "60dp"/>
<LinearLayout
    android:layout_width = "match_parent"
    android:layout_height = "wrap_content"
    android:orientation = "horizontal"
    android:layout_marginTop = "20dp"
    android:layout_marginLeft = "60dp"
    android:layout_marginRight = "60dp">
    <TextView
        android:layout_width = "wrap_content"
        android:layout_height = "wrap_content"
        android:text = "电    话"
        android:textSize = "15sp"
        android:textColor = "#000000"/>
    <EditText
        android:id = "@+id/et_phone_rg"
        android:layout_width = "match_parent"
        android:layout_height = "wrap_content"
```

```xml
            android:background = "@null"
            android:gravity = "center"
            android:textColor = "#000000"/>
</LinearLayout>
<View
    android:layout_width = "match_parent"
    android:layout_height = "2dp"
    android:background = "#3700b3"
    android:layout_marginRight = "60dp"
    android:layout_marginLeft = "60dp"/>
<Button
    android:id = "@+id/bt_ok_rg"
    android:layout_width = "match_parent"
    android:layout_height = "wrap_content"
    android:text = "注册"
    android:textSize = "20sp"
    android:textColor = "#ffffff"
    android:layout_marginLeft = "60dp"
    android:layout_marginRight = "60dp"
    android:background = "@drawable/btn_style_one"
    android:layout_marginTop = "40dp"/>
<Button
    android:id = "@+id/bt_cancel_rg"
    android:layout_width = "match_parent"
    android:layout_height = "wrap_content"
    android:text = "取消"
    android:textSize = "20sp"
    android:textColor = "#000000"
    android:layout_marginLeft = "60dp"
    android:layout_marginRight = "60dp"
    android:background = "@drawable/btn_style_two"
    android:layout_marginTop = "20dp"/>
</LinearLayout>
```

上述代码中,EditText 控件的属性 background="@null"表示 EditText 控件自带下方一条线取消显示。View 控件用来定义一条线,使用 layout_height 属性设置线的粗细,使用 background 属性设置线的颜色,增强页面的美观效果。

布局代码编写成功后,可以看到界面的预览效果,如图 8.21 所示。

(2) 创建数据库类文件。

因工程量较大,类文件较多,为了使程序文件条理清晰,创建 db 文件夹来存放数据库文件。方法为右击 com.example.zsfinacialmanage →New→Package→输入名称"db"(全称为 com.example.zsfinacialmanage.db)→OK,便可完成 db 文件夹的创建。然后创建数据库类文件,右击 db→New→Java Class→Name 中输入类名"MyDBHelper"→Superclass 中输入父类"SQLiteOpenHelper"(全称为 android.database.sqlite.SQLiteOpenHelper)→OK。在数据库类文件中输入如下代码。

图 8.21　注册界面布局设计效果

```java
public class MyDBHelper extends SQLiteOpenHelper {
    private static final String DBNAME = "financial.db"; //定义数据库名称
    private static final int VERSION = 1; //定义数据库的版本号
    public MyDBHelper(Context context) {
        super(context, DBNAME, null, VERSION);
    }
    //1.创建数据库
    @Override
    public void onCreate(SQLiteDatabase db) {
        //创建用户表
        db.execSQL("create table tb_userinfo(id integer primary key autoincrement,name varchar(10),pwd varchar(15),email varchar(50),phone varchar(11))");
    }
    //2.升级数据库
    @Override
    public void onUpgrade(SQLiteDatabase db, int oldVersion, int newVersion) {

    }
}
```

（3）注册界面功能的实现。双击打开 RegisterActivity.java 类文件，输入逻辑代码如下。

```java
public class RegisterActivity extends AppCompatActivity {
    //3.定义对象
    EditText et_name,et_pwd,et_email,et_phone;
    Button btn_register,btn_cancel;
    MyDBHelper mhelper;//创建一个数据库类文件
```

```java
SQLiteDatabase db;//创建一个可以操作的数据库对象
@Override
protected void onCreate(Bundle savedInstanceState) {
    super.onCreate(savedInstanceState);
    setContentView(R.layout.activity_register);
    //绑定控件
    initView();
    //"注册"按钮功能的实现
    btnRegister();
    //"取消"按钮功能的实现
    btnCancel();
}
//4.绑定控件
private void initView() {
    et_name = findViewById(R.id.et_name_rg);
    et_pwd = findViewById(R.id.et_pwd_rg);
    et_email = findViewById(R.id.et_email_rg);
    et_phone = findViewById(R.id.et_phone_rg);
    btn_register = findViewById(R.id.bt_ok_rg);
    btn_cancel = findViewById(R.id.bt_cancel_rg);
    mhelper = new MyDBHelper(RegisterActivity.this);
    db = mhelper.getWritableDatabase();
}
//5."注册"按钮功能的实现
private void btnRegister() {
    btn_register.setOnClickListener(new View.OnClickListener() {
        @Override
        public void onClick(View v) {
            ContentValues values = new ContentValues(); //创建一个对象,用来封装一行数据
            values.put("name",et_name.getText().toString());//将输入的用户名放到 name 列
            values.put("pwd",et_pwd.getText().toString());//将输入的密码放到 pwd 列
            values.put("email",et_email.getText().toString());//将输入的邮箱放到 email 列
            values.put("phone",et_phone.getText().toString());//将输入的电话放到 phone 列
            db.insert("tb_userinfo",null,values); //将封装好的一行数据保存到数据库的
                                                  //tb_userinfo 表中
            Toast.makeText(RegisterActivity.this,"注册成功",Toast.LENGTH_SHORT).show();
        }
    });
}
//6."取消"按钮功能的实现
private void btnCancel() {
    btn_cancel.setOnClickListener(new View.OnClickListener() {
        @Override
        public void onClick(View v) {
            Intent intent = new Intent(RegisterActivity.this, LoginActivity.class);
            startActivity(intent);
```

```
                finish();
            }
        });
    }
}
```

上述代码中,用户自定义三个方法:initView()方法用来对控件和数据库进行初始化;btnRegister()方法用来实现用户注册功能,方法内部,首先创建了一个 ContentValues 对象,用来组装一行数据,然后调用 put()方法,将每个 EditText 控件上输入的内容,放在用户表中的不同列;最后调用 insert()方法将封装好的一组数据保存到数据库的用户表中。btnCancel()方法用来完成"取消"按钮的功能,在"取消"按钮单击事件中完成了从注册界面到登录界面的跳转。

(4) 修改工程从注册界面启动。打开工程的配置文件 AndroidManifest.xml,修改注册界面 RegisterActivity 的开始节点和结束节点,将工程启动的四行代码放置在注册界面节点内部,代码如下。

```
<activity android:name=".LoginActivity"></activity>
<activity android:name=".RegisterActivity" >
    <intent-filter>
        <action android:name="android.intent.action.MAIN" />
        <category android:name="android.intent.category.LAUNCHER" />
    </intent-filter>
</activity>
<activity android:name=".WelcomeActivity" >
</activity>
<activity android:name=".MainActivity">
</activity>
```

(5) 运行程序,效果如图 8.22 所示。

(6) 查看保存到数据库 tb_userinfo 表中的用户信息。在模拟器中找到 RE 文件管理器,双击打开文件管理器,挂载为可读写模式,根据数据库的存储路径依次操作 data/data/com.example.zsfinacialmanage/databases/financial.db/tb_userinfo,其中,databases 是数据库,financial.db 是数据库的名称,tb_userinfo 是数据表名称。双击 tb_userinfo 将其打开,便可查看刚刚保存到数据表中的文件内容,如图 8.23 所示。

8.2.4 登录界面的设计与功能

微课视频

【任务描述】

在登录界面中,当用户名或密码为空时,弹出提示"用户名或密码不能为空"。如果用户输入了用户名和密码,获取输入的用户名和用户密码,与数据库用户表中的用户信息进行匹配,如果匹配不成功,则弹出提示"用户名或密码错误,请重新输入",如果匹配成功,则弹出提示"用户名和密码正确,欢迎登录",同时跳转到主界面。

项目8 我的第一桶金——理财通App的设计与实现 | 319

图8.22 注册界面的运行效果

图8.23 从RE文件管理器中查看用户表中注册的用户信息

【设计思路】

布局界面中的控件通过添加属性的方式达到用户需求,在功能上,单击"登录"按钮,将对输入的用户名和密码进行一系列判断,达到用户的需求。

【任务实施】

(1) 登录界面布局设计。

双击layout文件夹下的activity_login.xml文件,便可打开布局编辑器,修改布局类型为LinearLayout,添加方向属性为vertical。依次添加控件,代码如下。

```
<?xml version = "1.0" encoding = "utf - 8"?>
< LinearLayout xmlns:android = "http://schemas.android.com/apk/res/android"
    xmlns:app = "http://schemas.android.com/apk/res - auto"
    xmlns:tools = "http://schemas.android.com/tools"
    android:layout_width = "match_parent"
    android:layout_height = "match_parent"
    android:orientation = "vertical"
    android:background = "@drawable/loginbg"
    tools:context = ".LoginActivity">

    < LinearLayout
        android:layout_width = "match_parent"
        android:layout_height = "wrap_content"
        android:orientation = "horizontal"
```

```xml
            android:layout_marginTop = "260dp"
            android:layout_marginLeft = "60dp"
            android:layout_marginRight = "60dp">
        <TextView
            android:layout_width = "wrap_content"
            android:layout_height = "wrap_content"
            android:text = "用户名"
            android:textSize = "15sp"
            android:textColor = "#000000"/>
        <EditText
            android:id = "@+id/et_name_lg"
            android:layout_width = "match_parent"
            android:layout_height = "wrap_content"
            android:background = "@null"
            android:gravity = "center"
            android:textColor = "#000000"/>
    </LinearLayout>
    <View
        android:layout_width = "match_parent"
        android:layout_height = "2dp"
        android:background = "#3700b3"
        android:layout_marginRight = "60dp"
        android:layout_marginLeft = "60dp"/>
    <LinearLayout
        android:layout_width = "match_parent"
        android:layout_height = "wrap_content"
        android:orientation = "horizontal"
        android:layout_marginTop = "20dp"
        android:layout_marginLeft = "60dp"
        android:layout_marginRight = "60dp">
        <TextView
            android:layout_width = "wrap_content"
            android:layout_height = "wrap_content"
            android:text = "密    码"
            android:textSize = "15sp"
            android:textColor = "#000000"/>
        <EditText
            android:id = "@+id/et_pwd_lg"
            android:layout_width = "match_parent"
            android:layout_height = "wrap_content"
            android:background = "@null"
            android:inputType = "textPassword"
            android:gravity = "center"
            android:textColor = "#000000"/>
    </LinearLayout>
    <View
        android:layout_width = "match_parent"
        android:layout_height = "2dp"
        android:background = "#3700b3"
```

```xml
            android:layout_marginRight = "60dp"
            android:layout_marginLeft = "60dp"/>
    <Button
        android:id = "@+id/bt_newregister_lg"
        android:layout_width = "wrap_content"
        android:layout_height = "wrap_content"
        android:text = "新用户注册"
        android:textSize = "20sp"
        android:textColor = "#000000"
        android:layout_marginLeft = "60dp"
        android:layout_marginRight = "60dp"
        android:background = "@null"
        android:layout_marginTop = "20dp"
        android:layout_gravity = "right"/>
    <Button
        android:id = "@+id/bt_login_lg"
        android:layout_width = "match_parent"
        android:layout_height = "wrap_content"
        android:text = "登录"
        android:textSize = "20sp"
        android:textColor = "#ffffff"
        android:layout_marginLeft = "60dp"
        android:layout_marginRight = "60dp"
        android:background = "@drawable/btn_style_one"
        android:layout_marginTop = "40dp"/>
</LinearLayout>
```

上述代码中，输入用户名和密码的控件与注册界面类似，其下方 Button 控件属性 background="@null" 表示 Button 控件默认的灰色背景不显示。只显示控件上的文字，达到较美观的用户体验。

布局代码编写成功后，可以看到界面的预览效果，如图 8.24 所示。

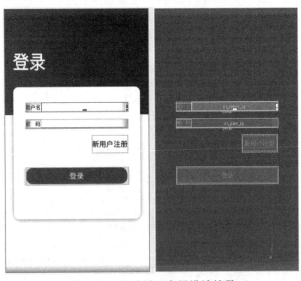

图 8.24　登录界面布局设计效果

（2）登录界面功能的实现。双击打开 LoginActivity.java 类文件，输入逻辑代码如下。

```java
public class LoginActivity extends AppCompatActivity {
    //1.定义对象
    private EditText et_name,et_pwd;
    private Button btn_newregister,btn_login;
    private MyDBHelper mhelper;
    private SQLiteDatabase db;
    @Override
    protected void onCreate(Bundle savedInstanceState) {
        super.onCreate(savedInstanceState);
        setContentView(R.layout.activity_login);
        //绑定控件
        initView();
        //"登录"按钮功能的实现
        btnLogin();
        //"新用户注册"按钮功能的实现
        btnNewRegister();
    }
    //2.绑定控件
    private void initView() {
        et_name = findViewById(R.id.et_name_lg);
        et_pwd = findViewById(R.id.et_pwd_lg);
        btn_newregister = findViewById(R.id.bt_newregister_lg);
        btn_login = findViewById(R.id.bt_login_lg);
        mhelper = new MyDBHelper(LoginActivity.this);
        db = mhelper.getWritableDatabase();
    }
    //3."登录"按钮功能的实现
    private void btnLogin() {
        btn_login.setOnClickListener(new View.OnClickListener() {
            @Override
            public void onClick(View v) {
                //首先:获取输入的用户名和密码
                String inputname = et_name.getText().toString();
                String inputpwd = et_pwd.getText().toString();
                //其次:对获取的用户名和密码进行判断
                if(inputname.equals("")||inputpwd.equals("")){ //用户名或密码为空
                    Toast.makeText(LoginActivity.this,"用户名或密码不能为空",Toast.LENGTH_SHORT).show();
                }else{//用户名或密码不为空时,再对输入的正确性进行判断
                    //根据输入的用户名和密码从数据库中查询
                    Cursor cursor = db.rawQuery("select * from tb_userinfo where name = ? and pwd = ?",new String[]{inputname,inputpwd});
                    //根据查询到的结果进行判断
                    if (cursor.moveToNext()){//查询到时
                        String getname = cursor.getString(cursor.getColumnIndex("name"));
                        String getpwd = cursor.getString(cursor.getColumnIndex("pwd"));
                        if(inputname.equalsIgnoreCase(getname)&&inputpwd.equalsIgnoreCase(getpwd)){
                            SharedPreferences.Editor editor = getSharedPreferences("userinfo",0).edit();
```

```java
                            editor.putString("username",inputname);
                            editor.putString("userpwd",inputpwd);
                            editor.commit();
        Toast.makeText(LoginActivity.this,"用户名和密码正确,欢迎登录",Toast.LENGTH_SHORT).show();
                            Intent intent = new Intent(LoginActivity.this, MainActivity.class);
                            startActivity(intent);
                            finish();
                        }
                    }else{//没有查询到结果时
        Toast.makeText(LoginActivity.this,"用户名或密码错误,请重新输入",Toast.LENGTH_SHORT).show();
                            et_name.setText("");
                            et_pwd.setText("");
                        }
                    }
                }
            });
        }
        //4."新用户注册"按钮功能的实现
        private void btnNewRegister() {
            btn_newregister.setOnClickListener(new View.OnClickListener() {
                @Override
                public void onClick(View v) {
                    Intent intent = new Intent(LoginActivity.this, RegisterActivity.class);
                    startActivity(intent);
                    finish();
                }
            });
        }
    }
```

上述代码中,用户自定义了以下三个方法。

initView()方法用来对控件和数据库进行初始化。

btnLogin()方法用来实现登录按钮的功能,方法内部,首先获取输入的用户名和密码,其次对获取的用户名和密码进行判断,用户名或密码为空,弹出提示"用户名或密码不能为空"。用户名或密码不为空时,再对输入的正确性进行判断,采用 rawQuery()方法从数据库中查询,查询结果放到了 cursor 结果集当中,当查询到结果时,取出查询结果集中的用户名和密码,与用户输入的用户名和密码进行匹配,如果不相同则提示"用户名或密码错误,请重新输入",如果相同则提示"用户名和密码正确,欢迎登录",同时利用 SharedPreferences 将用户名和密码以键值对的方式存储起来,方便系统设置页面获取到该用户名和修改密码。

btnNewRegister()方法是用户自定义方法,方法用来实现新用户注册按钮的功能,在新用户注册按钮单击事件内部完成了从登录页面到注册界面的跳转。

(3) 修改工程从欢迎界面启动。打开工程的配置文件 AndroidManifest.xml,将工程启动的四行代码放置在欢迎界面节点内部,代码如下。

```xml
<activity android:name = ".LoginActivity"></activity>
<activity android:name = ".RegisterActivity" >
</activity>
<activity android:name = ".WelcomeActivity" >
<intent - filter >
        <action android:name = "android.intent.action.MAIN" />
        <category android:name = "android.intent.category.LAUNCHER" />
</intent - filter >
</activity>
<activity android:name = ".MainActivity">
</activity>
```

（4）运行程序,出现欢迎界面,单击欢迎界面中的"知道了"按钮,跳转到登录界面,单击"新用户注册"按钮,跳转到新用户注册界面。这里用已经注册过的用户信息登录：用户名 admin,密码 123456。在登录界面中未输入用户名和密码时,单击"登录"按钮效果如图 8.25 所示；输入错误的用户名和密码时,单击"登录"按钮效果如图 8.26 所示；输入正确的用户名和密码时,单击"登录"按钮效果如图 8.27 所示。

图 8.25　未输入用户名或密码　　图 8.26　输入错误的用户名或密码　　图 8.27　正确的用户信息

微课视频

8.2.5　主界面的设计与功能

【任务描述】

主界面是整个作品展示内容的入口点。界面下方分别显示了不同功能区域,单击不同的区域,便可打开不同的页面：新增收入、收入明细、新增支出、支出明细、数据分析、系统设置。

【设计思路】

布局界面中的控件通过添加属性的方式达到用户需求，在功能上，单击"不同的图标"按钮，跳转到不同的界面，达到用户的需求。

【任务实施】

（1）主界面布局设计。

双击 layout 文件夹下的 activity_main.xml 文件，便可打开布局编辑器，修改布局类型为 LinearLayout，添加方向属性为 vertical，添加背景图片。依次添加控件，代码如下。

```xml
<?xml version = "1.0" encoding = "utf-8"?>
<LinearLayout xmlns:android = "http://schemas.android.com/apk/res/android"
    xmlns:app = "http://schemas.android.com/apk/res-auto"
    xmlns:tools = "http://schemas.android.com/tools"
    android:layout_width = "match_parent"
    android:layout_height = "match_parent"
    android:orientation = "vertical"
    android:background = "@drawable/welcomebg"
    tools:context = ".view.MainActivity">
    <ImageButton
        android:layout_width = "match_parent"
        android:layout_height = "250dp"
        android:background = "@drawable/main_top"/>
    <LinearLayout
        android:layout_width = "match_parent"
        android:layout_height = "wrap_content"
        android:orientation = "horizontal"
        android:layout_marginTop = "40dp"
        android:gravity = "center">
        <Button
            android:id = "@+id/bt_newincome_main"
            android:layout_width = "90dp"
            android:layout_height = "90dp"
            android:background = "@drawable/new_pay_in"
            android:layout_marginRight = "50dp"/>
        <Button
            android:id = "@+id/bt_incomedetail_main"
            android:layout_width = "90dp"
            android:layout_height = "90dp"
            android:background = "@drawable/my_pay_in"
            android:layout_marginRight = "50dp"/>
        <Button
            android:id = "@+id/bt_newpay_main"
            android:layout_width = "90dp"
            android:layout_height = "90dp"
            android:background = "@drawable/new_pay_out" />
    </LinearLayout>
    <LinearLayout
        android:layout_width = "match_parent"
        android:layout_height = "wrap_content"
```

```xml
            android:orientation = "horizontal"
            android:layout_marginTop = "20dp"
            android:gravity = "center">
            <TextView
                android:layout_width = "wrap_content"
                android:layout_height = "wrap_content"
                android:text = "新增收入"
                android:textSize = "15sp"
                android:textColor = "#ffffff"
                android:layout_marginRight = "80dp"/>
            <TextView
                android:layout_width = "wrap_content"
                android:layout_height = "wrap_content"
                android:text = "收入明细"
                android:textSize = "15sp"
                android:textColor = "#ffffff"
                android:layout_marginRight = "80dp"/>
            <TextView
                android:layout_width = "wrap_content"
                android:layout_height = "wrap_content"
                android:text = "新增支出"
                android:textSize = "15sp"
                android:textColor = "#ffffff"/>
    </LinearLayout>
    <LinearLayout
            android:layout_width = "match_parent"
            android:layout_height = "wrap_content"
            android:orientation = "horizontal"
            android:layout_marginTop = "40dp"
            android:gravity = "center">
            <Button
                android:id = "@+id/bt_paydetail_main"
                android:layout_width = "90dp"
                android:layout_height = "90dp"
                android:background = "@drawable/my_pay_out"
                android:layout_marginRight = "50dp"/>
            <Button
                android:id = "@+id/bt_dataanalyse_main"
                android:layout_width = "90dp"
                android:layout_height = "90dp"
                android:background = "@drawable/data_administration"
                android:layout_marginRight = "50dp"/>
            <Button
                android:id = "@+id/bt_syssetting_main"
                android:layout_width = "90dp"
                android:layout_height = "90dp"
                android:background = "@drawable/sys_setting" />
    </LinearLayout>
    <LinearLayout
```

```xml
    android:layout_width = "match_parent"
    android:layout_height = "wrap_content"
    android:orientation = "horizontal"
    android:layout_marginTop = "20dp"
    android:gravity = "center">
    <TextView
        android:layout_width = "wrap_content"
        android:layout_height = "wrap_content"
        android:text = "支出明细"
        android:textSize = "15sp"
        android:textColor = "#ffffff"
        android:layout_marginRight = "80dp"/>
    <TextView
        android:layout_width = "wrap_content"
        android:layout_height = "wrap_content"
        android:text = "数据分析"
        android:textSize = "15sp"
        android:textColor = "#ffffff"
        android:layout_marginRight = "80dp"/>
    <TextView
        android:layout_width = "wrap_content"
        android:layout_height = "wrap_content"
        android:text = "系统设置"
        android:textSize = "15sp"
        android:textColor = "#ffffff"/>
    </LinearLayout>
</LinearLayout>
```

布局代码编写成功后,可以看到界面的预览效果,如图 8.28 所示。

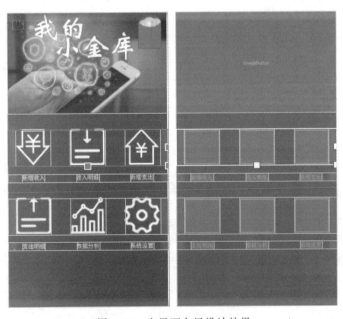

图 8.28 主界面布局设计效果

（2）创建 view 文件夹，存放 4 个页面的类文件。

因工程量较大，类文件较多，为了程序文件条理清晰，创建 view 文件夹来存放欢迎界面、注册界面、登录界面、主界面。方法为：右击 com. example. zsfinacialmanage→New→Package→输入名称"view"（全称为 com. example. zsfinacialmanage. view）→OK，便可完成 view 文件夹的创建。将目前已经完成的 4 个界面类文件分别拖放至 view 文件夹上，松开鼠标左键，单击 Refactor 按钮，此时 4 个类文件便转移到了 view 文件夹的内部。

（3）创建 activity 文件夹。

主界面中单击 6 个按钮，可以跳转到 6 个不同子功能界面。为了让工程类文件清晰明了，把主界面中单击 6 个按钮可以跳转的 6 个子功能界面放到一个文件夹里面，因此创建 activity 文件夹。方法为：右击 com. example. zsfinacialmanage→New→Package→输入名称"activity"（全称为 com. example. zsfinacialmanage. activity）→OK，便可完成 activity 文件夹的创建。

（4）新建 6 个子功能页面，放到 activity 文件夹内。

选中 activity 文件夹→单击鼠标右键→New→Activity→Empty Activity，会弹出一个创建活动的对话框，将活动命名为 NewInComeActivity，默认勾选 Generate Layout File 关联布局界面，布局界面名称为 activity_new_in_come，但不要勾选 Launcher Activity。单击 Finish 按钮，便可在工程中完成第一个子功能页面的添加。同理，添加第二个页面，类文件名称与关联的布局界面的名称如表 8.10 所示，6 个子功能页面被添加后的效果如图 8.29 所示。

表 8.10　主界面中 6 个按钮对应的 6 个子功能界面

类文件名	关联的布局界面	含义
NewInComeActivity	activity_new_in_come	新增收入
InComeDetailActivity	activity_in_come_detail	收入明细
NewPayActivity	activity_new_pay	新增支出
PayDetailActivity	activity_pay_detail	支出明细
DataAnalyseActivity	activity_data_analyse	数据分析
SysSettingActivity	activity_sys_setting	系统设置

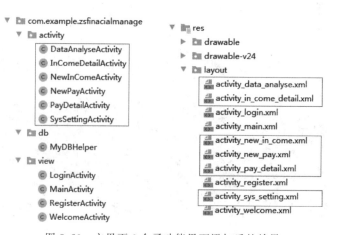

图 8.29　主界面 6 个子功能界面添加后的效果

(5) 主界面功能的实现。

双击打开 MainActivity.java 类文件,输入逻辑代码如下。

```java
public class MainActivity extends AppCompatActivity {
    //定义对象
    Button bt_newincome, bt_incomedetail, btn_newpay, btn_paydetail, bt_dataanalyse, btn_setting;
    @Override
    protected void onCreate(Bundle savedInstanceState) {
        super.onCreate(savedInstanceState);
        setContentView(R.layout.activity_main);
        //绑定控件
        initView();
        //按钮单击事件
        btnOnClick();
    }
    //绑定控件
    private void initView() {
        bt_newincome = findViewById(R.id.bt_newincome_main);
        bt_incomedetail = findViewById(R.id.bt_incomedetail_main);
        btn_newpay = findViewById(R.id.bt_newpay_main);
        btn_paydetail = findViewById(R.id.bt_paydetail_main);
        bt_dataanalyse = findViewById(R.id.bt_dataanalyse_main);
        btn_setting = findViewById(R.id.bt_syssetting_main);
    }
    //按钮单击事件
    private void btnOnClick() {
        bt_newincome.setOnClickListener(new View.OnClickListener() {
            @Override
            public void onClick(View v) {
                Intent intent = new Intent(MainActivity.this, NewInComeActivity.class);
                startActivity(intent); //跳转到新增收入页面
            }
        });
        bt_incomedetail.setOnClickListener(new View.OnClickListener() {
            @Override
            public void onClick(View v) {
                Intent intent = new Intent(MainActivity.this, InComeDetailActivity.class);
                startActivity(intent); //跳转到收入明细页面
            }
        });
        btn_newpay.setOnClickListener(new View.OnClickListener() {
            @Override
            public void onClick(View v) {
                Intent intent = new Intent(MainActivity.this, NewPayActivity.class);
                startActivity(intent); //跳转到新增支出页面
            }
        });
        btn_paydetail.setOnClickListener(new View.OnClickListener() {
```

```
            @Override
            public void onClick(View v) {
                Intent intent = new Intent(MainActivity.this, PayDetailActivity.class);
                startActivity(intent);  //跳转到支出明细页面
            }
        });
        bt_dataanalyse.setOnClickListener(new View.OnClickListener() {
            @Override
            public void onClick(View v) {
                Intent intent = new Intent(MainActivity.this, DataAnalyseActivity.class);
                startActivity(intent);  //跳转到数据分析页面
            }
        });
        btn_setting.setOnClickListener(new View.OnClickListener() {
            @Override
            public void onClick(View v) {
                Intent intent = new Intent(MainActivity.this, SysSettingActivity.class);
                startActivity(intent);  //跳转到系统设置页面
            }
        });
    }
}
```

(6) 运行程序,效果如图 8.30 所示。在主界面中,单击不同的按钮图标,便可打开不同功能的子页面。

微课视频

8.2.6 新增收入界面的设计与功能

【任务描述】

整个作品中跟收入相关的内容分别是新增收入、浏览收入、收入管理(修改或删除收入)、数据分析中的收入预览图等。其中,新增收入用来添加收入信息,收入浏览用来显示所有的收入信息,收入管理用来根据编号来修改或者删除收入信息,收入预览图以图表方式直观展示各类收入情况。本节主要学习新增收入。

【设计思路】

布局界面中的控件通过添加属性的方式满足用户需求,在功能上,单击"添加"按钮,将输入的信息保存到数据库的收入表中,满足用户的需求。

【任务实施】

(1) 新增收入界面布局设计。

双击 layout 文件夹下的 activity_new_in_come.xml 文件,便可打开布局编辑器,修改布局类型为 LinearLayout,添加方向属性为 vertical,依次添加控件,代码如下。

图 8.30 主界面效果图

```xml
<?xml version="1.0" encoding="utf-8"?>
<LinearLayout xmlns:android="http://schemas.android.com/apk/res/android"
    xmlns:app="http://schemas.android.com/apk/res-auto"
    xmlns:tools="http://schemas.android.com/tools"
    android:layout_width="match_parent"
    android:layout_height="match_parent"
    android:orientation="vertical"
    android:background="@drawable/new_income_bg"
    tools:context=".activity.NewInComeActivity">
    <LinearLayout
        android:layout_width="match_parent"
        android:layout_height="wrap_content"
        android:orientation="horizontal"
        android:layout_marginTop="220dp"
        android:layout_marginLeft="30dp"
        android:layout_marginRight="30dp">
        <TextView
            android:layout_width="wrap_content"
            android:layout_height="wrap_content"
            android:text="金    额:"
            android:textColor="#000000"
            android:textSize="20sp"
            android:textStyle="bold"/>
        <EditText
            android:id="@+id/et_money_newin"
            android:layout_width="match_parent"
            android:layout_height="wrap_content"
            android:hint="0.00"
            android:textColor="#000000"
            android:textSize="20sp"
            android:textStyle="bold"
            android:gravity="center"/>
    </LinearLayout>
    <LinearLayout
        android:layout_width="match_parent"
        android:layout_height="wrap_content"
        android:orientation="horizontal"
        android:layout_marginTop="10dp"
        android:layout_marginLeft="30dp"
        android:layout_marginRight="30dp">
        <TextView
            android:layout_width="wrap_content"
            android:layout_height="wrap_content"
            android:text="日    期:"
            android:textColor="#000000"
            android:textSize="20sp"
            android:textStyle="bold"/>
        <EditText
            android:id="@+id/et_time_newin"
```

```xml
            android:layout_width = "match_parent"
            android:layout_height = "wrap_content"
            android:hint = "2020 - 05 - 12"
            android:textColor = "#000000"
            android:textSize = "20sp"
            android:textStyle = "bold"
            android:gravity = "center"/>
    </LinearLayout>
    <LinearLayout
        android:layout_width = "match_parent"
        android:layout_height = "wrap_content"
        android:orientation = "horizontal"
        android:layout_marginTop = "10dp"
        android:layout_marginLeft = "30dp"
        android:layout_marginRight = "30dp">
        <TextView
            android:layout_width = "wrap_content"
            android:layout_height = "wrap_content"
            android:text = "类    型:"
            android:textColor = "#000000"
            android:textSize = "20sp"
            android:textStyle = "bold"/>
        <Spinner
            android:id = "@+id/sp_type_newin"
            android:layout_width = "match_parent"
            android:layout_height = "wrap_content"
            android:entries = "@array/incometype"/>
    </LinearLayout>
    <LinearLayout
            android:layout_width = "match_parent"
            android:layout_height = "wrap_content"
            android:orientation = "horizontal"
            android:layout_marginTop = "10dp"
            android:layout_marginLeft = "30dp"
            android:layout_marginRight = "30dp">
        <TextView
            android:layout_width = "wrap_content"
            android:layout_height = "wrap_content"
            android:text = "付款方:"
            android:textColor = "#000000"
            android:textSize = "20sp"
            android:textStyle = "bold"/>
        <EditText
            android:id = "@+id/et_payer_newin"
            android:layout_width = "match_parent"
            android:layout_height = "wrap_content"
            android:hint = "海明有限公司"
            android:textColor = "#000000"
            android:textSize = "20sp"
```

```xml
                android:textStyle = "bold"
                android:gravity = "center"/>
    </LinearLayout>
    <LinearLayout
        android:layout_width = "match_parent"
        android:layout_height = "wrap_content"
        android:orientation = "horizontal"
        android:layout_marginTop = "10dp"
        android:layout_marginLeft = "30dp"
        android:layout_marginRight = "30dp">
        <TextView
            android:layout_width = "wrap_content"
            android:layout_height = "wrap_content"
            android:text = "备    注:"
            android:textColor = "#000000"
            android:textSize = "20sp"
            android:textStyle = "bold"/>
        <EditText
            android:id = "@+id/et_remark_newin"
            android:layout_width = "match_parent"
            android:layout_height = "wrap_content"
            android:hint = "远程技术指导费"
            android:textColor = "#000000"
            android:textSize = "20sp"
            android:textStyle = "bold"
            android:gravity = "center"/>
    </LinearLayout>
<Button
    android:id = "@+id/bt_save_newin"
    android:layout_width = "match_parent"
    android:layout_height = "wrap_content"
    android:text = "保存"
    android:textSize = "20sp"
    android:textColor = "#ffffff"
    android:layout_marginLeft = "60dp"
    android:layout_marginRight = "60dp"
    android:background = "@drawable/btn_style_one"
    android:layout_marginTop = "40dp"/>
<Button
    android:id = "@+id/bt_cancel_newin"
    android:layout_width = "match_parent"
    android:layout_height = "wrap_content"
    android:text = "取消"
    android:textSize = "20sp"
    android:textColor = "#000000"
    android:layout_marginLeft = "60dp"
    android:layout_marginRight = "60dp"
    android:background = "@drawable/btn_style_two"
    android:layout_marginTop = "20dp"/>
</LinearLayout>
```

布局代码编写成功后,可以看到界面的预览效果,如图 8.31 所示。

图 8.31　新增收入布局界面效果

(2) 在工程中添加付款方 Spinner 控件的数据类型。

在工程中,依次展开 res→values,右击 values→New→Values Resource File,在弹出的对话框中输入资源文件名称"financialType",单击 OK 按钮,此时在 values 文件夹下便出现了刚刚创建的文件,如图 8.32 所示。

图 8.32　付款方 Spinner 控件数据类型

双击打开 financialType.xml 文件,输入如下代码。

```
<?xml version = "1.0" encoding = "utf-8"?>
<resources>
    <array name = "incometype">
        <item>---请选择---</item>
        <item>学习-奖金</item>
        <item>补助-奖金</item>
        <item>比赛-奖励</item>
        <item>业余-兼职</item>
        <item>基本-工资</item>
        <item>福利-分红</item>
        <item>加班-津贴</item>
        <item>其他</item>
    </array>

    <array name = "paytype">
        <item>---请选择---</item>
        <item>电影-娱乐</item>
```

```xml
        <item>美食-畅饮</item>
        <item>欢乐-购物</item>
        <item>手机-充值</item>
        <item>交通-出行</item>
        <item>教育-培训</item>
        <item>社交-礼仪</item>
        <item>生活-日用</item>
        <item>其他</item>
    </array>
</resources>
```

（3）在数据库类文件 MyDBHelper.java 中添加收入表和支出表，完善代码如下。

```java
public class MyDBHelper extends SQLiteOpenHelper {
    …//此处代码保持原来的样子,无修改,故此处省略
    //1.创建数据库
    @Override
    public void onCreate(SQLiteDatabase db) {
        //创建用户表
        db.execSQL("create table tb_userinfo(id integer primary key autoincrement, name varchar(10),pwd varchar(15),email varchar(50),phone varchar(11))");
        //创建收入表
        db.execSQL("create table in_come(id integer primary key autoincrement, inmoney double,intime varchar(20),intype varchar(30),inpayer varchar(100),inremark varchar(500))");
        //创建支出表
        db.execSQL("create table pay_out(id integer primary key autoincrement, outmoney double,outtime varchar(20),outtype varchar(30),outpayee varchar(100),outremark varchar(500))");
    }
    …//此处代码保持原来的样子,无修改,故此处省略
    }
```

（4）新增收入功能的实现，双击打开 NewInComeActivity.java 类文件，添加如下代码。

```java
public class NewInComeActivity extends AppCompatActivity {
    //1.定义对象
    private EditText et_money,et_time,et_payer,et_remark;
    private Spinner sp_type;
    private Button bt_sava,bt_cancel;
    private MyDBHelper mhelper;
    private SQLiteDatabase db;
    @Override
    protected void onCreate(Bundle savedInstanceState) {
        super.onCreate(savedInstanceState);
        setContentView(R.layout.activity_new_in_come);
        //绑定控件
        initView();
        //"保存"按钮功能的实现
        btnSave();
```

```java
            //"取消"按钮功能的实现
            btnCancel();
    }
    //2.绑定控件
    private void initView() {
        et_money = findViewById(R.id.et_money_newin);
        et_time = findViewById(R.id.et_time_newin);
        sp_type = findViewById(R.id.sp_type_newin);
        et_payer = findViewById(R.id.et_payer_newin);
        et_remark = findViewById(R.id.et_remark_newin);
        bt_sava = findViewById(R.id.bt_save_newin);
        bt_cancel = findViewById(R.id.bt_cancel_newin);
        mhelper = new MyDBHelper(NewInComeActivity.this);
        db = mhelper.getWritableDatabase();
    }
    //3."保存"按钮功能的实现
    private void btnSave() {
        bt_sava.setOnClickListener(new View.OnClickListener() {
            @Override
            public void onClick(View v) {
                //获取输入的内容保存到数据库的收入表中
                ContentValues values = new ContentValues();
                values.put("inmoney",et_money.getText().toString());
                values.put("intime",et_time.getText().toString());
                values.put("intype",sp_type.getSelectedItem().toString());
                values.put("inpayer",et_payer.getText().toString());
                values.put("inremark",et_remark.getText().toString());
                db.insert("in_come",null,values);
                Toast.makeText(NewInComeActivity.this,"保存成功",Toast.LENGTH_SHORT).show();
                //刷新本页面
                Intent intent = new Intent(NewInComeActivity.this,NewInComeActivity.class);
                startActivity(intent);
                finish();
            }
        });
    }
    //4."取消"按钮功能的实现
    private void btnCancel() {
        bt_cancel.setOnClickListener(new View.OnClickListener() {
            @Override
            public void onClick(View v) {
                Intent intent = new Intent(NewInComeActivity.this, MainActivity.class);
                startActivity(intent);
                finish();
            }
        });
    }
}
```

上述代码中,一共定义了三种方法,其中,initView()方法用来实现界面控件的初始化和数据库的初始化；btnSave()方法用来实现新增收入数据的保存；btnCancel()方法中实现了关闭本页面，返回主界面的功能。sp_type.getSelectedItem().toString()这行代码用来获取 Spinner 下拉列表框选择的选项内容，将其存放到收入表的 intype 列中，其他几行分别是获取输入控件输入的内容并存放到收入表不同的列，最后调用数据库的 insert()方法实现数据的保存。

（5）因数据库中数据表有更新，在用户表的基础上，又新增了收入表和支出表，如果直接运行程序，在新增收入界面操作时工程会报错，提示找不到收入表，因此需要将模拟器上的工程删除，重新安装即可。因此打开模拟器，选中 ZSFinacialManage 工程，长按鼠标左键，会出现一个垃圾桶，将工程拖放到垃圾桶上即可把工程删除。注册一个新的账号，并用新注册的账号登录，打开新增收入界面，输入多条数据并保存，效果如图 8.33 所示。打开 RE 文件管理器，查看收入表如图 8.34 所示。

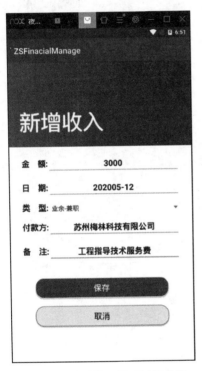

图 8.33　新增收入界面运行效果

图 8.34　查看收入表中的数据

8.2.7　新增支出界面的设计与功能

【任务描述】

整个作品中跟支出相关的内容分别是新增支出、浏览支出、支出管理（修改或删除支出）、数据分析中的支出预览图等。其中，新增支出用来添加支出信息，支出浏览用来显示所有的支出信息，支出管理用来根据编号修改或者删除支出信息，支出预览图以图表方式直观展示各类支出情况。本节主要学习新增支出。

微课视频

【设计思路】

布局界面中的控件通过添加属性的方式达到用户需求,在功能上,单击"添加"按钮,将输入的支出信息保存到数据库的支出表中,达到用户的需求。

【任务实施】

(1) 新增支出界面布局设计。

双击 layout 文件夹下的 activity_new_pay.xml 文件,便可打开布局编辑器,修改布局类型为 LinearLayout,添加方向属性为 vertical,界面中控件代码同新增收入,可以采用复制粘贴的方法来快速完成,修改支出界面中几个控件的 id 属性,其 id 如下。

```xml
<?xml version = "1.0" encoding = "utf-8"?>
<LinearLayout xmlns:android = "http://schemas.android.com/apk/res/android"
    xmlns:app = "http://schemas.android.com/apk/res-auto"
    xmlns:tools = "http://schemas.android.com/tools"
    android:layout_width = "match_parent"
    android:layout_height = "match_parent"
    android:orientation = "vertical"
    android:background = "@drawable/new_pay_bg"
    tools:context = ".activity.NewPayActivity">

<!--此处代码与新增收入雷同,可以采用复制粘贴的方法完成,其 id 有所区别,分别如下 -->

<!-- 金额控件的 id    android:id = "@ + id/et_money_newout" -->
    ...

<!-- 日期控件的 id    android:id = "@ + id/et_time_newout" -->
    ...

<!-- 支出类型控件的 id    android:id = "@ + id/sp_type_newout",支出类型:android:entries = "@array/paytype" -->
    ...

<!-- 收款方控件的 id    android:id = "@ + id/et_payer_newout" -->
    ...

<!-- 备注控件的 id    android:id = "@ + id/et_remark_newout" -->
    ...

<!-- "保存"按钮控件的 id    android:id = "@ + id/bt_save_newout" -->
    ...

<!-- "取消"按钮控件的 id    android:id = "@ + id/bt_cancel_newout" -->
    ...

</LinearLayout>
```

布局代码编写成功后,可以看到界面的预览效果,如图 8.35 所示。

项目8 我的第一桶金——理财通App的设计与实现 | 339

图 8.35 新增支出布局界面设计效果

（2）新增支出功能的实现，双击打开 NewPayActivity.java 类文件，其代码与新增收入代码类似，可采用复制粘贴的方法快速完成，注意修改绑定控件的 id。新增支出数据保存的核心代码如下，其他代码请自行修改完善。

```java
public class NewPayActivity extends AppCompatActivity {
    //1.定义对象
    EditText et_money,et_time,et_payer,et_remark;
    Spinner sp_type;
    Button bt_sava,bt_cancel;
    MyDBHelper mhelper;
    SQLiteDatabase db;
    @Override
    protected void onCreate(Bundle savedInstanceState) {
        super.onCreate(savedInstanceState);
        setContentView(R.layout.activity_new_pay);
        initView();  //绑定控件
        btnSave();   //"保存"按钮功能的实现
        btnCancel(); //"取消"按钮功能的实现
    }
    //2.绑定控件
    private void initView() {
        et_money = findViewById(R.id.et_money_newout);
        et_time = findViewById(R.id.et_time_newout);
        sp_type = findViewById(R.id.sp_type_newout);
        et_payer = findViewById(R.id.et_payer_newout);
        et_remark = findViewById(R.id.et_remark_newout);
        bt_sava = findViewById(R.id.bt_save_newout);
```

```java
        bt_cancel = findViewById(R.id.bt_cancel_newout);
        mhelper = new MyDBHelper(NewPayActivity.this);
        db = mhelper.getWritableDatabase();
    }
    //3."保存"按钮功能的实现
    private void btnSave() {
        bt_sava.setOnClickListener(new View.OnClickListener() {
            @Override
            public void onClick(View v) {
                //获取输入的内容保存到数据库的支出表中
                ContentValues values = new ContentValues();
                values.put("outmoney",et_money.getText().toString());
                values.put("outtime",et_time.getText().toString());
                values.put("outtype",sp_type.getSelectedItem().toString());
                values.put("outpayee",et_payer.getText().toString());
                values.put("outremark",et_remark.getText().toString());
                db.insert("pay_out",null,values);
                Toast.makeText(NewPayActivity.this,"保存成功", Toast.LENGTH_SHORT).show();
                //刷新本页面
                Intent intent = new Intent(NewPayActivity.this,NewPayActivity.class);
                startActivity(intent);
                finish();
            }
        });
    }
    //4."取消"按钮功能的实现
    private void btnCancel() {
        bt_cancel.setOnClickListener(new View.OnClickListener() {
            @Override
            public void onClick(View v) {
                Intent intent = new Intent(NewPayActivity.this, MainActivity.class);
                startActivity(intent);
                finish();
            }
        });
    }
}
```

（3）重新运行程序，打开新增支出界面，输入数据并保存，效果如图 8.36 所示。打开 RE 文件管理器，查看支出表，如图 8.37 所示。

微课视频

8.2.8 收入明细界面的设计与功能

【任务描述】

从收入表中获取所有的收入数据，动态展示收入信息。

【设计思路】

布局界面中的控件通过添加属性的方式达到用户需求，功能实现时，考虑到用户可能会添加很多的数据，所以可以考虑使用 RecyclerView 控件实现。主要包括以下几个操作。

项目8 我的第一桶金——理财通App的设计与实现 | 341

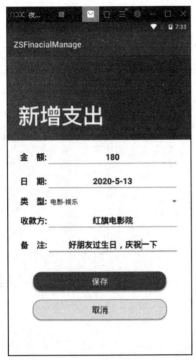

图 8.36 新增支出界面运行效果

图 8.37 查看支出表中的数据

(1) 数据:从数据库的收入表中获取数据,封装成一个类 IncomeBean,保存到动态数据当中。

(2) 子布局:根据要显示的样式自行设计子布局 recy_item_in。

(3) 适配器:采用自定义适配器 IncomeAdapter。

(4) 收入明细界面的设计与功能实现。

【任务实施】

(1) 添加 RecyclerView 的依赖。

打开工程 app 下方的 build.gradle 文件,添加 RecyclerView 的依赖,代码如下。

```
dependencies {
    implementation fileTree(dir: 'libs', include: ['*.jar'])
    implementation 'androidx.appcompat:appcompat:1.1.0'
    implementation 'androidx.constraintlayout:constraintlayout:1.1.3'
    testImplementation 'junit:junit:4.12'
    androidTestImplementation 'androidx.test.ext:junit:1.1.1'
    androidTestImplementation 'androidx.test.espresso:espresso-core:3.2.0'
    implementation 'com.android.support:recyclerview-v7:29.0.3'
}
```

Android 软件版本不同,所需要的依赖版本也不同,可以根据自己的 Android 软件版本,到网络上自行下载适合自己计算机的 RecyclerView 的依赖版本。

(2) 建立 IncomeBean 信息类。

因工程量较大,类文件较多,为了使程序文件条理清晰,创建 bean 文件夹来存放收入信息类文件。方法为:右击 com. example. zsfinacialmanage→New→Package→输入名称"bean"(com. example. zsfinacialmanage. bean)→OK,便可完成 bean 文件夹的创建。然后创建收入信息类文件,右击 bean→New→Java Class→Name 中输入类名"IncomeBean"→OK。输入如下代码。

```java
public class IncomeBean {
//首先:定义下面 6 个属性,属性下方的代码可通过快捷键自动生成
    private int id;
    private double money;
    private String time;
    private String type;
    private String payer;
    private String remark;
//鼠标定位到属性的下方,按 Alt + Insert 组合键,在弹出列表中选择 Constructor,将几个属性值全选,单击 OK,便可自动出现下方的代码
    public IncomeBean(int id, double money, String time, String type, String payer, String remark) {
        this.id = id;
        this.money = money;
        this.time = time;
        this.type = type;
        this.payer = payer;
        this.remark = remark;
    }
//鼠标定位到属性的下方,按 Alt + Insert 组合键,在弹出列表中选择 Getter and Setter,将几个属性值全选,单击 OK,便可自动出现下方的代码
    public int getId() {
        return id;
    }
    public void setId(int id) {
        this.id = id;
    }
    public double getMoney() {
        return money;
    }
    public void setMoney(double money) {
        this.money = money;
    }
    public String getTime() {
        return time;
    }
    public void setTime(String time) {
        this.time = time;
    }
    public String getType() {
        return type;
    }
    public void setType(String type) {
```

```java
        this.type = type;
    }
    public String getPayer() {
        return payer;
    }
    public void setPayer(String payer) {
        this.payer = payer;
    }
    public String getRemark() {
        return remark;
    }
    public void setRemark(String remark) {
        this.remark = remark;
    }
}
```

(3) 建立子布局 recy_item_in.xml 文件。

右击 layout→New→Layout Resource File→输入子布局名称"recy_item_in"→OK,子布局界面代码如下。

```xml
<?xml version = "1.0" encoding = "utf-8"?>
<LinearLayout
    xmlns:android = "http://schemas.android.com/apk/res/android"
    android:layout_width = "match_parent"
    android:layout_height = "wrap_content"
    android:orientation = "horizontal">
    <ImageView
        android:layout_width = "50dp"
        android:layout_height = "50dp"
        android:layout_gravity = "center_vertical"
        android:src = "@drawable/incomeicon"/>
    <LinearLayout
        android:layout_width = "wrap_content"
        android:layout_height = "wrap_content"
        android:orientation = "vertical"
        android:layout_weight = "1"
        android:padding = "5dp">
        <TextView
            android:id = "@+id/item_payer_in"
            android:layout_width = "wrap_content"
            android:layout_height = "wrap_content"
            android:text = "付款方"/>
        <TextView
            android:id = "@+id/item_type_in"
            android:layout_width = "wrap_content"
            android:layout_height = "wrap_content"
            android:text = "类型"/>
        <TextView
            android:id = "@+id/item_time_in"
```

```
                android:layout_width = "wrap_content"
                android:layout_height = "wrap_content"
                android:text = "时间"/>
            <TextView
                android:id = "@ + id/item_remark_in"
                android:layout_width = "wrap_content"
                android:layout_height = "wrap_content"
                android:text = "备注"/>
        </LinearLayout>
        <TextView
            android:id = "@ + id/item_money_in"
            android:layout_width = "wrap_content"
            android:layout_height = "wrap_content"
            android:text = "金额"
            android:textSize = "25sp"
            android:textColor = "#ff0000"
            android:textStyle = "bold"
            android:layout_gravity = "center_vertical"/>
</LinearLayout>
```

子布局代码编写成功后,可以看到界面的预览效果,如图 8.38 所示。

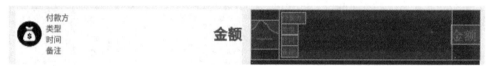

图 8.38 收入明细子布局 item 设计效果

(4) 自定义收入明细适配器 IncomeAdapter。

创建 adapter 文件夹,专门用来存放适配器。右击 com. example. zsfinacialmanage→New→Package→输入名称"adapter"(全称为 com. example. zsfinacialmanage. adapter)→OK,便可完成 adapter 文件夹的创建。然后创建收入明细适配器,右击 adapter→New→Java Class→Name 中输入类名"IncomeAdapter"→Superclass 中输入父类"RecyclerView. Adapter",自动出现 android. support. v7. widget. RecyclerView. Adapter→OK。在适配器类文件中输入如下代码。

```
//1.类文件后面添加泛型
//2.鼠标定位类文件行红色波浪线处,按 Alt + Enter 组合键:添加未实现的方法
//3.鼠标定位类文件行 ViewHolder 处,按 Alt + Enter 组合键:添加内部类
//4.鼠标定位界面最下方内部类 ViewHolder 处,添加 extends RecyclerView.ViewHolder
//5.鼠标定位界面最下方内部类 ViewHolder 红色波浪线处,按 Alt + Enter 组合键:添加构造方法
//6.定位两个对象:上下文环境和数组
//7.定位两个对象下方的空白处:按 Alt + Insert 组合键,添加适配器的构造方法
public class IncomeAdapter extends RecyclerView.Adapter<IncomeAdapter.ViewHolder> {
    Context mcontext;
    List<IncomeBean> arr2;
    public IncomeAdapter(Context mcontext, List<IncomeBean> arr2) {
```

```
        this.mcontext = mcontext;
        this.arr2 = arr2;
    }
    //用于创建 ViewHolder 实例
    @Override
    public IncomeAdapter.ViewHolder onCreateViewHolder(ViewGroup parent, int viewType) {
        View view = LayoutInflater.from(mcontext).inflate(R.layout.recy_item_in,parent,false);
        ViewHolder mholder = new ViewHolder(view);
        return mholder;
    }
    //对 RecyclerView 子项进行赋值
    @Override
    public void onBindViewHolder(IncomeAdapter.ViewHolder mholder, int position) {
        IncomeBean incomeBean = arr2.get(position);
        mholder.item_payer.setText("收款-来自" + incomeBean.getPayer());
        mholder.item_type.setText(incomeBean.getType());
        mholder.item_time.setText(incomeBean.getTime());
        mholder.item_remark.setText(incomeBean.getRemark());
        mholder.item_money.setText(" + " + incomeBean.getMoney());
    }
    //RecyclerView 一共有多少子项
    @Override
    public int getItemCount() {
        return arr2.size();
    }
    public class ViewHolder extends RecyclerView.ViewHolder{
        TextView item_payer,item_type,item_time,item_remark,item_money;
        public ViewHolder( View itemView) {
            super(itemView);
            item_payer = itemView.findViewById(R.id.item_payer_in);
            item_type = itemView.findViewById(R.id.item_type_in);
            item_time = itemView.findViewById(R.id.item_time_in);
            item_remark = itemView.findViewById(R.id.item_remark_in);
            item_money = itemView.findViewById(R.id.item_money_in);
        }
    }
}
```

上述代码中,可以使用快捷键 Alt+Enter 添加未实现的方法,使用 Alt+Insert 快捷键添加构造方法,提高书写代码的速度和技巧。

(5) 收入明细界面设计。

双击 layout 文件夹下的 activity_in_come_detail.xml 文件,便可打开布局编辑器,修改布局类型为 LinearLayout,添加 RecyclerView 控件,代码如下。

```
<?xml version = "1.0" encoding = "utf-8"?>
<LinearLayout xmlns:android = "http://schemas.android.com/apk/res/android"
    xmlns:app = "http://schemas.android.com/apk/res-auto"
```

```xml
    xmlns:tools = "http://schemas.android.com/tools"
    android:layout_width = "match_parent"
    android:layout_height = "match_parent"
    android:orientation = "vertical"
    android:background = "@drawable/in_detail_bg"
    tools:context = ".activity.InComeDetailActivity">
    <androidx.recyclerview.widget.RecyclerView
        android:id = "@+id/recy_view_indetail"
        android:layout_width = "match_parent"
        android:layout_height = "wrap_content"
        android:layout_marginTop = "120dp"
        android:layout_marginLeft = "30dp"
        android:layout_marginRight = "30dp"
        android:layout_marginBottom = "40dp"/>
</LinearLayout>
```

布局代码编写成功后,可以看到界面的预览效果,如图 8.39 所示。

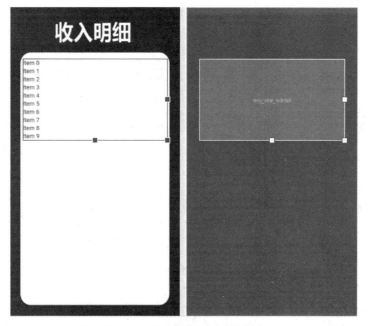

图 8.39 收入明细界面布局效果

(6)收入明细功能的实现。

双击打开 InComeDetailActivity.java 类文件,输入逻辑代码如下,代码并不是特别难以理解,在备忘录、音乐播放器、视频播放器中已多次书写类似的代码,请务必掌握相关内容。

```java
public class InComeDetailActivity extends AppCompatActivity {
    //1.定义对象
    RecyclerView recy_view;
    MyDBHelper mhelper;
    SQLiteDatabase db;
```

```java
        List < IncomeBean > arr1 = new ArrayList();
        @Override
        protected void onCreate(Bundle savedInstanceState) {
            super.onCreate(savedInstanceState);
            setContentView(R.layout.activity_in_come_detail);
            //绑定控件
            initView();
            //准备数据
            initData();
            //设计每一行的子布局
            //定义适配器:数据和子布局关联起来(桥梁的作用)
            IncomeAdapter adapter = new IncomeAdapter(InComeDetailActivity.this,arr1);
            //将适配器和布局管理器加载到控件中
StaggeredGridLayoutManager st = new StaggeredGridLayoutManager(StaggeredGridLayoutManager.
VERTICAL,1);
            recy_view.setLayoutManager(st);
            recy_view.setAdapter(adapter);
        }
        //2.绑定控件
        private void initView() {
            recy_view = findViewById(R.id.recy_view_indetail);
            mhelper = new MyDBHelper(InComeDetailActivity.this);
            db = mhelper.getWritableDatabase();
        }
        //3.准备数据
        private void initData() {
            //从数据库查询所有的新增收入信息,取出数据
            Cursor cursor = db.rawQuery("select * from in_come",null);
            while(cursor.moveToNext()){
                int myid = cursor.getInt(cursor.getColumnIndex("id"));
                double mymoney = cursor.getDouble(cursor.getColumnIndex("inmoney"));
                String mytime = cursor.getString(cursor.getColumnIndex("intime"));
                String mytype = cursor.getString(cursor.getColumnIndex("intype"));
                String mypayer = cursor.getString(cursor.getColumnIndex("inpayer"));
                String myremark = cursor.getString(cursor.getColumnIndex("inremark"));
            IncomeBean incomeBean = new IncomeBean( myid, mymoney, mytime, mytype, mypayer,
myremark);
                arr1.add(incomeBean);
            }
        }
    }
```

（7）重新运行程序,打开收入明细界面,效果如图8.40所示。也可以单击"新增收入"按钮,添加多条收入数据,再次运行程序,预览效果,如图8.41所示。

微课视频

8.2.9 支出明细界面的设计与功能

【任务描述】

从支出表中获取所有的支出数据,动态展示支出信息。

图 8.40 支出明细界面运行效果　　图 8.41 添加多条支出数据后的运行效果

【设计思路】

布局界面中的控件通过添加属性的方式达到用户需求,实现功能时,考虑到用户可能会添加很多的数据,所以可以考虑使用 RecyclerView 控件实现。主要包括以下几个操作。

(1) 数据:从数据库的支出中获取数据,封装成一个类 OutpayBean,保存到动态数据中。

(2) 子布局:根据要显示的样式自行设计子布局 recy_item_out。

(3) 适配器:采用自定义适配器 OutpayAdapter。

(4) 支出明细界面的设计与功能实现。

【任务实施】

(1) 建立 OutpayBean 信息类。

右击 db→New→Java Class→Name 中输入类名"OutpayBean"→OK。输入如下代码。

```
public class OutpayBean {
    //首先,定义下面 6 个属性,属性下方的代码可通过快捷键自动生成
    private int id;
    private double money;
    private String time;
    private String type;
    private String payee;
    private String remark;
    //鼠标定位到属性的下方,按 Alt + Insert 组合键,在弹出列表中选择 Constructor,将几个属性
    //值全选,单击 OK,便可自动出现下方的代码
```

```java
    public OutpayBean(int id, double money, String time, String type, String payee, String remark) {
        this.id = id;
        this.money = money;
        this.time = time;
        this.type = type;
        this.payee = payee;
        this.remark = remark;
    }

    //然后,鼠标定位到属性的下方,按 Alt + Insert 组合键,在弹出列表中选择 Getter and Setter,将几
    //个属性值全选,单击 OK,便可自动出现下方的代码,如果快捷键无法使用,可通过菜单 Code→
    //Generate 打开对话框
    …//这里有近五十行的代码请自行添加

}
```

(2) 建立子布局 recy_item_out.xml 文件。

右击 layout→New→Layout Resource File→输入子布局名称"recy_item_out"→OK,子布局界面代码同收入明细子布局代码类似,请根据下方提供的控件 id,自行修改。

```xml
<?xml version = "1.0" encoding = "utf - 8"?>
<LinearLayout
    xmlns:android = "http://schemas.android.com/apk/res/android"
    android:layout_width = "match_parent"
    android:layout_height = "wrap_content"
    android:orientation = "horizontal">

    <!-- 把收入明细子布局代码复制粘贴过来,重点修改控件 text 属性和 id -->

    <!-- 图片控件的 src 属性为:android:src = "@drawable/outpayicon" -->

    <!-- 收款方控件的 id 为:android:id = "@ + id/item_payee_out" -->

    <!-- 类型方控件的 id 为:android:id = "@ + id/item_type_out" -->

    <!-- 时间控件的 id 为:android:id = "@ + id/item_time_out" -->

    <!-- 备注控件的 id 为:android:id = "@ + id/item_remark_out" -->

    <!-- 金额控件的 id 为:android:id = "@ + id/item_money_out",文字颜色改为蓝色 -->
    …//这里有近五十行代码,请自行完成

</LinearLayout>
```

子布局代码编写成功后,可以看到界面的预览效果,如图 8.42 所示。

(3) 自定义支出明细适配器 OutpayAdapter。

右击 adapter→New→Java Class→Name 中输入类名"OutpayAdapter"→Superclass 中

图 8.42 支出明细子布局 item 设计效果

输入父类"RecyclerView. Adapter",自动出现 android. support. v7. widget. RecyclerView. Adapter→OK。在适配器类文件中输入如下代码。

```java
public class OutpayAdapter extends RecyclerView.Adapter<OutpayAdapter.ViewHolder>{
    Context mcontext;
    List<OutpayBean> arr2;
    public OutpayAdapter(Context mcontext, List<OutpayBean> arr2) {
        this.mcontext = mcontext;
        this.arr2 = arr2;
    }
    //用于创建 ViewHolder 实例
    @Override
    public OutpayAdapter.ViewHolder onCreateViewHolder(ViewGroup parent, int viewType) {
        View view = LayoutInflater.from(mcontext).inflate(R.layout.recy_item_out, parent, false);
        ViewHolder mholder = new ViewHolder(view);
        return mholder;
    }
    //对 RecyclerView 子项进行赋值
    @Override
    public void onBindViewHolder(OutpayAdapter.ViewHolder mholder, int position) {
      OutpayBean outpayBean = arr2.get(position);
        mholder.item_payee.setText("付款-给" + outpayBean.getPayee());
        mholder.item_type.setText(outpayBean.getType());
        mholder.item_time.setText(outpayBean.getTime());
        mholder.item_remark.setText(outpayBean.getRemark());
        mholder.item_money.setText(" - " + outpayBean.getMoney());
     }
    //RecyclerView 一共有多少子项
    @Override
    public int getItemCount() {
      return arr2.size();
    }
    public class ViewHolder extends  RecyclerView.ViewHolder{
        TextView item_payee,item_type,item_time,item_remark,item_money;
        public ViewHolder( View itemView) {
            super(itemView);
            item_payee = itemView.findViewById(R.id.item_payee_out);
            item_type = itemView.findViewById(R.id.item_type_out);
            item_time = itemView.findViewById(R.id.item_time_out);
            item_remark = itemView.findViewById(R.id.item_remark_out);
            item_money = itemView.findViewById(R.id.item_money_out);
        }
    }
}
```

上述代码中,与收入明细适配器类似,可采用复制粘贴的方法快速完成,提高书写代码的速度和技巧。但要注意有部分代码需要修改,请根据适配器的技术原理认真完成。

(4) 支出明细界面设计。

双击 layout 文件夹下的 activity_pay_detail.xml 文件,便可打开布局编辑器,代码如下。

```xml
<?xml version="1.0" encoding="utf-8"?>
<LinearLayout xmlns:android="http://schemas.android.com/apk/res/android"
    xmlns:app="http://schemas.android.com/apk/res-auto"
    xmlns:tools="http://schemas.android.com/tools"
    android:layout_width="match_parent"
    android:layout_height="match_parent"
    android:orientation="vertical"
    android:background="@drawable/pay_detail_bg"
    tools:context=".activity.PayDetailActivity">
    <androidx.recyclerview.widget.RecyclerView
        android:id="@+id/recy_view_outdetail"
        android:layout_width="match_parent"
        android:layout_height="wrap_content"
        android:layout_marginTop="120dp"
        android:layout_marginLeft="30dp"
        android:layout_marginRight="30dp"
        android:layout_marginBottom="40dp"/>
</LinearLayout>
```

修改布局类型为 LinearLayout,添加 RecyclerView 控件,其代码同收入明细界面,修改界面背景图片为 pay_detail_bg,同时修改 recyclerView 控件的 id 为 recy_view_outdetail 即可。布局代码编写成功后,可以看到界面的预览效果,如图 8.43 所示。

图 8.43 支出明细界面布局效果

(5) 支出明细功能的实现。

双击打开 PayDetailActivity.java 类文件，逻辑代码与收入明细类文件雷同，请自行书写。

```java
public class PayDetailActivity extends AppCompatActivity {
    //1.定义对象
    RecyclerView recy_view;
    MyDBHelper mhelper;
    SQLiteDatabase db;
    List<OutpayBean> arr1 = new ArrayList();
    @Override
    protected void onCreate(Bundle savedInstanceState) {
        super.onCreate(savedInstanceState);
        setContentView(R.layout.activity_pay_detail);
        //绑定控件
        initView();
        //准备数据
        initData();
        //设计每一行的子布局
        //创建适配器
        OutpayAdapter adapter = new OutpayAdapter(PayDetailActivity.this, arr1);
        //将适配器和布局管理器加载到控件当中
        StaggeredGridLayoutManager st = new StaggeredGridLayoutManager(StaggeredGridLayoutManager.VERTICAL, 1);
        recy_view.setLayoutManager(st);
        recy_view.setAdapter(adapter);
    }
    //2.绑定控件
    private void initView() {
        recy_view = findViewById(R.id.recy_view_outdetail);
        mhelper = new MyDBHelper(PayDetailActivity.this);
        db = mhelper.getWritableDatabase();
    }
    //3.准备数据
    private void initData() {
        //从数据库查询所有的新增收入信息，取出数据
        Cursor cursor = db.rawQuery("select * from pay_out", null);
        while(cursor.moveToNext()){
            int myid = cursor.getInt(cursor.getColumnIndex("id"));
            double mymoney = cursor.getDouble(cursor.getColumnIndex("outmoney"));
            String mytime = cursor.getString(cursor.getColumnIndex("outtime"));
            String mytype = cursor.getString(cursor.getColumnIndex("outtype"));
            String mypayer = cursor.getString(cursor.getColumnIndex("outpayee"));
            String myremark = cursor.getString(cursor.getColumnIndex("outremark"));
            OutpayBean outpayBean = new OutpayBean(myid, mymoney, mytime, mytype, mypayer, myremark);
            arr1.add(outpayBean);
        }
    }
}
```

（6）重新运行程序,打开支出明细界面,效果如图 8.44 所示。也可以单击"新增支出"按钮,添加多条支出数据,再次运行程序,预览效果,如图 8.45 所示。

图 8.44　收入明细界面运行效果　　图 8.45　添加多条收入数据后的运行效果

8.2.10　数据分析界面的设计与功能

【任务描述】

以图表方式动态展示支出表和收入表中各类型汇总结果。

【设计思路】

功能实现时,考虑到绘图的便利性和快捷性,所以采用 Android 的开源图标库 AndroidMPChart 绘制图表。

微课视频

【任务实施】

（1）导入 AndroidMPChart 依赖。

选中素材中的开源图标库 jar 包 MPAndroidChart-v3.0.3.jar,复制该文件,展开工程 app 目录,选择 libs 文件夹,右键粘贴,单击 OK 按钮,便可添加成功,如图 8.46 所示。选择该文件,右击,在弹出的快捷菜单中选择 Add As Library,如图 8.47 所示,单击 OK 按钮,便可将依赖成功地添加到本工程中。

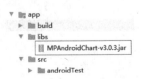 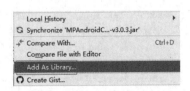

图 8.46　把 jar 包复制到工程中　　图 8.47　将 jar 包对应的依赖添加到本工程中

打开工程 app 下方的 build.gradle 文件,查看添加 AndroidMPChart 的依赖,代码如下。

```
dependencies {
    ...
    implementation 'com.android.support:recyclerview-v7:29.0.3'
    implementation files('libs/MPAndroidChart-v3.0.3.jar')
}
```

(2) 数据分析布局界面设计。

双击 layout 文件夹下的 activity_data_analyse.xml 文件将其打开,添加控件,代码如下。

```
<?xml version="1.0" encoding="utf-8"?>
<LinearLayout xmlns:android="http://schemas.android.com/apk/res/android"
    xmlns:app="http://schemas.android.com/apk/res-auto"
    xmlns:tools="http://schemas.android.com/tools"
    android:layout_width="match_parent"
    android:layout_height="match_parent"
    android:orientation="vertical"
    android:background="@drawable/welcomebg"
    tools:context=".activity.DataAnalyseActivity">
    <TextView
        android:layout_width="match_parent"
        android:layout_height="wrap_content"
        android:text="个人收入预览图"
        android:textColor="#ffffff"
        android:textStyle="bold"
        android:textSize="25sp"
        android:gravity="center"/>
    <com.github.mikephil.charting.charts.LineChart
        android:id="@+id/income_chart_data"
        android:layout_width="match_parent"
        android:layout_height="wrap_content"
        android:layout_weight="1"/>
    <View
        android:layout_width="match_parent"
        android:layout_height="2dp"
        android:background="#ffffff"/>
    <TextView
        android:layout_width="match_parent"
        android:layout_height="wrap_content"
        android:text="个人支出预览图"
        android:textColor="#ffffff"
        android:textStyle="bold"
        android:textSize="25sp"
        android:gravity="center"/>
    <com.github.mikephil.charting.charts.LineChart
        android:id="@+id/outpay_chart_data"
```

```
        android:layout_width = "match_parent"
        android:layout_height = "wrap_content"
        android:layout_weight = "1"/>
</LinearLayout>
```

布局代码编写成功后，可以看到界面的预览效果，如图8.48所示。

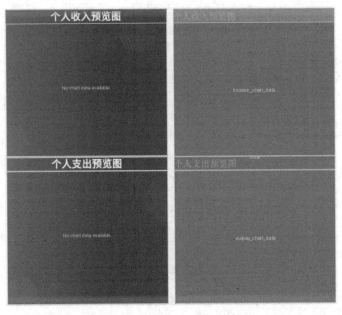

图 8.48　数据分析界面设计效果

（3）数据分析界面功能的实现。

双击打开 DataAnalyseActivity.java 类文件，输入实现数据分析的逻辑代码，如下。

```
public class DataAnalyseActivity extends AppCompatActivity {
    //1.定义对象
    LineChart income_chart,outpay_chart;
    MyDBHelper mhelper;
    SQLiteDatabase db;
    String[] indata = {"学习－奖金","补助－奖金","比赛－奖励","业余－兼职","基本－工资","福利－分红","加班－津贴","其他"};
    //收入类型数据统计的初始值
    int xxjjmoney = 0;
    int bzjjmoney = 0;
    int bsjlmoney = 0;
    int yyjzmoney = 0;
    int jbgzmoney = 0;
    int flfhmoney = 0;
    int jbjtmoney = 0;
    int qtmoney = 0;
    String[] outdata = {"电影－娱乐","美食－畅饮","欢乐－购物","手机－充值","交通－出行","教育－培训","社交－礼仪","生活－日用","其他"};
```

```java
//收入类型数据统计的初始值
int dyylmoney = 0;
int mscymoney = 0;
int hlgwmoney = 0;
int sjczmoney = 0;
int jtcxmoney = 0;
int jypxmoney = 0;
int sjlymoney = 0;
int shrymoney = 0;
int othermoney = 0;
@Override
protected void onCreate(Bundle savedInstanceState) {
    super.onCreate(savedInstanceState);
    setContentView(R.layout.activity_data_analyse);
    //绑定控件
    initView();
    //收入汇总分析
    inComeData();
    //支出汇总分析
    outComeData();
}
//2.绑定控件
private void initView() {
    income_chart = findViewById(R.id.income_chart_data);
    outpay_chart = findViewById(R.id.outpay_chart_data);
    mhelper = new MyDBHelper(DataAnalyseActivity.this);
    db = mhelper.getWritableDatabase();
}
//3.收入汇总分析
private void inComeData() {
    //第一部分:获取数据
    Cursor cursor = db.rawQuery("select * from in_come",null);
    while(cursor.moveToNext()){
        Double mymoney = cursor.getDouble(cursor.getColumnIndex("inmoney"));
        String mytype = cursor.getString(cursor.getColumnIndex("intype"));
        if(mytype.equals("学习-奖金")){
            xxjjmoney += mymoney;
        }else if(mytype.equals("补助-奖金")){
            bsjlmoney += mymoney;
        }else if(mytype.equals("比赛-奖励")){
            bzjjmoney += mymoney;
        }else if(mytype.equals("业余-兼职")){
            yyjzmoney += mymoney;
        }else if(mytype.equals("基本-工资")){
            jbgzmoney += mymoney;
        }else if(mytype.equals("福利-分红")){
            flfhmoney += mymoney;
        }else if(mytype.equals("加班-津贴")){
            jbjtmoney += mymoney;
        }else if(mytype.equals("其他")){
```

```java
            qtmoney += mymoney;
        }
    }
    //第二部分:LineChart 图表初始化设置——XY 轴的设置
    XAxis xAxis = income_chart.getXAxis();  //获取此图表的 X 轴轴线
    YAxis yAxisleft = income_chart.getAxisLeft();  //获取此图表的 Y 轴左侧轴线
    YAxis yAxisright = income_chart.getAxisRight();  //获取此图表的 Y 轴右侧轴线
    xAxis.setPosition(XAxis.XAxisPosition.BOTTOM);  //设置 X 轴线的位置为底部
    yAxisleft.setAxisMinimum(0f);  //保证 Y 轴从 0 开始,不然会上移一点儿
    yAxisright.setAxisMinimum(0f);
    xAxis.setValueFormatter(new IAxisValueFormatter() {  //X 轴自定义标签的设置
        @Override
        public String getFormattedValue(float v, AxisBase axisBase) {
            return indata[(int) v];
        }
    });
    //第三部分:LineDataSet 曲线初始化设置
    List<Entry> inentries = new ArrayList<>();  //Y 轴的数据
    inentries.add(new Entry(0,xxjjmoney));
    inentries.add(new Entry(1,bzjjmoney));
    inentries.add(new Entry(2,bsjlmoney));
    inentries.add(new Entry(3,yyjzmoney));
    inentries.add(new Entry(4,jbgzmoney));
    inentries.add(new Entry(5,flfhmoney));
    inentries.add(new Entry(6,jbjtmoney));
    inentries.add(new Entry(7,qtmoney));
    LineDataSet lineDataSet = new LineDataSet(inentries,"金额");  //代表一条线,"金额"是
                                                                //曲线名称
    lineDataSet.setValueTextSize(25);  //曲线上文字的大小
    lineDataSet.setValueTextColor(Color.WHITE);  //曲线上文字的颜色
    lineDataSet.setDrawFilled(true);  //设置折线图填充
    //第四部分:曲线展示
    LineData data = new LineData(lineDataSet);  //创建 LineData 对象,属于 LineChart 折线
                                                //图的数据集合
    income_chart.setData(data);  //添加到图表中
}
//4.支出汇总分析
private void outComeData() {
    //第一部分:获取数据
    Cursor cursor = db.rawQuery("select * from pay_out",null);
    while(cursor.moveToNext()){
        Double mymoney = cursor.getDouble(cursor.getColumnIndex("outmoney"));
        String mytype = cursor.getString(cursor.getColumnIndex("outtype"));
        if(mytype.equals("电影-娱乐")){
            dyylmoney += mymoney;
        }else if(mytype.equals("美食-畅饮")){
            mscymoney += mymoney;
        }else if(mytype.equals("欢乐-购物")){
            hlgwmoney += mymoney;
```

```java
        }else if(mytype.equals("手机-充值")){
            sjczmoney += mymoney;
        }else if(mytype.equals("交通-出行")){
            jtcxmoney += mymoney;
        }else if(mytype.equals("教育-培训")){
            jypxmoney += mymoney;
        }else if(mytype.equals("社交-礼仪")){
            sjlymoney += mymoney;
        }else if(mytype.equals("生活-日用")){
            shrymoney += mymoney;
        }else if(mytype.equals("其他")){
            othermoney += mymoney;
        }
    }
    //第二部分:LineChart图表初始化设置——XY轴的设置
    XAxis xAxis = outpay_chart.getXAxis(); //获取此图表的X轴轴线
    YAxis yAxisleft = outpay_chart.getAxisLeft(); //获取此图表的Y轴左侧轴线
    YAxis yAxisright = outpay_chart.getAxisRight(); //获取此图表的Y轴右侧轴线
    xAxis.setPosition(XAxis.XAxisPosition.BOTTOM); //设置X轴线的位置为底部
    yAxisleft.setAxisMinimum(0f); //保证Y轴从0开始,不然会上移一点儿
    yAxisright.setAxisMinimum(0f);
    xAxis.setValueFormatter(new IAxisValueFormatter() {//X轴自定义标签的设置
        @Override
        public String getFormattedValue(float v, AxisBase axisBase) {
            return outdata[(int) v];
        }
    });
    //第三部分:LineDataSet曲线初始化设置
    List<Entry> outentries = new ArrayList<>(); //Y轴的数据
    outentries.add(new Entry(0,dyylmoney));
    outentries.add(new Entry(1,mscymoney));
    outentries.add(new Entry(2,hlgwmoney));
    outentries.add(new Entry(3,sjczmoney));
    outentries.add(new Entry(4,jtcxmoney));
    outentries.add(new Entry(5,jypxmoney));
    outentries.add(new Entry(6,sjlymoney));
    outentries.add(new Entry(7,shrymoney));
    outentries.add(new Entry(8,othermoney));
    LineDataSet lineDataSet = new LineDataSet(outentries,"金额"); //代表一条线,"金额"
                                                                  //是曲线名称
    lineDataSet.setValueTextSize(25); //曲线上文字的大小
    lineDataSet.setValueTextColor(Color.WHITE); //曲线上文字的颜色
    lineDataSet.setDrawFilled(true); //设置折线图填充
    //第四部分:曲线展示
    LineData data = new LineData(lineDataSet); //创建LineData对象 属于LineChart折线图
                                                //的数据集合
    outpay_chart.setData(data); //添加到图表中
    }
}
```

上述代码中,定义了两个折线图对象,income_chart 用来显示收入统计图,outpay_chart 对象用来显示支出统计图。这些图中的数据是从数据库的收入表和支出表中获取的,因此定义了数据库帮助类对象 mhelper 和数据库对象 db。此外,为了显示图形的需要,定义了两个字符串数组,用来存放收入类型和支出类型常量数据。统计图显示结果时,会求出各分类的数据总和并显示出来,连成一条折线,因此根据收入类型不同定义 8 个变量,设置变量初始值为 0。同理,根据支出类型不同,定义 9 个变量,初始值也为 0。

initView()方法中,用来实现控件绑定和数据库的初始化操作。

inComeData()方法用来实现收入数据的汇总分析统计图的绘制。首先从数据查询所有的数据,通过 while 循环,依次取出每条数据的收入类型及收入金额,然后根据收入类型进行判断,将该类型的收入金额累加放到预定义的该类型的变量中,经过多次循环之后,就求得了不同收入类型的数据总和。接下来的任务将各类型的计算总和以折线图的形式显示出来。

首先对 LineChart 图表进行初始化设置,即 XY 轴的设置。在折线图中有几十种方法,可通过百度详细了解折线图的各种参数。其中,getXAxis()方法可以获取图表的 X 轴轴线,getAxisLeft()方法获取图表的 Y 轴左侧轴线,getAxisRight()方法获取图表的 Y 轴右侧轴线。xAxis. setPosition (XAxis. XAxisPosition. BOTTOM)设置 X 轴线的位置为底部,调用 Y 轴的 setAxisMinimum(0f)方法,取值为 0f,表示 Y 轴刻度从坐标原点(0,0)开始,不然会上移一点。然后调用 X 轴的 setValueFormatter()方法自定义 X 轴标签描述,将输入类型的数组放到统计图的 X 轴当作分类坐标。

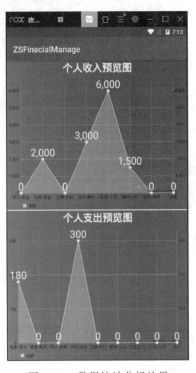

图 8.49　数据统计分析效果

其次,将各种类型的求和数据显示到图表上,即曲线初始化设置。曲线上每个点都有对应的 X 轴和 Y 轴两个坐标,通过 inentries. add(new Entry(0,xxjjmoney))方法将每个点的坐标添加到动态数组中。调用 new LineDataSet(inentries,"金额")方法将点连成一条线,曲线的名称为"金额"。调用 setValueTextSize(25)方法设置曲线上文字的大小为 25,调用 setValueTextColor (Color. WHITE)方法设置曲线上文字的颜色为白色,调用 setDrawFilled(true)方法设置曲线为填充模式,最后将曲线添加到图标中,从而使曲线显示出来。

outComeData()方法实现支出数据的汇总分析统计图的绘制,其代码含义与收入预览图类似。

(4) 运行程序,效果如图 8.49 所示。在新增收入和新增支出中给各种类型添加数据,再次运行效果如图 8.50 所示。修改折线图的参数,运行后效果如图 8.51 所示。

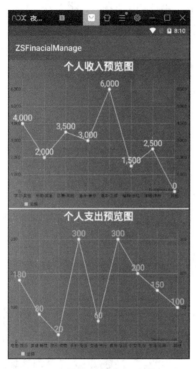 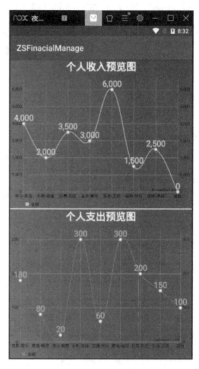

图 8.50 增加数据和设置参数后效果　　图 8.51 调整线条参数后效果

```
lineDataSet.setValueTextSize(20);                              //曲线上文字的大小
lineDataSet.setValueTextColor(Color.GREEN);                    //曲线上文字的颜色
lineDataSet.setDrawFilled(false);                              //设置折线图填充
lineDataSet.setMode(LineDataSet.Mode.CUBIC_BEZIER);            //设置线条为曲线
lineDataSet.setColor(Color.RED);                               //设置曲线的颜色
```

微课视频

8.2.11　系统设置界面的设计与功能

【任务描述】

登录用户名显示到页面的左上角位置,并且该页面可以实现密码修改功能。

【设计思路】

从数据库中读取用户表中的数据,调用数据库的 update()方法实现对用户表中当前登录用户密码的修改。

【任务实施】

(1) 系统设置界面布局界面的设计。

```
<?xml version = "1.0" encoding = "utf - 8"?>
<LinearLayout xmlns:android = "http://schemas.android.com/apk/res/android"
    xmlns:app = "http://schemas.android.com/apk/res - auto"
    xmlns:tools = "http://schemas.android.com/tools"
    android:layout_width = "match_parent"
    android:layout_height = "match_parent"
```

```xml
    android:orientation = "vertical"
    android:background = "@drawable/welcomebg"
    tools:context = ".activity.SysSettingActivity">
    <LinearLayout
        android:layout_width = "match_parent"
        android:layout_height = "wrap_content"
        android:orientation = "horizontal"
        android:layout_marginLeft = "10dp">
        <TextView
            android:layout_width = "wrap_content"
            android:layout_height = "wrap_content"
            android:text = "当前用户:"
            android:textColor = "#ffffff"/>
        <TextView
            android:id = "@+id/txt_name_sys"
            android:layout_width = "wrap_content"
            android:layout_height = "wrap_content"
            android:textColor = "#ffffff"/>
    </LinearLayout>
    <TextView
        android:layout_width = "match_parent"
        android:layout_height = "wrap_content"
        android:text = "用户密码修改"
        android:textSize = "30sp"
        android:textStyle = "bold"
        android:textColor = "#ffffff"
        android:layout_margin = "80dp"
        android:gravity = "center"/>
    <LinearLayout
        android:layout_width = "match_parent"
        android:layout_height = "wrap_content"
        android:orientation = "horizontal"
        android:layout_marginLeft = "20dp"
        android:layout_marginRight = "20dp">
        <TextView
            android:layout_width = "wrap_content"
            android:layout_height = "wrap_content"
            android:text = "原始密码:"
            android:textColor = "#ffffff"/>
        <EditText
            android:id = "@+id/et_ypwd_sys"
            android:layout_width = "match_parent"
            android:layout_height = "wrap_content"
            android:hint = "请输入原密码"
            android:textColor = "#ffffff"/>
    </LinearLayout>
    <LinearLayout
        android:layout_width = "match_parent"
        android:layout_height = "wrap_content"
```

```xml
        android:orientation = "horizontal"
        android:layout_marginLeft = "20dp"
        android:layout_marginRight = "20dp">
        <TextView
            android:layout_width = "wrap_content"
            android:layout_height = "wrap_content"
            android:text = "新 密 码:"
            android:textColor = "#ffffff"/>
        <EditText
            android:id = "@+id/et_xpwd_sys"
            android:layout_width = "match_parent"
            android:layout_height = "wrap_content"
            android:hint = "请输入新密码"
            android:textColor = "#ffffff"/>
    </LinearLayout>
    <LinearLayout
        android:layout_width = "match_parent"
        android:layout_height = "wrap_content"
        android:orientation = "horizontal"
        android:layout_marginLeft = "20dp"
        android:layout_marginRight = "20dp">
        <TextView
            android:layout_width = "wrap_content"
            android:layout_height = "wrap_content"
            android:text = "再次输入:"
            android:textColor = "#ffffff"/>
        <EditText
            android:id = "@+id/et_zxpwd_sys"
            android:layout_width = "match_parent"
            android:layout_height = "wrap_content"
            android:hint = "请再次输入新密码"
            android:textColor = "#ffffff"/>
    </LinearLayout>
    <LinearLayout
        android:layout_width = "match_parent"
        android:layout_height = "wrap_content"
        android:orientation = "horizontal"
        android:layout_marginLeft = "20dp"
        android:layout_marginRight = "20dp"
        android:layout_marginTop = "80dp">
        <Button
            android:id = "@+id/bt_modify_sys"
            android:layout_width = "wrap_content"
            android:layout_height = "wrap_content"
            android:text = "确认修改"
            android:textColor = "#000000"
            android:background = "@drawable/btn_style_two"
            android:layout_marginRight = "40dp"
            android:layout_weight = "1"/>
```

```xml
<Button
    android:id = "@+id/bt_cancel_sys"
    android:layout_width = "wrap_content"
    android:layout_height = "wrap_content"
    android:text = "取消"
    android:textColor = "#000000"
    android:background = "@drawable/btn_style_two"
    android:layout_weight = "1"/>
</LinearLayout>
</LinearLayout>
```

上述布局代码编辑完成后,布局设计效果如图 8.52 所示。

图 8.52 系统设置界面布局界面效果

(2) 系统设置界面功能设计。

```java
public class SysSettingActivity extends AppCompatActivity {
    //1.定义对象
    private TextView txt_user; //创建一个显示用户名的文本对象
    private EditText et_ypwd, et_xpwd, et_zxpwd; //创建三个 EditText 对象
    private Button bt_modify, bt_cancel; //创建两个 Button 对象
    private MyDBHelper mhelper;
    private SQLiteDatabase db;
    private String name;
    private String pwd;
    @Override
    protected void onCreate(Bundle savedInstanceState) {
        super.onCreate(savedInstanceState);
        setContentView(R.layout.activity_sys_setting);
        //绑定控件
        initView();
```

```java
            //显示当前登录的用户名
            displayInfo();
            //"确认修改"按钮功能
            btnModify();
            //"取消"按钮功能
            btncancel();
        }
        //2.绑定控件
        private void initView() {
            txt_user = findViewById(R.id.txt_name_sys);
            et_ypwd = findViewById(R.id.et_ypwd_sys);
            et_xpwd = findViewById(R.id.et_xpwd_sys);
            et_zxpwd = findViewById(R.id.et_zxpwd_sys);
            bt_modify = findViewById(R.id.bt_modify_sys);
            bt_cancel = findViewById(R.id.bt_cancel_sys);
            mhelper = new MyDBHelper(SysSettingActivity.this);
            db = mhelper.getWritableDatabase();
        }
        //3.显示当前登录的用户名
        private void displayInfo() {
            name = getSharedPreferences("userinfo",0).getString("username","");
            pwd = getSharedPreferences("userinfo",0).getString("userpwd","");
            txt_user.setText(name);
        }
        //4."修改"按钮功能
        private void btnModify() {
            bt_modify.setOnClickListener(new View.OnClickListener() {
                @Override
                public void onClick(View v) {
                    //获取三个输入框中的内容
                    String ypwd = et_ypwd.getText().toString(); //获取输入的原密码
                    String xpwd = et_xpwd.getText().toString(); //获取输入的新密码
                    String zxpwd = et_zxpwd.getText().toString(); //获取第二次输入的新密码
                    //对每个密码进行逻辑判断
                    if(ypwd.equals("")){
Toast.makeText(SysSettingActivity.this,"请输入原始密码",Toast.LENGTH_SHORT).show();
                    }else if(!ypwd.equalsIgnoreCase(pwd)){
Toast.makeText(SysSettingActivity.this,"输入的密码与原密码不一致",Toast.LENGTH_SHORT).show();
                    }else if(xpwd.equals("")){
Toast.makeText(SysSettingActivity.this,"请输入新密码",Toast.LENGTH_SHORT).show();
                    }else if(xpwd.equalsIgnoreCase(ypwd)){
Toast.makeText(SysSettingActivity.this,"所输入的新密码与原密码不能相同",Toast.LENGTH_SHORT).show();
                    }else if(zxpwd.equals("")){
Toast.makeText(SysSettingActivity.this,"请再次输入新密码",Toast.LENGTH_SHORT).show();
                    }else if(!zxpwd.equalsIgnoreCase(xpwd)){
Toast.makeText(SysSettingActivity.this,"两次输入的新密码不一致",Toast.LENGTH_SHORT).show();
                    }else{
                        ContentValues values = new ContentValues();
                        values.put("pwd",xpwd);
```

```
                    db.update("tb_userinfo",values,"name = ?",new String[]{name});
Toast.makeText(SysSettingActivity.this, "密码修改成功", Toast.LENGTH_SHORT ).show();
                    Intent intent = new Intent(SysSettingActivity.this, LoginActivity.class);
                    startActivity(intent);
                    finish();
                }
            }
        });
    }
    //5."取消"按钮功能
    private void btncancel() {
        bt_cancel.setOnClickListener(new View.OnClickListener() {
            @Override
            public void onClick(View v) {
                Intent intent = new Intent(SysSettingActivity.this, MainActivity.class);
                startActivity(intent);
                finish();
            }
        });
    }
}
```

上述代码中，自定义 initView()方法用来实现控件绑定和数据库初始化。自定义 displayInfo()方法用来实现用户名的显示。首先采用 SharedPreferences 数据存储技术，调用 getSharedPreferences()方法，从 userinfo 文件中取 username 键中存放的值，获取的值即为用户名。同时获取 pwd 键中存放的值，获取的值即为密码。调用 txt_user.setText(name)方法，将获取的用户名显示到页面左上角的文本控件中。

自定义 btnModify()方法用来实现密码的修改功能。首先获取三个输入框中的内容，然后对每个密码进行逻辑判断，如果原密码为空，提示"请输入原始密码"；如果原密码与从 SharedPreference 中获取的密码不一致，提示"输入的密码与原密码不一致"；如果新密码为空，提示"请输入新密码"；如果所输入的新密码与原密码相同，则提示"所输入的新密码与原密码不能相同"；如果再次输入新密码为空，则提示"请再次输入新密码"；如果再次输入的新密码与第一次输入的新密码不同，则提示"两次输入的新密码不一致"；在上面三个密码输入都正确且符合要求的情况下，调用数据库的 update()方法，根据当前登录的用户名，完成密码的更新。同时打开登录界面，用更改后的新密码重新登录。

(3) 运行程序，效果如图 8.53 所示。

图 8.53 系统设置界面运行效果

微课视频

8.2.12 收入管理界面的设计与功能

【任务描述】

对已经添加的收入数据进行修改和删除。

【设计思路】

在收入明细界面上,单击某条数据,便可跳转到收入管理界面,显示收入信息;重新修改收入信息后,单击"修改"按钮,便可完成收入数据的修改;单击"删除"按钮,从数据库收入表当中删除该条数据。

【任务实施】

(1) 建立 other 文件夹。

因工程量较大,类文件较多,为了使程序文件条理清晰,创建 other 文件夹来存放收入管理页面和支出管理页面。方法为:右击 com.example.zsfinacialmanage→New→Package→输入名称"other"(com.example.zsfinacialmanage.other)→OK,便可完成 other 文件夹的创建。然后

图 8.54 收入管理页面与支出管理页面的创建

创建收入管理页面,右击 other→New→Activity→Empty Activity,会弹出一个创建活动的对话框,将活动命名为 InManageActivity,默认勾选 Generate Layout File 关联布局界面,布局界面名称为 activity_in_manage,但不要勾选 Launcher Activity。单击 Finish 按钮,便可在工程中完成收入管理页面的添加。同理,添加支出管理页面,类文件名称为 OutManageActivity,其对应的布局界面名称为 activity_out_manage。完成效果如图 8.54 所示。

(2) 将 IncomeBean 收入信息类设置为实现序列化接口。

展开工程中的 bean 文件夹,打开收入信息类 IncomeBean 文件,在类文件名的右侧添加 implements Serializable,便可实现序列化接口,代码如下,其他代码保持不变。

```
public class IncomeBean implements Serializable {
    …//内部所有代码保持不变
}
```

(3) 在 IncomeAdapter 适配器中增加单击某一行跳转的代码。

展开工程中的 adapter 文件夹,打开收入信息适配器 IncomeAdapter 类文件,修改代码如下,其他代码保持不变。

```
//对 RecyclerView 子项进行赋值
@Override
public void onBindViewHolder(IncomeAdapter.ViewHolder mholder, int position) {
    final IncomeBean incomeBean = arr2.get(position);
    mholder.item_payer.setText("收款 - 来自" + incomeBean.getPayer());
    mholder.item_type.setText(incomeBean.getType());
    mholder.item_time.setText(incomeBean.getTime());
    mholder.item_remark.setText(incomeBean.getRemark());
    mholder.item_money.setText(" +" + incomeBean.getMoney());
    //完善:单击某一个条目,跳转到收入管理页面
```

```
    mholder.itemView.setOnClickListener(new View.OnClickListener() {
        @Override
        public void onClick(View v) {
            //跳转到收入管理页面的代码
            Intent intent = new Intent(mcontext, InManageActivity.class);
            intent.putExtra("seri", incomeBean);
            mcontext.startActivity(intent);
            ((Activity)mcontext).finish();
        }
    });
}
```

(4) 收入管理页面布局界面设计。

双击打开 activity_in_manage.xml 布局界面,其界面外观与新增收入界面类似,请参考新增收入布局界面设计自行完成,其界面背景图片为 inmanagebg.png,界面中控件 id 如表 8.11 所示。

表 8.11 收入管理界面背景图片和界面中控件 id

控件类型	控件含义	id 号取值
EditText	输入修改后的金额	android:id="@+id/et_money_inmag"
EditText	输入修改后的日期	android:id="@+id/et_time_inmag"
Spinner	选择修改后的收入类型	android:id="@+id/sp_type_inmag"
EditText	输入修改后的付款方	android:id="@+id/et_payer_inmag"
EditText	输入修改后的备注	android:id="@+id/et_remark_inmag"
Button	"修改"按钮(注意按钮上的文字)	android:id="@+id/bt_modify_inmag"
Button	"删除"按钮(注意按钮上的文字)	android:id="@+id/bt_delete_inmag"

布局界面代码编辑完成后,便可预览界面的设计效果,如图 8.55 所示。

图 8.55 收入管理界面布局设计效果图

(5) 收入管理页面的功能设计。

双击打开 InManageActivity.java 类文件,编辑代码如下。

```java
public class InManageActivity extends AppCompatActivity {
    //1.定义对象
    private EditText et_money,et_time,et_payer,et_remark;
    private Spinner sp_type;
    private Button btn_modify,btn_delete;
    private MyDBHelper mhelper;
    private SQLiteDatabase db;
    private IncomeBean incomeBean;
    @Override
    protected void onCreate(Bundle savedInstanceState) {
        super.onCreate(savedInstanceState);
        setContentView(R.layout.activity_in_manage);
        //绑定控件
        initView();
        //获取单击的那条数据并显示出来
        getDataDisplay();
        //"修改"按钮功能的实现
        btnModidfy();
        //"删除"按钮功能的实现
        btnDelete();
    }
//2.绑定控件
private void initView() {
    et_money = findViewById(R.id.et_money_inmag);
    et_time = findViewById(R.id.et_time_inmag);
    sp_type = findViewById(R.id.sp_type_inmag);
    et_payer = findViewById(R.id.et_payer_inmag);
    et_remark = findViewById(R.id.et_remark_inmag);
    btn_modify = findViewById(R.id.bt_modify_inmag);
    btn_delete = findViewById(R.id.bt_delete_inmag);
    mhelper = new MyDBHelper(InManageActivity.this);
    db = mhelper.getWritableDatabase();
}
//3.获取单击的那条数据并显示出来
private void getDataDisplay() {
incomeBean = (IncomeBean) getIntent().getSerializableExtra("seri");
et_money.setText(incomeBean.getMoney() + "");
et_time.setText(incomeBean.getTime());
    if (incomeBean.getType().equals("学习-奖金")){
        sp_type.setSelection(1);
    }else if (incomeBean.getType().equals("补助-奖金")) {
        sp_type.setSelection(2);
    }else if (incomeBean.getType().equals("比赛-奖励")){
        sp_type.setSelection(3);
    }else if (incomeBean.getType().equals("业余-兼职")){
        sp_type.setSelection(4);
    }else if (incomeBean.getType().equals("基本-工资")){
        sp_type.setSelection(5);
```

```java
        }else if (incomeBean.getType().equals("福利-分红")){
            sp_type.setSelection(6);
        }else if (incomeBean.getType().equals("加班-津贴")){
            sp_type.setSelection(7);
        } else if(incomeBean.getType().equals("其他")){
            sp_type.setSelection(8);
        }else {
            sp_type.setSelection(0);
        }
    et_payer.setText(incomeBean.getPayer());
    et_remark.setText(incomeBean.getRemark());
}
//4."修改"按钮功能的实现
private void btnModidfy() {
    btn_modify.setOnClickListener(new View.OnClickListener() {
        @Override
        public void onClick(View v) {
            //创建一个对象,封装一行数据
            ContentValues values = new ContentValues();
            values.put("inmoney",et_money.getText().toString());
            values.put("intime",et_time.getText().toString());
            values.put("intype",sp_type.getSelectedItem().toString());
            values.put("inpayer",et_payer.getText().toString());
            values.put("inremark",et_remark.getText().toString());
            //把该行数据更新到收入表中
            db.update("in_come",values,"id = ?",new String[]{incomeBean.getId() + ""});
            Toast.makeText(InManageActivity.this,"修改成功",Toast.LENGTH_SHORT).show();
            //关闭本页面,重新打开收入明细界面,即可查询修改后的结果
            Intent intent = new Intent(InManageActivity.this, InCoMeDetailActivity.class);
            startActivity(intent); //执行 Intent 操作
            finish(); //退出当前程序,或关闭当前页面
        }
    });
}
//5."删除"按钮功能的实现
private void btnDelete() {
    btn_delete.setOnClickListener(new View.OnClickListener() {
        @Override
        public void onClick(View v) {
            //从数据库中删除该条记录即可
            db.delete("in_come","id = ?",new String[]{incomeBean.getId() + ""});
            Toast.makeText(InManageActivity.this,"删除成功",Toast.LENGTH_SHORT).show();
            //关闭本页面,重新打开收入明细界面,即可查看删除后的结果
            Intent intent = new Intent(InManageActivity.this, InCoMeDetailActivity.class);
            startActivity(intent); //执行 Intent 操作
            finish(); //退出当前程序,或关闭当前页面
        }
    });
}
}
```

上述代码中，通过 getIntent().getSerializableExtra("seri")方法获取收入明细页单击的那条数据，因为每条数据传递时是传递 IncomeBean 类的二进制序列，因此页面也先接受传递过来二进制序列并将其转换为 IncomeBean 类，之后调用类的 get()方法获取该条数据本身的信息，显示到对应的控件上。

btnModidfy()方法用来实现数据的修改功能，在该方法的内部，首先构建一个 ContentValues 对象，代表要修改的那条数据，然后调用 put()方法将获取修改后的数据放到不同的列中，最后调用数据库的 update()方法根据该行数据的 id，更新该条数据的内容。完成更新后跳转到收入明细界面，将修改后的数据显示出来。btnDelete()方法用来实现数据的删除，删除过程中调用数据库的 delete()方法，删除完成后跳转到收入明细界面，发现删除后的数据列表中已经不存在了。

（6）运行程序，效果如图 8.56 和图 8.57 所示。

图 8.56　修改收入　　　　　　　　图 8.57　删除收入

微课视频

8.2.13　支出管理界面的设计与功能

【任务描述】

对已经添加的支出数据进行修改和删除。

【设计思路】

在支出明细界面上，单击某条数据，便可跳转到支出管理界面；显示支出信息；重新输入支出信息后，单击"修改"按钮，便可完成支出数据的修改；单击"删除"按钮，从数据库支出表当中删除该条数据。

【任务实施】

（1）将 OutpayBean 收入信息类设置为实现序列化接口。

展开工程中的 bean 文件夹，打开收入信息类 OutpayBean 文件，在类文件名的右侧添加 implements Serializable，便可实现序列化接口，代码如下，其他代码保持不变。

```java
public class OutpayBean implements Serializable {
    …//内部所有代码保持不变
}
```

（2）在 OutpayAdapter 适配器中增加单击某一行跳转的代码。

展开工程中的 adapter 文件夹，打开收入信息适配器 OutpayAdapter 类文件，修改代码如下，其他代码保持不变。

```java
//对 RecyclerView 子项进行赋值
@Override
public void onBindViewHolder(OutpayAdapter.ViewHolder mholder, int position) {
    final OutpayBean outpayBean = arr2.get(position);
    mholder.item_payee.setText("付款-给" + outpayBean.getPayer());
    mholder.item_type.setText(outpayBean.getType());
    mholder.item_time.setText(outpayBean.getTime());
    mholder.item_remark.setText(outpayBean.getRemark());
    mholder.item_money.setText("-" + outpayBean.getMoney());
    //完善:单击某一个条目,跳转到收入管理页面
    mholder.itemView.setOnClickListener(new View.OnClickListener() {
        @Override
        public void onClick(View v) {
            //跳转到收入管理页面的代码
            Intent intent = new Intent(mcontext, OutManageActivity.class);
            intent.putExtra("sero", outpayBean);
            mcontext.startActivity(intent);
            ((Activity)mcontext).finish();
        }
    });
}
```

（3）支出管理页面布局界面设计。

双击打开 activity_out_manage.xml 布局界面，其界面外观与新增支出界面类似，请参考新增支出布局界面设计自行完成，其界面背景图片为 outmanagebg.png，界面中控件 id 如表 8.12 所示。

表 8.12　支出管理界面背景图片和界面中控件 id

控件类型	控件含义	id 号取值
EditText	输入修改后的金额	android:id="@+id/et_money_outmag"
EditText	输入修改后的日期	android:id="@+id/et_time_outmag"
Spinner	选择修改后的支出类型	android:id="@+id/sp_type_outmag"
EditText	输入修改后的收款方	android:id="@+id/et_payer_outmag"
EditText	输入修改后的备注	android:id="@+id/et_remark_outmag"
Button	"修改"按钮(注意按钮上的文字)	android:id="@+id/bt_modify_outmag"
Button	"删除"按钮(注意按钮上的文字)	android:id="@+id/bt_delete_outmag"

布局界面代码编辑完成后，便可预览界面的设计效果，如图 8.58 所示。

图 8.58　支出管理界面布局设计效果图

（4）支出管理页面的功能设计。

双击打开 OutManageActivity.java 类文件，编辑代码如下。

```java
public class OutManageActivity extends AppCompatActivity {
    //1.定义对象
    private EditText et_money,et_time,et_payer,et_remark;
    private Spinner sp_type;
    private Button btn_modify,btn_delete;
    private MyDBHelper mhelper;
    private SQLiteDatabase db;
    private OutpayBean outpayBean;
    @Override
    protected void onCreate(Bundle savedInstanceState) {
        super.onCreate(savedInstanceState);
        setContentView(R.layout.activity_out_manage);
        //绑定控件
        initView();
        //获取单击的那条数据并显示出来
        getDataDisplay();
        //"修改"按钮功能的实现
        btnModidfy();
        //"删除"按钮功能的实现
        btnDelete();
    }
    //2.绑定控件
    private void initView() {
        et_money = findViewById(R.id.et_money_outmag);
        et_time = findViewById(R.id.et_time_outmag);
        sp_type = findViewById(R.id.sp_type_outmag);
```

```java
        et_payer = findViewById(R.id.et_payer_outmag);
        et_remark = findViewById(R.id.et_remark_outmag);
        btn_modify = findViewById(R.id.bt_modify_outmag);
        btn_delete = findViewById(R.id.bt_delete_outmag);
        mhelper = new MyDBHelper(OutManageActivity.this);
        db = mhelper.getWritableDatabase();
    }
    //3.获取单击的那条数据并显示出来
    private void getDataDisplay() {
        outpayBean = (OutpayBean) getIntent().getSerializableExtra("sero");
        et_money.setText(outpayBean.getMoney() + "");
        et_time.setText(outpayBean.getTime());
        if (outpayBean.getType().equals("电影-娱乐")){
            sp_type.setSelection(1);
        }else if (outpayBean.getType().equals("美食-畅饮")) {
            sp_type.setSelection(2);
        }else if (outpayBean.getType().equals("欢乐-购物")){
            sp_type.setSelection(3);
        }else if (outpayBean.getType().equals("手机-充值")){
            sp_type.setSelection(4);
        }else if (outpayBean.getType().equals("交通-出行")){
            sp_type.setSelection(5);
        }else if (outpayBean.getType().equals("教育-培训")){
            sp_type.setSelection(6);
        }else if (outpayBean.getType().equals("社交-礼仪")){
            sp_type.setSelection(7);
        } else if(outpayBean.getType().equals("生活-日用")){
            sp_type.setSelection(8);
        }else if(outpayBean.getType().equals("其他")){
            sp_type.setSelection(9);
        }else {
            sp_type.setSelection(0);
        }
        et_payer.setText(outpayBean.getPayee());
        et_remark.setText(outpayBean.getRemark());
    }
    //4."修改"按钮功能的实现
    private void btnModidfy() {
        btn_modify.setOnClickListener(new View.OnClickListener() {
            @Override
            public void onClick(View v) {
                //创建一个对象,封装一行数据
                ContentValues values = new ContentValues();
                values.put("outmoney",et_money.getText().toString());
                values.put("outtime",et_time.getText().toString());
                values.put("outtype",sp_type.getSelectedItem().toString());
                values.put("outpayee",et_payer.getText().toString());
                values.put("outremark",et_remark.getText().toString());
                //把该行数据更新到支出表中
                db.update("pay_out",values,"id = ?",new String[]{outpayBean.getId() + ""});
```

```
                Toast.makeText (OutManageActivity.this,"修改成功",Toast.LENGTH_SHORT ).show();
                //关闭本页面,重新打开支出明细界面,即可查询修改后的结果
                Intent intent = new Intent(OutManageActivity.this, PayDetailActivity.class);
                startActivity(intent); //执行 Intent 操作
                finish(); //退出当前程序,或关闭当前页面
            }
        });
    }
    //5."删除"按钮功能的实现
    private void btnDelete() {
        btn_delete.setOnClickListener(new View.OnClickListener() {
            @Override
            public void onClick(View v) {
                //从数据库中删除条记录即可
                db.delete("pay_out","id = ?",new String[]{outpayBean.getId() + ""});
                Toast.makeText (OutManageActivity.this,"删除成功",Toast.LENGTH_SHORT ).show();
                //关闭本页面,重新打开支出明细界面,即可删除后的结果
                Intent intent = new Intent(OutManageActivity.this, PayDetailActivity.class);
                startActivity(intent); //执行 Intent 操作
                finish(); //退出当前程序,或关闭当前页面
            }
        });
    }
}
```

代码含义同收入修改和删除,这里不再详解,运行效果如图 8.59 和图 8.60 所示。

图 8.59　修改支出

图 8.60　删除支出

项目小结

本项目内容紧紧围绕 Android 当中的 SQLite 数据库存储,通过理财通案例详细而全面地讲解了数据的增、删、改、查操作,Android 中复杂界面设计、RecyclerView 滚动列表、自定义适配器、SharedPreference 存储、SQLite 数据库存储的综合运用。让学生对大型 App 项目有一个全面了解和认识,培养学生大型 App 的设计与开发能力。

习题

完善理财通 App,要求如下。
(1) 实现首次登录从欢迎页面启动,二次登录时直接打开主界面。
(2) 在收入明细界面,增加 TextView 控件,显示收入总金额。
(3) 在支出明细界面,增加 TextView 控件,显示支出总金额。
(4) 完善数据分析功能,本页面采用滚动布局,页面下方添加图表控件,实现按照日期统计金额并绘制图表。

项目 9

我的第一道菜——菜谱App的设计与实现

【教学导航】

学习目标	(1) 学会 ViewPager 控件的使用。 (2) 理解 HTTP 协议。 (3) 理解并掌握 Java 中的输入/输出流操作。 (4) 学会使用 HttpURLConnection 访问网络。 (5) 学会使用第三方网络通信库 OKHttp 访问网络。 (6) 学会如何解析 JSON 格式的数据。
教学方法	任务驱动法、理论实践一体化、探究学习法、分组讨论法
课时建议	8 课时

9.1 ViewPager 控件

在 Android App 中,所有的用户界面元素都是由 View 和 ViewGroup 的对象构成的。在 Android 程序中,View 类是最基本的一个 UI(User Interface,用户界面)类,一个 View 通常占用屏幕上的一个矩形区域,并负责绘图及事件处理。而 ViewGroup 是一个特殊的 View 类,它继承于 android.view.View。一个 ViewGroup 对象是一个 Android.view.ViewGroup 的实例,它的功能就是装载和管理下一层的 View 对象和 ViewGroup 对象。因此 ViewGroup 可以理解成是一个用于存放 View 和 ViewGroup 对象的布局容器。

ViewPager(android.support.v4.view.ViewPager)是 Android 扩展包 v4 包中的类,这个类可以让用户左右切换当前的 View,实现滑动切换的效果。因为 ViewPager 类直接继承了 ViewGroup 类,所以它也是一个容器类,可以在其中添加多个 View 类。通过在 App 中使用 ViewPager 控件,可以做很多事情,从最简单的广告页轮播到页面菜单等。那如何使用它呢?与 ListView 类似,也需要一个适配器,它就是 PagerAdapter。

在实际使用过程中，ViewPager 又经常和 Fragment 一起使用，并且还专门提供了 FragmentPagerAdapter 和 FragmentStatePagerAdapter 类供 Fragment 中的 ViewPager 使用。

9.1.1 使用场景

在 App 设计中，如果涉及左右滑动功能，如下方导航条、上方导航条、图片滑动、翻页功能时，则可以直接使用 ViewPager 进行开发，ViewPager 自带滑动和翻页效果，可以用 ViewPager 自身的 Adapter 进行开发，方便很多。

9.1.2 ViewPager 的基本用法

微课视频

ViewPager 中包含 3 个视图，并且这 3 个视图可以通过左右滑动来进行切换。这里先创建一个 ViewPagerTest 的工程。既然 ViewPager 需要包含 3 个视图，就先准备 3 个页面的布局，每个页面居中显示一个图片。

（1）在主布局文件中加入 ViewPager 控件（activity_main.xml 文件）。

```
<?xml version = "1.0" encoding = "utf - 8"?>
< LinearLayout xmlns:android = "http://schemas.android.com/apk/res/android"
    xmlns:tools = "http://schemas.android.com/tools"
    android:layout_width = "match_parent"
    android:layout_height = "match_parent">
< android.support.v4.view.ViewPager
    android:id = "@ + id/viewpager"
    android:layout_width = "match_parent"
    android:layout_height = "wrap_content"
/>
</ LinearLayout >
```

这里需要注意的是，ViewPager 引入的时候必须写完整 android.support.v4.view. ViewPager，如果按照一般控件那样直接写 ViewPager，程序就会报错，从报错的描述中可以看出是因为无法在 android.view 下面找到 ViewPager 这个类，主要涉及兼容包的问题。

（2）新建三个 Layout，用于滑动切换的视图。

这里分别把它们命名为 page1.xml、page2.xml、page3.xml，并且每个布局中只是简单的居中显示了一张图片，在实际应用中，可以往布局中添加各种控件。

page1.xml 布局文件：

```
<?xml version = "1.0" encoding = "utf - 8"?>
< RelativeLayout xmlns:android = "http://schemas.android.com/apk/res/android"
    android:layout_width = "match_parent"
    android:layout_height = "match_parent" >
    < ImageView
        android:layout_width = "wrap_content"
        android:layout_height = "wrap_content"
        android:layout_centerInParent = "true"
        android:src = "@drawable/view1" />
</RelativeLayout >
```

page2.xml 布局文件:

```xml
<?xml version = "1.0" encoding = "utf-8"?>
<RelativeLayout xmlns:android = "http://schemas.android.com/apk/res/android"
    android:layout_width = "match_parent"
    android:layout_height = "match_parent" >
<ImageView
        android:layout_width = "wrap_content"
        android:layout_height = "wrap_content"
        android:layout_centerInParent = "true"
        android:src = "@drawable/view2" />
</RelativeLayout>
```

page3.xml 布局文件:

```xml
<?xml version = "1.0" encoding = "utf-8"?>
<RelativeLayout xmlns:android = "http://schemas.android.com/apk/res/android"
    android:layout_width = "match_parent"
    android:layout_height = "match_parent" >
<ImageView
        android:layout_width = "wrap_content"
        android:layout_height = "wrap_content"
        android:layout_centerInParent = "true"
        android:src = "@drawable/view3" />
</RelativeLayout>
```

(3) 将分页显示的 View 装入数组。

```
private View view1, view2, view3;                       //ViewPager 包含的 3 个页面
private List<View> viewList; //ViewPager 包含的页面列表,一般给 Adapter 传递的是一个 List 数组
…
LayoutInflater inflater = getLayoutInflater();          //布局渲染器
view1 = inflater.inflate(R.layout.page1, null);         //对应 page1 的布局
view2 = inflater.inflate(R.layout.page2, null);         //对应 page2 的布局
view3 = inflater.inflate(R.layout.page3, null);         //对应 page3 的布局
viewList = new ArrayList<View>();
viewList.add(view1);
viewList.add(view2);
viewList.add(view3);
```

上面的过程就是将资源与变量联系起来布局,最后将实例化的 view1、view2、view3 添加到 viewList 列表中。

(4) 定义适配器来管理要显示的页面。

和 ListView 等控件使用相同,ViewPager 控件也需要设置适配器来完成页面和数据的绑定,而 PagerAdapter 就是用于"将多个页面填充到 ViewPager"的一个基类适配器,大多数情况下,我们更倾向于自定义一个具体的适配器,并让它继承 PagerAdapter。

在继承 PagerAdapter 基类之后,需要在程序中重写四个方法。具体如表 9.1 所示。

表 9.1 继承 PagerAdapter 基类重写的四个方法

函　　数	说　　明
instantiateItem(ViewGroup container, int position)	instantiateItem()做了两件事：①将当前视图添加到 container 中；②返回当前 View
destroyItem(ViewGroup container, int position, Object object)	从当前 container 中删除指定位置(position)的 View
getCount()	返回要滑动的 View 的个数
isViewFromObject(View arg0, Object arg1)	判断显示的是否是同一张图片，这里将两个参数相比较返回即可

其中，getCount( )、isViewFromObject( )方法是在继承之后自动重写的，然而 instantiateItem()、destroyItem()这两个方法则需要在进行操作时手动重写。

接着，新建一个类 MyPagerAdapter，并继承基类适配器 PagerAdapter，并重写相应的四个方法。

```java
public class MyPagerAdapter extends PagerAdapter {
    private List<View> pageList;
    public MyPagerAdapter(List<View> viewList){
        //接收从 Activity 传递过来的页面列表
        this.pageList = viewList;
    }
    @Override
    public Object instantiateItem(ViewGroup container, int position) {
        //将当前视图添加到 container 中
        container.addView(pageList.get(position));
        //返回当前的 View 视图
        return pageList.get(position);
    }
    @Override
    public void destroyItem(ViewGroup container, int position, Object object) {
        //将当前位置的 View 移除
        container.removeView(pageList.get(position));
    }
    @Override
    public int getCount() {
        //返回要展示的图片数量
        return pageList.size();
    }
    @Override
    public boolean isViewFromObject(View arg0, Object arg1) {
        return arg0 == arg1;
    }
}
```

(5) 将 ViewPager 控件与适配器进行绑定。

最后需要完成的工作就是在 Activity 文件中将页面中的 ViewPager 控件和自定义的适配器进行绑定。

```
private ViewPager viewPager;              //定义 ViewPager 对象
private MyPagerAdapter myPagerAdapter;    //定义自定义适配器对象
…
viewPager = findViewById(R.id.viewpager);
myPagerAdapter = new MyPagerAdapter(viewList);
viewPager.setAdapter(myPagerAdapter);     //ViewPager 控件和适配器绑定
```

运行程序,最终实现效果,如图 9.1 所示。

图 9.1　ViewPager 实现效果

微课视频

9.1.3　ViewPager 结合 Fragment 的使用

前面介绍了 ViewPager 的基本使用方法,通过自定义适配器来管理和显示滑动的界面,但在实际使用过程中,项目中的 ViewPager 会和 Fragment 同时出现,每一个 ViewPager 的页面就是一个 Fragment。例如,我们平时使用的微信主界面就是两者结合使用的效果,而且这种使用方法也是 Android 官方极力推荐的做法。

Android 还提供了专门的适配器来让 ViewPager 与 Fragment 一起工作,它们分别为 FragmentPagerAdapter 与 FragmentStatePagerAdapter。对比基类适配器 PagerAdapter,FragmentPagerAdapter 和 FragmentStatePagerAdapter 更专注于每一页 Fragment 的情况,并且这两个子类适配器在使用上也是有区别的。FragmentPagerAdapter 适用于有限个静态 Fragment 页面的管理,而 FragmentStatePagerAdapter 适用于处理大量的页面切换。因为 FragmentStatePagerAdapter 中 Fragment 实例在 destroyItem 的时候被真正释放,所以 FragmentStatePagerAdapter 节省了内存开销。而 FragmentPagerAdapter 中的 Fragment 实例在 destroyItem 的时候并没有真正释放,每一个 Fragment 页面都会一直被保存在 Fragment Manager 中,以便用户可以随时取用,所以 FragmentPagerAdapter 消耗更多的内存,好处是效率相对会更高一些。

首先,创建一个 FragmentTest 的工程。跟前面的做法不同的是,这里 ViewPager 所包含的 3 个视图将通过 Fragment 来实现。

(1) 在主布局文件中加入 ViewPager 控件(activity_main.xml 文件)。

```xml
<?xml version = "1.0" encoding = "utf-8"?>
<LinearLayout xmlns:android = "http://schemas.android.com/apk/res/android"
    xmlns:tools = "http://schemas.android.com/tools"
    android:layout_width = "match_parent"
    android:layout_height = "match_parent">
<android.support.v4.view.ViewPager
    android:id = "@+id/viewpager"
    android:layout_width = "match_parent"
    android:layout_height = "wrap_content"
/>
</LinearLayout>
```

同样需要注意的是,ViewPager 引入的时候必须写完整 android.support.v4.view.ViewPager。

(2) 创建三个 Fragment 类。

① 第一个 Fragment 类,命名为 Fragment1,其所对应的布局文件命名为 layout1.xml。为了让大家更好地理解 Fragment 的用法,这里滑动的 3 个页面内容仍然采用图片居中显示。

```xml
<?xml version = "1.0" encoding = "utf-8"?>
<RelativeLayout xmlns:android = "http://schemas.android.com/apk/res/android"
    android:layout_width = "match_parent"
    android:layout_height = "match_parent" >
    <ImageView
        android:layout_width = "wrap_content"
        android:layout_height = "wrap_content"
        android:layout_centerInParent = "true"
        android:src = "@drawable/view1" />
</RelativeLayout>
```

接着需要创建 Fragment,并加载其所对应的布局文件。要创建一个 Fragment,必须创建一个 Fragment 的子类。Fragment 类的代码看起来很像 Activity,它与 Activity 一样都有回调函数,例如 onCreate()、onstart()、onPause()和 onstop()。一般情况下,在创建 Fragment 时至少需要实现以下几个生命周期方法,具体如表 9.2 所示。

表 9.2 创建 Fragment 时需实现的生命周期方法

函　　数	说　　明
onCreate()	在创建 Fragment 时系统会调用此方法。可实现初始化想要在 Fragment 中保持的那些必要组件,当 Fragment 处于暂停或者停止状态之后可重新启用它们
onCreateView()	第一次为 Fragment 绘制用户界面时系统会调用此方法。并返回所绘出的 Fragment 的根 View。如果 Fragment 没有用户界面可以返回空值
onPause()	用户离开 Fragment 时系统会调用该函数。在当前用户会话结束之前,通常要在这里提交任何应该持久化的变化

创建第一个 Fragment 类,命名为 Fragment1,并继承 Fragment 基类,同时实现 onCreateView()方法,绘制第一个滑动的用户界面。

```java
public class Fragment1 extends Fragment {
    @Override
    public View onCreateView(LayoutInflater inflater, ViewGroup container,
            Bundle savedInstanceState) {
        View view = inflater.inflate(R.layout.layout1, container, false);
        return view;
    }
}
```

② 第二个 Fragment 类,命名为 Fragment2,其所对应的布局文件命名为 layout2.xml。

```xml
<?xml version = "1.0" encoding = "utf-8"?>
<RelativeLayout xmlns:android = "http://schemas.android.com/apk/res/android"
    android:layout_width = "match_parent"
    android:layout_height = "match_parent">
    <ImageView
        android:layout_width = "wrap_content"
        android:layout_height = "wrap_content"
        android:layout_centerInParent = "true"
        android:src = "@drawable/view2" />
</RelativeLayout>
```

创建第二个 Fragment 类,命名为 Fragment2,并继承 Fragment 基类,同时实现 onCreateView()方法,绘制第二个滑动的用户界面。

```java
public class Fragment2 extends Fragment {
    @Override
    public View onCreateView(LayoutInflater inflater, ViewGroup container,
            Bundle savedInstanceState) {
        View view = inflater.inflate(R.layout.layout2, container, false);
        return view;
    }
}
```

③ 第三个 Fragment 类,命名为 Fragment3,其所对应的布局文件命名为 layout3.xml。

```xml
<?xml version = "1.0" encoding = "utf-8"?>
<RelativeLayout xmlns:android = "http://schemas.android.com/apk/res/android"
    android:layout_width = "match_parent"
    android:layout_height = "match_parent">
    <ImageView
        android:layout_width = "wrap_content"
        android:layout_height = "wrap_content"
        android:layout_centerInParent = "true"
        android:src = "@drawable/view3" />
</RelativeLayout>
```

创建第三个 Fragment 类，命名为 Fragment3，并继承 Fragment 基类，同时实现 onCreateView()方法，绘制第三个滑动的用户界面。

```java
public class Fragment3 extends Fragment {
    @Override
    public View onCreateView(LayoutInflater inflater, ViewGroup container,
            Bundle savedInstanceState) {
        View view = inflater.inflate(R.layout.layout3, container, false);
        return view;
    }
}
```

（3）定义 Fragment 适配器。

新建一个类 FragAdapter。和前面不同的是，这次不是继承基类适配器 PagerAdapter，而是继承了 Fragment 专门的适配器 FragmentPagerAdapter，并重写相应的方法。对于 FragmentPagerAdapter 的派生类，只重写 getItem(int)和 getCount()就可以了。

```java
public class FragAdapter extends FragmentPagerAdapter {
    private List<Fragment> mFragments;
    public FragAdapter(FragmentManager fm, List<Fragment> fragments) {
        super(fm);
        mFragments = fragments;
    }
    @Override
    public Fragment getItem(int arg0) {
        return mFragments.get(arg0);
    }
    @Override
    public int getCount() {
        return mFragments.size();
    }
}
```

需要说明的是，在 FragAdapter 类的构造函数中，申请了一个 Fragment 的 List 对象，用于保存滑动的 Fragment 对象，并在构造函数中初始化。在 getItem(int arg0)中，根据传来的参数 arg0，来返回当前要显示的 Fragment，而 getCount()方法返回用于滑动的 Fragment 总数。

（4）将 Fragment 添加到 Activity 页面。

```java
public class MainActivity extends FragmentActivity {
    @Override
    protected void onCreate(Bundle savedInstanceState) {
        super.onCreate(savedInstanceState);
        setContentView(R.layout.activity_main);
        //构造 Fragment 列表
        List<Fragment> fragments = new ArrayList<Fragment>();
        //将前面定义的三个 Fragment 类对应的实例加入 Fragment 列表
```

```
            fragments.add(new Fragment1());
            fragments.add(new Fragment2());
            fragments.add(new Fragment3());
    //构造适配器
            FragAdapter adapter = new FragAdapter(getSupportFragmentManager(), fragments);
    //设定适配器
            ViewPager vp = (ViewPager)findViewById(R.id.viewpager);
            vp.setAdapter(adapter);
        }
    }
```

需要注意的是,这里的 Activity 派生自 FragmentActivity,因为只有 FragmentActivity 才能内嵌 Fragment 页面,普通 Activity 无法实现。另外,在整个程序中,涉及 Fragment 的地方都需要导入 android.support.v4.app.Fragment 包。

9.2 使用 HTTP 访问网络

HTTP(Hyper Text Transfer Protocol)即超文本传输协议,它规定了浏览器和万维网服务器之间相互通信的规则。当客户端与服务器建立连接后,向服务器发送的请求,称为 HTTP 请求。服务器端接收到请求后会做出响应,并返回一些数据给客户端,称为 HTTP 响应。其实一个浏览器的基本工作原理也是如此,当我们在浏览器中打开百度主页,此时就会向百度的服务器发起一条 HTTP 请求,接着服务器分析出我们想要访问的是百度的首页,于是会把该网页的 HTML 代码进行返回,然后客户端再对返回的 HTML 代码进行解析和处理,最终将页面展示出来。这个请求和响应的过程实际上就是 HTTP 通信的过程。

微课视频

9.2.1 使用 HttpURLConnection 访问网络

在 Android 程序开发中,也经常需要与服务器交互数据,过去,Android 上发送 HTTP 请求一般有两种方式:HttpURLConnection 和 HttpClient。不过由于 HttpClient 存在 API 数量过多、扩展困难等缺点,在 Android 6.0 系统中,HttpClient 的功能已经被完全移除了,也标志着此功能被正式弃用,此时就需要用到 HttpURLConnection 对象了。

HttpURLConnection 是 Java 的标准类,继承自 URLConnection,可用于发送 HTTP 请求和获取 HTTP 响应。由于该类是一个抽象类,不能直接实例化对象。因而需要使用 URL 的 openConnection()方法创建具体的实例。

同时考虑到网络操作会遇到不可预期的延迟,从而造成 UI 卡顿的情况,为了避免造成不好的用户体验,从 Android 4.0 以后,在主线程中的 HTTP 请求,运行时都会报 ANRs ("应用没有响应")的错误,因此,在进行网络请求的时候,需要单独开辟一个子线程,然后等到数据返回成功后再刷新 UI。

下面具体介绍 HttpURLConnection 的使用步骤。

(1) 获得 HttpURLConnection 实例。

在使用 HttpURLConnection 访问网络时,首先需要获取到 HttpURLConnection 的实例,一般只需 new 出一个 URL 对象,并传入目标的网络地址,然后调用 openConnection()

方法,可以使用下面的代码。

```
URL url = new URL("https://www.baidu.com");
HttpURLConnection connection = (HttpURLConnection) url.openConnection();
```

(2) 设置请求属性。

通过 openConnection()方法创建的 HttpURLConnection 对象,并没有真正执行连接操作,只是创建了一个新的实例,在正式连接操作前,往往还需要设置一些属性。例如,连接超时的时间和请求方式等。

HttpURLConnection 常用的属性设置如表 9.3 所示。

表 9.3 HttpURLConnection 常用连接属性

函数	说明
setRequestMethod()	设置请求参数,主要有两种方式: GET 请求、POST 请求
setConnectTimeOut()	设置连接超时时间
setReadTimeOut()	设置读取超时时间
setRequestProperty()	设置请求头参数,主要是添加 HTTP 请求 HEAD 中的一些参数
setDoOutput()	设置是否向 HttpURLConnection 输出,对于 POST 请求,参数要放在 http 正文中,因此需要设为 true,默认情况下为 false
setDoInput()	设置是否从 HttpURLConnection 读入,默认情况下为 true

其中,URL 请求通常有两种方式: GET 请求和 POST 请求。GET 请求的数据会附在 URL 之后,通过?来拼接所传的参数,参数之间以 & 相连。请求中发送的参数如果是字母或者数字则按照原样发送,空格则转换为+,中文或者其他字符则按照 base64 位加密,得到类似%E4%BD%A0%E5%A5%BD 的形式,其中,%xx 为该符号以十六进制表示的 ASCII 码。POST 请求的参数不是放在 URL 字符串里面,而是放在 HTTP 请求的正文内,请求的参数被封装起来以流的形式发送给服务端。

两者主要的区别: GET 请求参数直接拼接在 URL 后面,隐私性较差,长度可能会受限制,GET 请求能够被缓存,一般的请求默认为 GET,而 POST 请求通过实体内容传参隐私性相对较好,大小也没有限制,并且 POST 请求不能被缓存下来,需要声明采用 POST 方式。因此,GET 常用于向服务器索取数据的一种请求,而 POST 则是用于向服务器提交数据的一种请求。

(3) 调用 connect()连接远程资源。

```
connection.connect();
```

通过 connect()方法就与服务器建立了 Socket 连接,而连接以后,连接属性就不可以再修改了,但是可以查询服务器返回的头信息(head information)了。需要注意的是,在发起 POST 请求时,查询服务器消息要在写完所有要传输的数据以后,如果先调用了 getResponseCode()或者 getResponseMessage()方法,那么后面是不能再向 outputStream 写消息的,否则程序会报错。HttpURLConnection 在 URLConnection 的基础上也提供了一些方法,用于查询服务器的响应结果,常用的查询函数如表 9.4 所示。

表 9.4 服务器响应函数

函　　数	说　　明
getResponseCode()	获取服务器的响应代码
getResponseMessage()	获取服务器的响应消息

例如,对响应码进行判断:

```
if(connection. getResponseCode()!= 200)
    throw new RuntimeException("请求 url 失败!");
```

(4) 利用 getInputStream()访问资源数据、利用 getOutputStream()传输 POST 消息。

使用 getInputStream()方法只是得到一个流对象,并不是数据,不过可以从流中读出数据,但是需要注意从这个流对象中只能读取一次数据,第二次读取时将会得到空数据。

使用 getOutputStream()方法是用来传输 POST 消息,设置 GET 方式时不会使用该方法,该方法得到的是一个输出流,该输出流中保存的是发送给服务器端的数据。

那么,如何从输入流中读取数据,向输出流中写入参数呢?这里简单介绍下 Java 语言中的输入输出操作。

按照流的方向,可以分为输入流(InputStream)与输出流(OutputStream)。它们两者之间的区别,如表 9.5 所示。

表 9.5　Java 中的输入输出流

名　　称	说　　明
输入流(InputStream)	只能读取数据,不能写入数据
输出流(OutputStream)	只能写入数据,不能读取数据

在实际运用中,很多人往往分不清楚输入流与输出流的使用场合,这里通过一个图形来表述它们的概念,如图 9.2 所示。

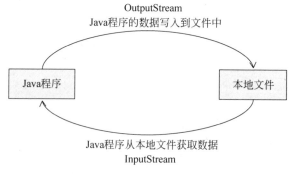

图 9.2　Java 中的输入输出流

从图 9.2 可以看出,输入与输出是一个相对的概念,数据写入文件,对于程序来说是输出流,对文件来说是输入流。但 Java 中的输入输出流是以程序作为中心,所以从程序写入数据到其他位置,则是输出流,将数据读入程序中则是输入流。在编写程序过程中,可以简单地这样理解:读取数据就是输入流,写入数据就是输出流。

按照处理的数据单位分为字节流和字符流。具体如表9.6所示。字节流可以处理所有数据文件，若处理的是纯文本数据，则建议使用字符流。

表9.6 字节流与字符流

名称	说 明
字节流	操作的数据单元是8位的字节。InputStream、OutputStream作为抽象基类
字符流	操作的数据单元是字符。以Writer、Reader作为抽象基类

同时，根据I/O流中数据的来源不同，可以将IO流分为三类数据源，分别如下。
① 基于磁盘文件：FileInputStream、FileOutputSteam、FileReader、FileWriter。
② 基于内存：ByteArrayInputStream、ByteArrayOutputStream。
③ 基于网络：SocketInputStream、SocketOutputStream。

按照流是否直接与特定的地方（例如磁盘、内存、设备等）相连，又可以分为节点流和处理流两类，分别如下。
① 节点流：可以从或向一个特定的地方（节点）读写数据，例如FileReader。
② 处理流：是对一个已存在的流的连接和封装，通过所封装的流的功能调用实现数据读写。例如，BufferedReader处理流的构造方法总是要带一个其他的流对象作参数。

Java中常用的节点流如表9.7所示。

表9.7 Java中常用的节点流

分类	字节输入流	字节输出流	字符输入流	字符输出流
抽象基类	InputStream	OutputStream	Reader	Writer
访问文件	FileInputStream	FileOutputStream	FileReader	FileWriter
访问数组	ByteArrayInputStream	ByteArrayOutputStream	CharArrayReader	CharArrayWriter
访问管道	PipedInputStream	PipedOutputStream	PipedReader	PipedWriter
访问字符串			StringReader	StringWriter

Java中常用的处理流如表9.8所示。

表9.8 Java中常用的处理流

分类	字节输入流	字节输出流	字符输入流	字符输出流
抽象基类	InputStream	OutputStream	Reader	Writer
缓冲流	BufferedInputStream	BufferedOutputStream	BufferedReader	BufferedWriter
转换流			InputStreamReader	OutputStreamWriter
数据流	DataInputStream	DataOutputStream		

处理流的优势在于在程序中通过调用节点流或其他处理流，以便达到更加灵活方便地读写各种类型的数据，所以通常由处理流执行I/O操作。在处理流中，缓冲流的使用最为频繁，主要是因为缓冲流提供了一个缓冲区，能够提高输入/输出的执行效率，减少同节点的频繁操作，同时它还增加了一些新的方法，例如，BufferedReader中的readLine()方法用于读取一行字符串，BufferedWriter中的newLine()方法用于写入一个行分隔符。另外，对于输出的缓冲流BufferedWriter和BufferedOutputStream，写出的数据会先在内存中缓存，然

后使用 flush() 方法将内存中的数据立刻写出。

这里通过使用 BufferedReader 缓冲流为例介绍下,如何从输入流中读取网络数据到字符串中,可以通过以下几个步骤来完成。

① 通过 getInputStream() 方法获取响应流。

② 构建 BufferedReader 对象。

③ 构建 StringBuilder 对象,用来接收 BufferedReader 中的数据。

在 Android 程序中,通过 getInputStream() 获取到输入流后,就可以通过上述的几个步骤来读取输入流中的数据了。

```
//得到响应流
InputStream in = connection.getInputStream();
//构建 BufferedReader 对象
BufferedReader reader = new BufferedReader(new InputStreamReader( in ));
//构建 StringBuilder 对象
StringBuilder sb = new StringBuilder();
String line = null;
while ((line = reader.readLine()) != null) {
    sb.append(line);
}
String reponse = sb.toString();
return reponse;
```

接着,再通过使用 BufferedWriter 缓冲流为例介绍下,如何往输出流中写入数据,可以通过以下几个步骤来完成。

① 通过 getOutputStream() 获取输出流。

② 构建 BufferedWriter 对象。

③ 调用输出流对象的写数据方法,并刷新缓冲区。

在 Android 程序中,通过 getOutputStream() 获取到输出流后,就可以通过上述的几个步骤来往输出流中写入数据了。

```
//获取输出流
OutputStream os = connection.getOutputStream();.
//构建 BufferedWriter 对象
BufferedWriter writer = new BufferedWriter(new OutputStreamWriter(os, "UTF-8");
//调用 write()方法完成写数据
writer.write("这里填写要写入的数据");
//刷新缓冲区
writer.flush();
```

需要注意的是,HTTP 传输的消息要使用 UTF-8 编码,英文字母、数字和部分符号保持不变,空格编码成'+',其他字符编码成"%XY"形式的字节序列,特别是中文字符,不能直接传输,所以在传输前可以考虑调用 URLEncoder.encode(string, "UTF-8")方法。

另外,在使用完节点流或处理流之后,一定要在程序结束前关闭相应的流操作,一般情况下流的关闭顺序是"先打开的后关闭,后打开的先关闭"。

(5) 关闭 HttpURLConnection。

所有的操作全部完成后，就可以调用 disconnect()方法将这个 HTTP 连接关闭掉，可以采用如下的方法。

```
if(connection!= null)
    connection.disconnect();
```

到这个地方是不是就可以顺利地从网络上获取数据了呢？答案是否定的。在开始运行程序前，千万别忘了要声明一下网络权限。修改 AndroidManifest.xml 中的代码，如下：

```
< manifest xmlns:android = "http://schemas.android.com/apk/res/android"
    package = "com.example.httpurlconnectiontest">
    < uses - permission android:name = "android.permission.INTERNET"/>
    ...
</manifest >
```

9.2.2 使用 OkHttp 访问网络

微课视频

当然并不是只能使用 HttpURLConnection，完全没有任何其他选择，事实上在开源盛行的今天，有许多出色的网络通信库都可以替代原生的 HttpURLConnection，而其中 OkHttp 无疑是做得最出色的一个。

OkHttp 是由鼎鼎大名的 Square 公司开发的，这个公司在开源事业上面贡献颇多，除了 OkHttp 之外，还开发了像 Picasso、Retrofit 等著名的开源项目。OkHttp 不仅在接口封装上面做得简单易用，就连在底层实现上也是自成一派，比起原生的 HttpURLConnection，可以说是有过之而无不及，现在已经成为广大 Android 开发者首选的网络通信库。OkHttp 的项目主页地址是 https://github.com/square/okhttp。

OkHttp 有 2.0 版本和 3.0 的版本，目前常用的是 OkHttp 3.0。它支持 Android 5.0 及以上版本的 Android 平台，对于 Java，支持 JDK 1.8 及以上。OkHttp 是一款优秀的 HTTP 框架，它支持 get 请求和 post 请求，支持基于 HTTP 的文件上传和下载，支持加载图片，支持下载文件透明的 GZIP 压缩，支持响应缓存避免重复的网络请求，支持使用连接池来降低响应延迟问题。

在使用 OkHttp 之前，需要先在项目中添加 OkHttp 库的依赖。编辑 app/build.gradle 文件，在 dependencies 闭包中添加如下内容。

```
dependencies {
    implementation fileTree(dir: 'libs', include: ['*.jar'])
    implementation 'com.android.support:appcompat-v7:28.0.0'
    implementation 'com.android.support.constraint:constraint-layout:1.1.3'
    testImplementation 'junit:junit:4.12'
    androidTestImplementation 'com.android.support.test:runner:1.0.2'
    androidTestImplementation 'com.android.support.test.espresso:espresso-core:3.0.2'
    implementation 'com.squareup.okhttp3:okhttp:4.4.0'
}
```

添加上述依赖会自动下载两个库,一个是OkHttp库,一个是Okio库,后者是前者的通信基础。目前,OkHttp的最新版本是4.4.0,可以访问OkHttp的项目主页来查看目前OkHttp的最新版本。

OkHttp的使用非常简单,它的请求/响应API使用构造器模式builders来设计,它支持阻塞式的同步请求和带回调的异步请求。下面来看一下OkHttp的具体用法。

(1)首先需要创建一个OkHttpClient的实例。

```
OkHttpClient client = new OkHttpClient();
```

(2)接着需要创建一个Request对象。

```
Request request = new Request.Builder().build();
```

上述代码只是创建了一个空的Request对象,并没有什么实际作用,在实际编程中,往往需要在build()方法前连缀很多其他方法来丰富这个Request对象。例如,可以通过url()方法来设置目标的网络地址,通过get()、post()方法来设置HTTP请求方式等。

```
Request request = new Request.Builder()
    .url("http://www.baidu.com")
    .get()       //默认就是GET请求,可以省略
    .build();
```

(3)调用OkHttpClient的newCall()方法来创建一个Call对象。

```
Call call = client.newCall(request);
```

(4)发送请求并获取服务器返回的数据。

这里可以根据不同的需求,采用同步请求或异步请求的方式来完成,它们在使用上略有不同,具体如下:

同步请求:执行请求的操作是阻塞式,直到HTTP响应返回。同一时刻只能有一个任务发起,synchronized关键字锁住了整个代码,那么如果当前OkhttpClient已经执行了一个同步任务,如果这个任务没有释放锁,那么新发起的请求将被阻塞,直到当前任务释放锁,它对应OkHttp中的execute()方法。

异步请求:执行的是非阻塞式的请求,它的执行结果一般都是通过接口回调的方式告知调用者。同一时刻可以发起多个请求,因为异步请求每一个都是在一个独立的线程,由两个队列管理,并且synchronized只锁住了代码校验是否执行的部分。它对应OkHttp中的enqueue()方法。

下面通过代码的方式,具体介绍下如何采用同步请求、异步请求的方式来发起HTTP请求并获得响应的数据。

1)同步GET请求

因为同步请求的方式会阻塞调用线程,所以在Android中应放在子线程中执行,否则有可能会引起ANR异常,并且在Android 3.0以后已经不允许在主线程中进行网络操作了。

```java
new Thread(new Runnable() {
    @Override
    public void run() {
        try {
            Response response = call.execute();
            String responseData = response.body().string();
            Log.d(TAG, "run: " + responseData);
        } catch (IOException e) {
            e.printStackTrace();
        }
    }
}).start();
```

通过调用它的 execute() 方法来发送请求并获取服务器返回的数据,其中,response 对象就是服务器返回的数据了。

2) 异步 GET 请求

通过 Call 对象的 enqueue(Callback)方法来提交异步请求,异步发起的请求会被加入到 Dispatcher(异步任务分发器)中的 runningAsyncCalls 双端队列中通过线程池来执行。

```java
call.enqueue(new Callback() {
    @Override
    public void onFailure(Call call, IOException e) {
        Log.d(TAG, "onFailure: ");
    }
    @Override
    public void onResponse(Call call, Response response) throws IOException {
            String responseData = response.body().string();
        Log.d(TAG, "onResponse: " + responseData);
    }
});
```

其中,onResponse 回调的参数是 response,一般情况下:

- 获得返回的字符串,通过 response.body().string()获取。
- 获得返回的二进制字节数组,则调用 response.body().bytes(),通过二进制字节数组,可以转换为 Bitmap 图片资源。
- 获得返回的 inputStream,则调用 response.body().byteStream;。这里支持大文件下载,有 inputStream 可以通过 IO 的方式写文件。

需要注意的是,这个 onResponse 执行的线程并不是 UI 线程,如果希望用服务器返回的数据去操作控件,还是需要使用 handler 操作。

```java
public void onResponse(final Response response) throws IOException
{
        final String res = response.body().string();
        runOnUiThread(new Runnable()
        {
```

```
            public void run()
            {
              TextView 控件对象.setText(res);
            }
        });
    }
```

如果是发起一条 POST 请求会比 GET 请求稍微复杂一点儿,在构造 Request 对象时,需要多构造一个 RequestBody 对象,用它来携带要提交的数据。在构造 RequestBody 时需要指定 MediaType,用于描述请求/响应 body 的内容类型。RequestBody 的几种构造方式如下:

- create(String content,MediaType contentType)　　　　//提交 String 类型数据
- create(byte[] content,MediaType contentType)　　　　//提交 byte[]类型数据
- create(File file,MediaType contentType)　　　　　　　//提交 File 类型数据。

其中,contentType 主要包括以下几种数据格式,如表 9.9 所示。

表 9.9　contentType 常见的数据格式

数据格式	说　　明
multipart/form-data	以表单形式提交,既可以上传键值对,也可以上传文件
application/x-www-from-urlencoded	以键值对的数据格式提交
raw	选择 text,则请求头是:text/plain 选择 javascript,则请求头是:application/javascript 选择 json,则请求头是:application/json 选择 html,则请求头是:text/html 选择 application/xml,则请求头是:application/xml
binary	相当于 Content-Type:application/octet-stream,只可以上传二进制数据,通常用来上传文件,由于没有键值,一次只能上传一个文件

下面通过代码的方式,介绍如何向服务器提交不同类型的数据。

1) POST 方式提交 String

字符串类型的数据在 POST 请求中比较常见,还有一种比较常见的是 JSON 字符串,如果提交的是 JSON 字符串,需要指定 MediaType 的类型为 application/json 格式。

```
//指定 MediaType 的类型
MediaType mediaType = MediaType.parse("text/plain; charset = utf - 8");
//设置提交字符串数据
String requestContent = "Hello World!";
//构建 RequestBody 对象
RequestBody requestBody = RequestBody.create(requestContent, mediaType);
//发起 HTTP 请求
OkHttpClient client = new OkHttpClient();
//构建 Request 对象
Request request = new Request.Builder()
        .url("https://api.github.com/markdown/raw")       //设置服务器地址
        .post(requestBody)                                //需要提交给服务器的数据
```

```
            .build();
Call call = client.newCall(request);
call.enqueue(new Callback() {
@Override
    public void onFailure(Call call, IOException e) {
            Log.d(TAG, "onFailure: " + e.getMessage());
    }
@Override
public void onResponse(Call call,Response response) throws IOException {
Log.d(TAG, response.protocol() + " " + response.code() + " " + response.message());
            Headers headers = response.headers();
            for (int i = 0; i < headers.size(); i++) {
                    Log.d(TAG, headers.name(i) + ":" + headers.value(i));
            }
            Log.d(TAG, "onResponse: " + response.body().string());
            }
    });
```

2) POST 方式提交文件

在某些应用中,有可能需要向服务器提交文件等数据格式,此时,除了要修改 MediaType 类型,还需要开启访问外部数据的相关权限等操作。

```
< uses - permission android:name = "android.permission.READ_EXTERNAL_STORAGE"/>
< uses - permission android:name = "android.permission.WRITE_EXTERNAL_STORAGE"/>
```

另外,从表 9.9 中发现 multipart/form-data 和 application/octet-stream 两种类型都可以用来上传文件,在 HTTP 的请求体中,它们的数据组织方式不一样,具体如下:

(1) multipart/form-data
- 既可以提交普通键值对,也可以提交文件键值对。
- HTTP 规范中的 Content-Type 不包含此类型,只能用在 POST 提交方式下,属于 HTTP 客户端的扩展。
- 通常在浏览器表单中,或者 HTTP 客户端中使用。

(2) application/octet-stream
- 只能提交二进制,而且只能提交一个二进制,如果提交文件的话,只能提交一个文件,后台接收参数只能有一个,而且只能是流或者是字节数组。
- 属于 HTTP 规范中 Content-Type 的一种。
- 很少使用。

因此,在提交文件的时候,Content-Type 应该设为 multipart/form-data 类型。另外,在上传文件时还需要用到 MuiltipartBody,它是 RequestBody 的一个子类,提交表单时需要利用这个类来构建一个 RequestBody 对象。

```
//要上传的文件
File file = new File(Environment.getExternalStorageDirectory() + "/Pictures","测试.mp4");
if (!file.exists()) {
```

```
            Log.d(TAG,"找不到该文件");
            return;
}
MediaType mediaType = MediaType.parse("multipart/form-data");
//通过 new MultipartBody.Builder()创建 requestBody 对象
RequestBody requestBody = new MultipartBody.Builder()
            //设置类型是表单
            .setType(MultipartBody.FORM)
            //添加数据
            .addFormDataPart("file",file.getName(),RequestBody.create(file,mediaType))
            .build();
OkHttpClient client = new OkHttpClient();
//创建 request 对象,并将 requestBody 作为 post()方法的参数传入
Request request = new Request.Builder()
            .url("http://requestbin.net/r/1m9w5g01")
            .post((requestBody))
            .build();
...
//(以下和提交 String 操作相同)
```

3) POST 方式提交表单

提交普通键值对类型的表单时,Content-Type 默认为 application/x-www-form-urlencoded, 在 OkHttp 中可以使用 FormBody.Builder 去提交该种类表单。通过 FormBody 添加多个 String 键值对,最后为 Request 添加 post()方法并传入 FormBody。

```
RequestBody requestBody = new FormBody.Builder()
        .add("username","admin")
        .add("password","123456")
        .build();
```

然后在 Request.Builder 中调用 post()方法,并将 RequestBody 对象传入。

```
Request request = new Request.Builder()
        .url("http://www.baidu.com")
        .post(requestBody)
        .build();
```

接下来的操作就和前面的 POST 请求一样了,调用 enqueue()方法来发送请求并获取服务器返回的数据。

9.2.3 网络访问框架的封装

微课视频

1. HttpURLConnection 的封装

相信读者已经掌握了 HttpURLConnection 和 OkHttp 的用法,知道了如何发起 HTTP 请求,以及如何解析服务器返回的数据,但也许会发现,在实际运用中,之前的写法并不是很好。因为一个应用程序很可能会在许多地方都使用到网络功能,而发送 HTTP 请求的代码

基本都是相同的，如果每次都去编写一遍发送 HTTP 请求的代码，这显然是非常差劲的做法，最好的做法是将 HTTP 请求进行封装。

通常情况下会将这些通用的网络操作提取到一个公共的类里，并提供一个静态方法，当想要发起网络请求的时候，只需简单地调用一下这个方法即可。例如，使用如下的写法：

```java
public class HttpUtil {
    //通过 sendHttpRequest()方法来发送 HTTP 请求，返回数据类型为 String
    public static String sendHttpRequest(String address) {
        HttpURLConnection connection = null;
        try {
            URL url = new URL(address);
            connection = (HttpURLConnection)url.openConnection();
            connection.setRequestMethod("GET");
            connection.setConnectTimeout(8000);
            connection.setReadTimeout(8000);
            connection.setDoInput(true);
            connection.setDoOutput(true);
            InputStream in = connection.getInputStream();
            BufferedReader reader = new BufferedReader(new InputStreamReader(in));
            StringBuilder response = new StringBuilder();
            String line;
            while( (line = reader.readLine()) != null){
                response.append(line);
            }
            return response.toString();
        }catch(Exception e){
            e.printStackTrace();
            return e.getMessage();
        }finally{
            if(connection != null){
                connection.disconnect();
            }
        }
    }
}
```

以后每当需要发起一个 HTTP 请求的时候就可以这样来写：

```java
String address = "http://www.baidu.com";   //服务器地址
String response = HttpUtil.sendHttpRequest(address);
```

在获取到服务器响应的数据后，就可以对它进行解析和处理了。但是需要注意的是，网络请求通常都是属于耗时的操作，而 sendHttpRequest()方法的内部并没有开启子线程，这样就有可能导致在调用 sendHttpRequest()方法的时候使得主线程被阻塞住，从而造成程序崩溃。

可能有人会说，这非常简单，在 sendHttpRequest()方法内部开启一个线程不就解决这个问题了么。其实没有想象中那么容易，因为如果在 sendHttpRequest()方法中开启了一个线程来发起 HTTP 请求，那么服务器响应的数据是无法进行返回的，所有的耗时逻辑都是在子线程里进行的，sendHttpRequest()方法会在服务器还未来得及响应的时候就已经执行结束了,当然也就无法返回响应的数据了。

那么遇到这种情况时应该怎么办呢?其实解决方法并不难,只需要使用 Java 的回调机制就可以了,下面一起来学习下回调机制到底是如何使用的。

首先需要定义一个接口,例如,将它命名成 HttpCallbackListener,代码如下:

```java
public interface HttpCallbackListener {
    void onFinish(String response);
    void onError(Exception e);
}
```

可以看到,在接口中定义了两个方法:onFinish()方法表示当服务器成功响应请求的时候调用,onError()方法表示当进行网络操作出现错误的时候调用。这两个方法都带有参数,onFinish()方法中的参数代表着服务器返回的数据,而 onError()方法中的参数记录着错误的详细信息。

接着修改 HttpUtil 中的代码,如下:

```java
public class HttpUtil {
    public static String sendHttpRequest(final String address , final HttpCallbackListener listener) {
        new Thread(new Runnable() {
        @Override
            public void run() {
                HttpURLConnection connection = null;
                try {
                    URL url = new URL(address);
                    connection = (HttpURLConnection)url.openConnection();
                    connection.setRequestMethod("GET");
                    connection.setConnectTimeout(8000);
                    connection.setReadTimeout(8000);
                    connection.setDoInput(true);
                    connection.setDoOutput(true);
                    InputStream in = connection.getInputStream();
                    BufferedReader reader = new BufferedReader(new InputStreamReader(in));
                    StringBuilder response = new StringBuilder();
                    String line;
                    while( (line = reader.readLine()) != null){
                        response.append(line);
                    }
                    if(listener != null) {
                        //回调 onFinish()方法
                    listener.onFinish(response.toString());
                    }
                }catch(Exception e){
                    if(listener != null) {
                        //回调 onError()方法
                        listener.onError(e);
                    }
                }finally{
```

```
                if(connection != null){
                    connection.disconnect();
                }
            }
        }
    }).start();
}
```

首先给 sendHttpRequest() 方法添加了一个 HttpCallbackListener 参数，并在方法的内部开启一个子线程，然后在子线程中去执行具体的网络操作。注意，子线程中是无法通过 return 语句来返回数据的，因此这里将服务器响应的数据传入了 HttpCallbackListener 的 onFinish() 方法中，如果出现异常就将异常原因传入到 onError() 方法中。

通过改写 sendHttpRequest() 方法后，现在该方法需要接收两个参数了，因此在调用它的时候还需要将 HttpCallbackListener 的实例传入，操作如下：

```
HttpUtil.sendHttpRequest(address, new HttpCallbackListener() {
    @Override
    public void onFinish(String response) {
        //在这里接收并解析从服务器返回的数据
    }
    @Override
    public void onError(Exception e) {
        //在这里对异常情况进行处理
    }
});
```

这样的话，当服务器成功响应的时候，就可以在 onFinish() 方法里对响应数据进行处理了。类似地，如果出现了异常，就可以在 onError() 方法里对异常情况进行处理。如此一来，就巧妙地利用回调机制将响应数据成功返回给调用方了。

2. OkHttp 的封装

通过上面的操作，不难发现使用 HttpURLConnection 的写法总体来说还是比较复杂的，那么使用 OkHttp 会变得简单吗？答案是肯定的，而且要简单得多。

下面在 HttpUtil 中加入一个 sendOkHttpRequest() 方法，具体如下：

微课视频

```
public class HttpUtil{
    …
    public static void sendOkHttpRequest(String address,okhttp3.Callback callback){
        OkHttpClient client = new OkHttpClient();
        Request request = new Request.Builder()
                    .url(address)
                    .build();
        client.newCall(request).enqueue(callback);
    }
}
```

可以看到，这里将 sendOkHttpRequest() 方法中的回调接口修改为 okhttp3.Callback 参数，因为 OkHttp 库中自带了一个回调接口 Callback，类似于前面编写的 HttpCallbackListener，通过调用 enqueue() 方法，其内部不仅已经帮我们开好了子线程，同时还会在子线程中去执行 HTTP 请求，最后将请求的结果回调到 Callback 接口中。

以后每当需要发起一条 HTTP 请求的时候，就可以这样来写：

```
HttpUtil.sendOkHttpRequest("https://www.baidu.com",new okhttp3.Callback(){
    @Override
    public void onResponse(Call call, Response response) throws IOException {
    //得到服务器返回的具体内容
        String responseData = response.body().string();
    }
    @Override
    public void onFailure(Call call , IOException e){
        //在这里对异常情况进行处理
    }
});
```

OkHttp 的接口设计确实非常人性化，它将一些常用的功能进行了很好的封装，使得我们只需编写少量的代码就能完成较为复杂的网络操作。

但是仍然需要注意的是，不管是使用 HttpURLConnection 还是 OkHttp，最终的回调接口都还是在子线程中执行的，因此不可以在这里执行任何的 UI 操作，除非借助 runOnUiThread() 方法来进行线程转换，或者是通过 Handler 来传递消息给主线程。

9.3 解析 JSON 数据格式

通常情况下，每个需要访问网络的应用程序都会有一个自己的服务器，我们可以向服务器提交数据，也可以从服务器上获取数据。不过这个时候就出现了一个问题，这些数据到底要以什么样的格式在网络上传输呢？随便传递一段文本肯定是不行的，因为另一方根本就不会知道这段文本的用途是什么。因此，一般都会在网络上传输一些格式化后的数据，这种数据会有一定的结构规格和语义，当另一方收到数据消息之后就可以按照相同的结构规格进行解析，从而取出他想要的那部分内容。

在网络上传输数据时最常用的格式有两种：XML 和 JSON。相比 XML，JSON 的主要优势在于它的体积更小，在网络上传输的时候可以更省流量。因此，本节重点讲解 JSON 格式数据的解析。

JSON(JavaScript Object Notation)是一种轻量级的数据交换格式，它可以将 JavaScript 对象中表示的一组数据转换为字符串，然后就可以在网络或者程序之间轻松地传递这个字符串，并在需要的时候将它还原为各编程语言所支持的数据格式。例如，在 PHP 中，可以将 JSON 还原为数组或者一个基本对象。在用到 AJAX 时，如果需要用到数组传值，此时就可以用 JSON 将数组转换为字符串来传递。虽然 JSON 数据看起来有点儿古怪，但是 JavaScript 能很容易地解释它，因此也使得 JSON 已经成为理想的数据交换语言。JSON 的建构主要有以下两种结构。

(1)"名称/值"对的集合。不同的语言中,它被理解为对象(object)、记录(record)、结构(struct)、字典(dictionary)、哈希表(hash table)、有键列表(keyed list)或者关联数组(associative array)。

(2)值的有序列表。在大部分语言中,它被理解为数组(array)。

正是因为 JSON 的两种建构结构,在实际使用中 JSON 格式数据主要表现为以下两种形式:对象和数组。

(1)对象。

对象是一个无序的"名称/值"对的集合。一个对象以"{"(左括号)开始,以"}"(右括号)结束。每个"名称"后跟一个":"(冒号),多个"名称/值"对之间使用","(逗号)来分隔。例如,采用 JSON 对象形式来描述一位学生的信息,可以表示为:

```
{"id":1, "name":"zhangsan","sex":"M","age":19}
```

(2)数组。

数组是值(value)的有序集合。一个数组以"["(左中括号)开始,以"]"(右中括号)结束。值之间使用","(逗号)分隔。例如,采用 JSON 数组形式来描述一组学生的信息,可以表示为:

```
[{"id":1, "name":"zhangsan","sex":"M","age":19},
{"id":2, "name":"lisi","sex":"F","age":18},
{"id":3, "name":"wangwu","sex":"M","age":20}]
```

另外,不管是对象还是数组的表示形式,其中值(value)可以是双引号括起来的字符串(string)、数值(number)、true、false、null、对象(object)或者数组(array)。因此,JSON 格式数据也表示成很复杂的形式,例如,利用 JSON 格式来描述中国部分省份信息,可以表示为:

```
{
    "name": "中国",
    "province": [
        {
            "name": "黑龙江",
            "cities": {
                "city": ["哈尔滨", "大庆"]
            }
        },
        {
            "name": "广东",
            "cities": {
                "city": ["广州", "深圳", "珠海"]
            }
        },
        {
            "name": "台湾",
            "cities": {
                "city": ["台北", "高雄"]
```

```
                }
            },
            {
                "name":"新疆",
                "cities": {
                        "city":["乌鲁木齐"]
                        }
            }]
        }
```

JSON 格式取代 XML 给网络传输带来了很大的便利,但是却没有了 XML 的一目了然,尤其是 JSON 数据很长的时候,会使我们陷入烦琐复杂的数据节点查找中。这时可以使用在线工具 BeJson、SoJson 帮助程序员,新接触 JSON 格式的人员更加快速地了解 JSON 的结构,更快地精确定位 JSON 格式错误。

微课视频

9.3.1 使用 JSONObject

了解了 JSON 数据的格式,那么如何对 JSON 格式的数据进行解析呢?其实,解析 JSON 数据也有很多种方法,可以使用官方提供的 JSONObject,也可以使用 Google 的开源库 GSON。另外,一些第三方的开源库如 Jackson、FastJSON 等也非常不错。本节中重点学习前两种解析方式的用法。

先来简单了解下 JSONObject(系统自带的类)类中的方法,其中,用方框标出来的方法就是后期在手动解析 JSON 格式数据时最常用的方法,具体如图 9.3 所示。

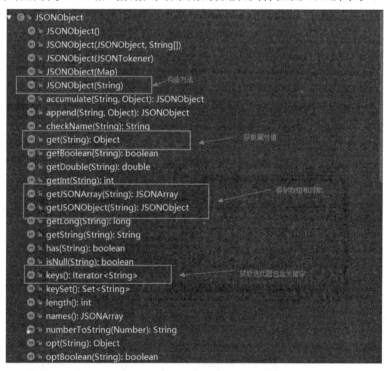

图 9.3 JSONObject 类提供的方法

相关函数介绍如下。

（1）JSONObject getJSONObject(String key)：如果 JSONObjct 对象中的 value 是一个 JSONObject 对象，即根据 key 获取对应的 JSONObject 对象。

（2）JSONArray getJSONArray(String key)：如果 JSONObject 对象中的 value 是一个 JSONObject 数组，即根据 key 获取对应的 JSONObject 数组。

（3）Object get(String key)：根据 key 值获取 JSONObject 对象中对应的 value 值，获取到的值是 Object 类型，需要手动转换为需要的数据类型。

JSONObject 对象既可以很方便地转换成字符串，也可以很方便地把其他对象转换成 JSONObject 对象。在介绍如何通过 JSONObject 解析 JSON 格式的数据之前，先来了解下如何来构造一个 JSONObject 对象。

方式一：通过 JSONObject 对象生成 JSON 数据格式

```java
JSONObject jsonObject = new JSONObject();
jsonObject.put("id", 1);
jsonObject.put("name", "张三");
jsonObject.put("sex","M");
jsonObject.put("age", 19);
jsonObject.put("major", new String[] {"程序设计","人工智能"});
System.out.println(jsonObject.toString());
```

方式二：通过 HashMap 数据结构生成。

```java
HashMap<String,Object> hashMap = new HashMap<>();
hashMap.put("id", 1);
hashMap.put("name", "张三");
hashMap.put("sex","M");
hashMap.put("age", 19);
hashMap.put("major", new String[] {"程序设计","人工智能"});
System.out.println(new JSONObject(hashMap).toString());
```

方式三：通过实体对象生成。

```java
User user = new User();
user.put("id", 1);
user.put("name", "张三");
user.put("sex","M");
user.put("age", 19);
user.put("major", new String[] {"程序设计","人工智能"});
System.out.println(new JSONObject(user));
```

从上述几种构造方法中，可以看出 JSONObject 只是一种数据结构，可以理解为 JSON 格式的数据结构(key-value 结构)，可以使用 put()方法给 JSON 对象添加元素。因此，在解析 JSONObject 对象时，可以采用 get()方法根据 key 的值，来得到相应的 value 值。

假设从服务器上返回的结果是一个 JSON 格式的字符串，内容如下：

```
{"id":1, "name":"zhangsan","sex":"M","age":19}
```

接下来，采用 JSONObject 类的相应方法来进行解析，可以采用如下的代码。

```java
public void jsonToJava(String str) {
    //采用 JSONObject 类解析数据
    JSONObject jsonObject = new JSONObject(str);
    //key 为"id"的字段，其 value 值为整型数,采用 getInt()方法
    int id = jsonObject.getInt("id");
    //key 为"name"的字段，其 value 值为字符串,采用 getString()方法
    String name = jsonObject.getString("name");
    String sex = jsonObject.getString("sex");
    int age = jsonObject.getInt("age");
    System.out.println("id = " + id + "\t" + "name = " + name + "\t" + "sex = " + sex + "\t" + "age = " + age);
}
```

但是，有时候从服务器上获取的数据结构非常复杂，不仅包含 JSON 对象，还包含 JSON 数组等，此时就给解析操作带来一定的难度。假设服务器返回的数据格式如下：

```
{
    "data": {
    "city": "深圳",
    "temphigh": "25",
    "templow": "19",
    "updatetime": "2020 - 03 - 04 13:23:00",
    "tempnow": "24",
    "sendibletemp": "27",
    "winddirect": "东北风",
    "windpower": "2 级",
    "humidity": "42",
    "sunrise": "06:29",
    "sunset": "17:45",
    "weather": "多云",
    "week": "星期三",
    "nl": null,
    "date": "2020 - 03 - 04",
    "index": [
      {
        "name": "化妆指数",
        "level": "控油",
        "msg": "建议用露质面霜打底,水质无油粉底霜,透明粉饼,粉质胭脂."
      },
      {
        "name": "感冒指数",
        "level": "易发",
        "msg": "感冒容易发生,少去人群密集的场所有利于降低感冒的几率."
      },
      {
        "name": "洗车指数",
        "level": "不宜",
```

```
      "msg": "雨(雪)水和泥水会弄脏您的爱车,不适宜清洗车辆."
    },
    {
      "name": "穿衣指数",
      "level": "舒适",
      "msg": "白天温度适中,但早晚凉,易穿脱的便携外套很实用."
    },
    {
      "name": "紫外线强度指数",
      "level": "弱",
      "msg": "辐射较弱,涂擦 SPF12－15、PA＋护肤品."
    },
    {
      "name": "运动指数",
      "level": "不适宜",
      "msg": "受到阵雨天气的影响,不宜在户外运动."
    }
  ],
  "pm25": {
    "aqi": 0,
    "co": 8,
    "o3": 42,
    "pm10": 63,
    "pm2_5": 64,
    "quality": "良",
    "so2": 4,
    "no2": 11,
    "updatetime": "2020－03－04 13:00:00"
  }
},
"status": 0,
"msg": "ok"
}
```

假设目前希望能从返回的结果数据中,获取当前深圳的天气信息等情况。从返回的数据中可以发现,当前的天气信息存放在名称为"data"的区域内,故首先需要将返回的数据传到 JSONObject 类型的 jsonObject 对象中,再通过 getJSONObject()方法获取到 data 域下的数据,可以采用如下操作:

```
//根据返回的字符串创建 JSONObject 对象
JSONObject jsonObject = new JSONObject(str);
JSONObject jsonData = jsonObject.getJSONObject("data");
//然后,再通过 getString()方法进行读值操作
String jsonTemplow = jsonData.getString("templow");
String jsonTempHigh = jsonData.getString("temphigh");
String jsonWeather = jsonData.getString("weather");
String jsonTempnow = jsonData.getString("tempnow");
String jsonWinddirect = jsonData.getString("winddirect");
```

```
String jsonWindpower = jsonData.getString("windpower");
String jsonHumidity = jsonData.getString("humidity");
```

另外,从返回的数据中发现,除了深圳当前的天气信息,还包含深圳当前的各项指数提醒,假设现在希望能够提取这些提醒信息,该如何操作呢? 这些信息全部被存放在名为 "index" 的 JSON 数组中,这就需要用到 JSONArray 类的相关操作了。

首先,将返回的字符串 str 数据传到 JSONObject 类型的 jsonObject 中,然后通过 getJSONObject()方法获取到 data 下的数据,最后通过 getJSONArray()方法获取到 index 下的数据,可以采用如下的代码:

```
//根据返回的字符串创建JSONObject对象
JSONObject jsonObject = new JSONObject(str);
JSONObject jsonData = jsonObject.getJSONObject("data");
JSONArray jsonArray = jsonData.getJSONArray("index");
```

接着再把 jsonArray 中的数据按分类进行解析,存放在不同的 ArrayList < String >数组,分别命名为 names、levels、msgs 等,然后通过 for 循环进行数据的存储。

```
List < String > names = new ArrayList < String >();
List < String > levels = new ArrayList < String >();
List < String > msgs = new ArrayList < String >();
for (int i = 0;i < jsonArray.length();i++){
    JSONObject partIndex = jsonArray.getJSONObject(i);
    String name = partIndex.getString("name");
    names.add(name);
    String level = partIndex.getString("level");
    levels.add(level);
    String msg = partIndex.getString("msg");
    msgs.add(msg);
}
```

通过上述操作,提醒信息就被成功解析了,并且相应的数据分别存放到了 names、levels、msgs 等 3 个数组中,后续可以根据用户的需求来进行处理了。回到 Android 编程中,需要修改 MainActivity 中的代码,操作如下:

```
public class MainActivity extends AppCompatActivity {
    @Override
    protected void onCreate(Bundle savedInstanceState) {
        super.onCreate(savedInstanceState);
        setContentView(R.layout.activity_main);
        new Thread(new Runnable() {
            @Override
            public void run() {
                OkHttpClient client = new OkHttpClient();
                Request request = new Request.Builder()
                        .url("http://v.juhe.cn/weather/index?format = 2&cityname = 苏州 &key = 49cbb077da729143d90374821fdaac84 ")
```

```java
                    .build();
            Call call = client.newCall(request);
            call.enqueue(new Callback() {
                @Override
                public void onFailure(Call call, IOException e) {
                }
                @Override
                public void onResponse(Call call, Response response) throws IOException {
                    String responseData = response.body().string();
                    try {
                        //下面解析的代码也可以放在一个自定义函数中完成
                        JSONObject jsonObject = new JSONObject(responseData);
                        JSONObject jsonData = jsonObject.getJSONObject("result");
                        JSONArray jsonArray = jsonData.getJSONArray("future");
                        for (int i = 0; i < jsonArray.length(); i++) {
                            JSONObject partFuture = jsonArray.getJSONObject(i);
                            String temperature = partFuture.getString("temperature");
                            String weather = partFuture.getString("weather");
                            String wind = partFuture.getString("wind");
                            String week = partFuture.getString("week");
                            String date = partFuture.getString("date");
                            Log.d("MainActivity", "temperature:" + temperature + " " + "weather:" + weather + " " + "wind:" + wind + " " + "week:" + week + " " + "date:" + date);
                        }
                    } catch (JSONException e) {
                    }
                }
            });
        }).start();
    }
}
```

程序中访问的是"聚合数据"网站提供的天气信息接口数据,因此首先需要将 HTTP 请求的地址修改为"http://v.juhe.cn/weather/index?format=2&cityname=城市名&key=您申请的 KEY",然后在得到了服务器返回的数据后,在 onResponse()方法中进行数据的解析。通过定义一个 JSONArray 对象,将接口中返回的未来7天苏州天气信息存入该数组,然后循环遍历 JSONArray,从中取出的每一个元素都是一个 JSONObject 对象,每个 JSONObject 对象中又包含 temperature、weather、wind、week、date 这些数据,接下来只需要调用 getString()方法将这些数据取出并打印在控制台。运行该程序,结果如图 9.4 所示。

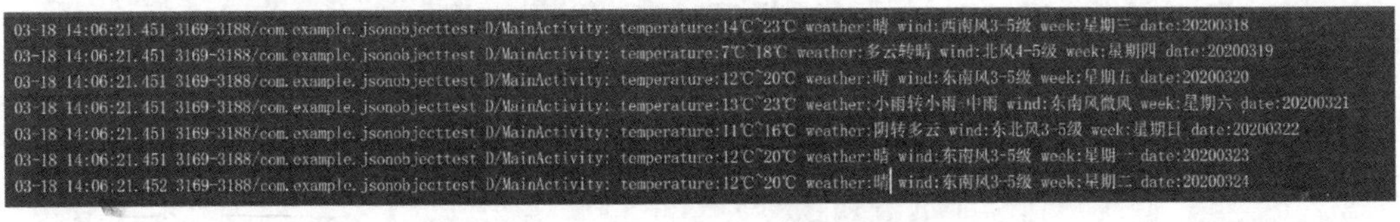

图 9.4 使用 JSONObject 解析天气数据

9.3.2 使用GSON

如果认为使用JSONObject来解析JSON数据已经非常简单了,那就太容易满足了。Google提供的GSON开源库可以让解析JSON数据的工作简单到让你不敢想象的地步。

不过GSON并没有被添加到Android官方的API中,因此如果想要使用这个功能的话,就必须要在项目中添加GSON库的依赖。编辑app/build.gradle文件,在dependencies闭包中添加如下内容。

```
dependencies {
    implementation fileTree(dir: 'libs', include: ['*.jar'])
    implementation 'com.android.support:appcompat-v7:28.0.0'
    implementation 'com.android.support.constraint:constraint-layout:1.1.3'
    testImplementation 'junit:junit:4.12'
    androidTestImplementation 'com.android.support.test:runner:1.0.2'
    androidTestImplementation 'com.android.support.test.espresso:espresso-core:3.0.2'
    implementation 'com.google.code.gson:gson:2.8.6'
}
```

GSON的项目主页地址是https://github.com/google/gson,目前最新的GSON开源库的版本是2.8.6,在实际运用时可以访问GSON的项目主页来查看当前最新的版本。

那么GSON库究竟神奇在哪里呢? GSON提供了fromJson()和toJson()两个直接用于解析和生成JSON数据的方法,前者实现反序列化,后者实现了序列化,同时每个方法都提供了重载方法。

- Serialization:序列化,将Java对象转换成JSON字符串。
- Deserialization:反序列化,JSON字符串转换成Java对象。

通过使用GSON开源库,可以将一段JSON格式的字符串自动映射成一个Java对象,从而不需要再手动去编写代码来进行解析了。例如,有一段JSON格式的数据如下。

```
{"name":"张三","age":20}
```

这时可以定义一个Person类,并加入name和age这两个字段,然后只需要简单地调用如下代码就可以将JSON字符串自动解析成一个Person对象。

(1) Person类的定义。

```
public class Person {
    public String name;
    public int age;
    //构造方法
    Person(String name, int age){
        this.name = name;
        this.age = age; }
}
```

（2）解析 JSON 字符串。

```
Gson gson = new Gson();                                    //创建 GSON 对象
String jsonString = "{\"name\":\"张三\",\"age\":20}";      //双引号需进行转义
Person person = gson.fromJson(jsonString, Person.class);
                    //通过 fromJson()方法,将 jsonString 字符串转换成 Person 类的对象
```

（3）生成 JSON 字符串。

同样,也可以将一个 Person 类的对象,进行序列化,生成一个 JSON 格式的字符串,以便于在网络上进行传输。

```
Gson gson = new Gson();
Person person = new Person("张三",20);
String jsonString = gson.toJson(person); //得到{"name":"张三","age":20}
```

如果需要解析的是一段 JSON 数组,解析的过程会稍微麻烦一点儿,需要借助 TypeToken 将期望解析成的数据类型传入到 fromJson()方法中,如下。

```
List<Person> people = gson.fromJson(jsonString,new TypeToken<List<Person>>(){}.getType());
```

为什么要在 TypeToken 类的构造方法后加上一个{}呢？其实这里 new 生成的是一个匿名类的对象,该匿名类继承自 TypeToken 类,这里的{}代表的是匿名类的内部,这里什么都没有写,因为只需要用到父类的一个 public 方法而已。

匿名内部类常用在监听里面,前面学习过给按钮设置监听事件,在 setOnClickListener()方法中需要接收一个 OnClickListener 类型的对象,而 OnClickListener 是一个接口,不能直接采用 new 生成对象,因此需要自定义一个类实现 OnClickListener 接口,并实现该接口的抽象方法,然后再将自定义类的对象作为参数传递给 setOnClickListener()方法,但那样太复杂,更多时候直接在参数中自定义实现类,并创建自定义类的对象。

```
button = findViewById(R.id.btn);
button.setOnClickListener(new View.OnClickListener() {
            @Override
            public void onClick(View v) {
            }
});
```

GSON 开源库基本的用法就是这样,非常简单！下面采用 GSON 开源库来对"聚合数据"网站获取天气信息的程序进行修改。先来看下该接口返回的天气信息格式,如下。

```
{
    "resultcode":"200",
    "reason":"successed!",
    "result":{
        "sk":{
            "temp":"18",
            "wind_direction":"北风",
            "wind_strength":"5 级",
```

```json
            "humidity":"53%",
            "time":"09:40"
        },
        "today":{
            "temperature":"6℃～18℃",
            "weather":"阴转晴",
            "weather_id":{
                "fa":"02",
                "fb":"00"
            },
            "wind":"北风3-5级",
            "week":"星期四",
            "city":"苏州",
            "date_y":"2020年03月19日",
            "dressing_index":"较舒适",
            "dressing_advice":"建议着薄外套、开衫、牛仔衫裤等服装.年老体弱者应适当添加衣物,宜着夹克衫、薄毛衣等.",
            "uv_index":"弱",
            "comfort_index":"",
            "wash_index":"较不宜",
            "travel_index":"较适宜",
            "exercise_index":"较适宜",
            "drying_index":""
        },
        "future":[
            {
                "temperature":"6℃～18℃",
                "weather":"阴转晴",
                "weather_id":{
                    "fa":"02",
                    "fb":"00"
                },
                "wind":"北风3-5级",
                "week":"星期四",
                "date":"20200319"
            },
            {
                "temperature":"11℃～20℃",
                "weather":"阴转多云",
                "weather_id":{
                    "fa":"02",
                    "fb":"01"
                },
                "wind":"东南风3-5级",
                "week":"星期五",
                "date":"20200320"
            },
            {
                "temperature":"14℃～25℃",
```

```
        "weather":"阴转中雨",
        "weather_id":{
            "fa":"02",
            "fb":"08"
        },
        "wind":"南风微风",
        "week":"星期六",
        "date":"20200321"
    },
    {
        "temperature":"9℃ ～20℃ ",
        "weather":"阴",
        "weather_id":{
            "fa":"02",
            "fb":"02"
        },
        "wind":"东北风微风",
        "week":"星期日",
        "date":"20200322"
    },
    {
        "temperature":"9℃ ～18℃ ",
        "weather":"阴转小雨",
        "weather_id":{
            "fa":"02",
            "fb":"07"
        },
        "wind":"东南风微风",
        "week":"星期一",
        "date":"20200323"
    },
    {
        "temperature":"14℃ ～25℃ ",
        "weather":"阴转中雨",
        "weather_id":{
            "fa":"02",
            "fb":"08"
        },
        "wind":"南风微风",
        "week":"星期二",
        "date":"20200324"
    },
    {
        "temperature":"9℃ ～20℃ ",
        "weather":"阴",
        "weather_id":{
            "fa":"02",
            "fb":"02"
        },
```

```
                "wind":"东北风微风",
                "week":"星期三",
                "date":"20200325"
            }
        ]
    },
    "error_code":0
}
```

从接口返回的数据格式可以发现，苏州未来 7 天的天气信息情况存放在"future"的区域内，而且数据格式是一个 JSON 数组。因此，为获取此部分的返回数据，首先需要定义一个 ResponseClass 类，用于将返回的 JSON 字符串转换成一个 Java 类的对象，对应的 ResponseClass 类定义如下。

（1）ResponseClass 类：

```
public class ResponseClass {
    private String resultcode;
    private String reason;
    private ResultClass result;
    private int error_code;
    public ResultClass getResult() {
        return result;
    }
    public void setResult(ResultClass result) {
        this.result = result;
    }
}
```

在 ResponseClass 类的定义中，暂时只设置了"result"区域数据的 getter()和 setter()方法，同时将该部分数据以 ResultClass 类的形式返回。

（2）ResultClass 类：

```
public class ResultClass {
    private SkClass sk;
    private TodayClass today;
    private List<MyWeather> future;
    public void setFuture(List<MyWeather> future) {
        this.future = future;
    }
    public List<MyWeather> getFuture() {
        return future;
    }
}
```

在 ResultClass 类的定义中，将"sk"区域返回的数据定义成 SkClass 类，将"today"区域返回的数据定义成 TodayClass 类，将"future"区域返回的数据定义成 List 数组，数组中每个元素的结构又被定义成 MyWeather 类，这里暂时只考虑解析苏州未来 7 天的天气信息，故可以先把 SkClass 类、TodayClass 类定义成一个空类。

（3）MyWeather 类：

```java
public class MyWeather {
    private String temperature;
    private String weather;
    private String wind;
    private String week;
    private String date;
    public void setTemperature(String temperature) {
        this.temperature = temperature;
    }
    public void setWeather(String weather) {
        this.weather = weather;
    }
    public void setWind(String wind) {
        this.wind = wind;
    }
    public void setWeek(String week) {
        this.week = week;
    }
    public void setDate(String date) {
        this.date = date;
    }
    public String getTemperature() {
        return temperature;
    }
    public String getWeather() {
        return weather;
    }
    public String getWind() {
        return wind;
    }
    public String getWeek() {
        return week;
    }
    public String getDate() {
        return date;
    }
}
```

MyWeather 类被定义成存放未来每天苏州的天气信息类，里面包含 temperature、weather、wind、week、date 等数据，同时为每个属性设置了相应的 getter() 和 setter() 方法。

（4）修改 MainActivity.java 文件：

```java
public class MainActivity extends AppCompatActivity {
    @Override
    protected void onCreate(Bundle savedInstanceState) {
        super.onCreate(savedInstanceState);
        setContentView(R.layout.activity_main);
```

```java
new Thread(new Runnable() {
    @Override
    public void run() {
        OkHttpClient client = new OkHttpClient();
        Request request = new Request.Builder()
                .url("http://v.juhe.cn/weather/index?format=2&cityname=苏州&key=ccfadb3491e63c1c3556f3ceb19e1237")
                .build();
        Call call = client.newCall(request);
        call.enqueue(new Callback() {
            @Override
            public void onFailure(Call call, IOException e) {
            }
            @Override
            public void onResponse(Call call, Response response) throws IOException {
                String responseData = response.body().string();
                Gson gson = new Gson();
                ResponseClass responseClass = gson.fromJson(responseData, new TypeToken<ResponseClass>(){}.getType());
                ResultClass resultClass = responseClass.getResult();
                List<MyWeather> weatherList = resultClass.getFuture();
                for (int i = 0; i < weatherList.size(); i++)
                {
                    MyWeather myWeather = weatherList.get(i);
                    Log.d("MainActivity", "temperature:" + myWeather.getTemperature() + " " + "weather:" + myWeather.getWeather() + " " + "wind:" + myWeather.getWind() + " " + "week:" + myWeather.getWeek() + " " + "date:" + myWeather.getDate());
                }
            }
        });
    }
}).start();
```

通过 gson.fromJson() 方法,将服务器返回的 JSON 字符串转换成 ResponseClass 类的对象,然后再调用 getter() 方法得到 resultClass 对象,最终得到 weatherList 数组对象。接着采用 for 循环的方式,将苏州未来 7 天内每天的天气信息显示到控制台。运行上述程序后,得到如图 9.5 所示运行结果。

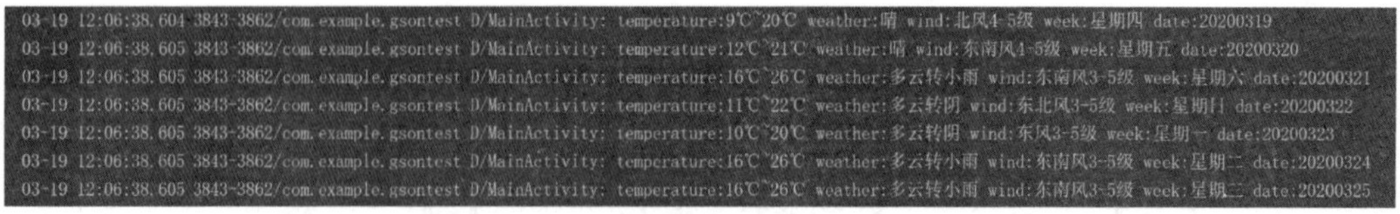

图 9.5 使用 GSON 解析天气数据

在采用 GSON 开源库解析 JSON 数据时,需要注意以下几点。

(1) 在碰到复杂的 JSON 数据时,如何定义待解析的类? 其实很简单,看到 JSON 结构里面有{}就定义一个类,看到[]就定义一个List,最后只剩下最简单的如String、int等基本类型的操作了。

(2) 内部嵌套的类,在定义时请使用public static class className{}框架。

(3) 类内部的属性名,必须与JSON串里面的Key的名称保持一致。

9.4 实战案例——菜谱App的设计与实现

本环节将综合运用前面所学理论知识,进行实际案例开发。项目中采用ViewPager控件制作菜谱App的引导界面,同时利用OkHttp开源库和GSON解析库从第三方网络API接口获取菜谱数据,并使用RecyclerView控件将菜谱分类和菜谱列表进行完美展现。另外,考虑到用户的使用习惯,项目中还将介绍如何使用SQLite数据库进行菜谱数据的收藏等操作。

9.4.1 菜谱App引导界面设计

微课视频

【任务描述】

实现三个子页面的滑屏界面,同时在页面底部添加一个导航栏,分别设置为"主页""收藏""个人中心"三个栏目。

【设计思路】

滑屏页面由两个部分组成:用来装载碎片的ViewPager和底部导航栏。底部导航栏由RadioGroup和RadioButton组成,可以实现导航栏中的按钮同时只能有一个处于按下状态。

【任务实施】

1. 创建Android工程

在Android Studio集成开发环境中创建一个空界面的工程,工程名为Cookbook。

2. 设计布局界面

双击layout文件夹下的activity_main.xml文件,便可打开布局编辑器,将根布局修改为LinearLayout,修改布局方向为vertical,并删除默认生成的TextView控件。

```xml
<?xml version = "1.0" encoding = "utf-8"?>
<LinearLayout xmlns:android = "http://schemas.android.com/apk/res/android"
    xmlns:tools = "http://schemas.android.com/tools"
    android:layout_width = "match_parent"
    android:layout_height = "match_parent"
    android:orientation = "vertical"
    tools:context = ".MainActivity">
</LinearLayout>
```

在根布局中依次添加以下所需控件:①ViewPager控件,并设置其权重为1;②View控件,用来分隔的空白;③横向排列的线性布局,用于放置底部导航栏,底部导航栏的RadioGroup和RadioButton控件。

```xml
<android.support.v4.view.ViewPager
    android:id="@+id/view_pager"
    android:layout_width="match_parent"
    android:layout_height="wrap_content"
    android:layout_weight="1"/>
<View
    android:layout_width="match_parent"
    android:layout_height="10dp"
    android:background="#000000"/>
<LinearLayout
    android:layout_width="match_parent"
    android:layout_height="60dp"
    android:orientation="horizontal"
    android:background="#e5a">
    <RadioGroup
        android:id="@+id/radio_group"
        android:layout_width="match_parent"
        android:layout_height="wrap_content"
        android:orientation="horizontal"
        android:weightSum="3">
        <RadioButton
            android:id="@+id/radio_one"
            android:layout_width="wrap_content"
            android:layout_height="wrap_content"
            android:button="@null"
            android:drawableTop="@drawable/a"
            android:layout_weight="1"/>
        <RadioButton
            android:id="@+id/radio_two"
            android:layout_width="wrap_content"
            android:layout_height="wrap_content"
            android:button="@null"
            android:drawableTop="@drawable/b"
            android:layout_weight="1"/>
        <RadioButton
            android:id="@+id/radio_three"
            android:layout_width="wrap_content"
            android:layout_height="wrap_content"
            android:button="@null"
            android:drawableTop="@drawable/c"
            android:layout_weight="1"/>
    </RadioGroup>
</LinearLayout>
```

用于导航栏图标的 drawable 资源需要设置"按下"和"未按下"两种状态,在 res/drawable 文件夹中放入图片资源 homepage.png、homepage2.png、collection.png、collection2.png、personalcenter.png、personalcenter2.png。新建 a.xml,使用 android:state_checked 选择器分别指定 homepage 和 homepage2 作为两种状态的资源图片。

```xml
<?xml version = "1.0" encoding = "utf-8"?>
<selector xmlns:android = "http://schemas.android.com/apk/res/android">
<item android:drawable = "@drawable/homepage" android:state_checked = "false"/>
<item android:drawable = "@drawable/homepage2" android:state_checked = "true"/>
</selector>
```

新建 b.xml,使用 android:state_checked 选择器分别指定 collection 和 collection2 作为两种状态的资源图片。

```xml
<?xml version = "1.0" encoding = "utf-8"?>
<selector xmlns:android = "http://schemas.android.com/apk/res/android">
<item android:drawable = "@drawable/collection" android:state_checked = "false"/>
<item android:drawable = "@drawable/collection2" android:state_checked = "true"/>
</selector>
```

新建 c.xml,使用 android:state_checked 选择器分别指定 personalcenter 和 personalcenter2 作为两种状态的资源图片。

```xml
<?xml version = "1.0" encoding = "utf-8"?>
<selector xmlns:android = "http://schemas.android.com/apk/res/android">
<item android:drawable = "@drawable/personalcenter" android:state_checked = "false"/>
<item android:drawable = "@drawable/personalcenter2" android:state_checked = "true"/>
</selector>
```

布局代码编写成功后,可以看到界面的预览效果,如图 9.6 所示。

图 9.6 导航栏效果图

3. 逻辑代码的实现

(1) 在 MainActivity 中,创建相应的控件。

```
private ViewPager view_pager;        //创建 ViewPager 控件
private RadioButton radio_one;       //创建"主页"按钮
```

```
private RadioButton radio_two;        //创建"收藏"按钮
private RadioButton radio_three;      //创建"个人中心"按钮
private RadioGroup radio_group;       //创建单选按钮组
```

(2) 创建 initView()方法,并在该方法中绑定控件。

```
private void initView() {
    view_pager = (ViewPager) findViewById(R.id.view_pager);
    radio_one = (RadioButton) findViewById(R.id.radio_one);
    radio_two = (RadioButton) findViewById(R.id.radio_two);
    radio_three = (RadioButton) findViewById(R.id.radio_three);
    radio_group = (RadioGroup) findViewById(R.id.radio_group);
}
```

(3) 创建 OneFragment、TwoFragment、ThreeFragment 三个碎片类,分别代表第一、第二、第三个滑屏界面。

```
public class OneFragment extends Fragment {
    @Override
    public View onCreateView ( LayoutInflater inflater, ViewGroup container, Bundle savedInstanceState) {
        TextView textView = new TextView(container.getContext());
        textView.setText("第一个页面---占位");
        return textView;
    }
}
public class TwoFragment extends Fragment {
    @Override
    public View onCreateView ( LayoutInflater inflater, ViewGroup container, Bundle savedInstanceState) {
        TextView textView = new TextView(container.getContext());
        textView.setText("第二个页面---占位");
        return textView;
    }
}
public class ThreeFragment extends Fragment {
    @Override
    public View onCreateView ( LayoutInflater inflater, ViewGroup container, Bundle savedInstanceState) {
        TextView textView = new TextView(container.getContext());
        textView.setText("第三个页面---占位");
        return textView;
    }
}
```

(4) 在 MainActivity 类中添加实例变量 onefragment、twofragment、threefragment 用于存放碎片,fragarrlist 用于存放碎片列表,position 用于记录当前所在的页面,默认值为 0 即选中第一个页面。

```
private Fragment onefragment,twofragment,threefragment;
private List<Fragment> fragarrlist = new ArrayList<>();
private int positon = 0;
```

(5)创建 MyFragmentAdapter 类,用于指定 ViewPager 的适配器。

```java
public class MyFragmentAdapter extends FragmentPagerAdapter {
    private List<Fragment> fragmentList2;
    public MyFragmentAdapter(FragmentManager fm ,List<Fragment> fragmentList2) {
        super(fm);
        this.fragmentList2 = fragmentList2;
    }
    //根据位置返回当前的碎片
    @Override
    public Fragment getItem(int i) {
        return fragmentList2.get(i);
    }
    //碎片的总数
    @Override
    public int getCount() {
        return fragmentList2.size();
    }
}
```

(6)在 MainActivity 类中创建 initData()方法,并在该方法中创建各个碎片类的实例并通过自定义的适配器添加到 ViewPager 控件。

```java
private void initData() {
//将碎片实例化,存放到动态数组
        Fragment onefragment =  new OneFragment();
        Fragment twofragment =  new TwoFragment();
        Fragment threefragment =  new ThreeFragment();
        fragarrlist.add(onefragment);
        fragarrlist.add(twofragment);
        fragarrlist.add(threefragment);
        //创建自定义适配器的实例
        MyFragmentAdapter adapter = new MyFragmentAdapter(this.getSupportFragmentManager(), fragarrlist);
            //为 ViewPager 控件绑定适配器
        view_pager.setAdapter(adapter);
        ((RadioButton)radio_group.getChildAt(positon)).setChecked(true);
}
```

(7)在 MainActivity 类中创建 initMove()方法,用于 ViewPager 滑屏时底部导航栏选中对应的按钮。

```java
private void initMove() {
   view_pager.addOnPageChangeListener(new ViewPager.OnPageChangeListener() {
```

```java
    @Override
     public void onPageScrolled(int i, float v, int i1) {
    }
    @Override
    public void onPageSelected(int i) {
        ((RadioButton)radio_group.getChildAt(i)).setChecked(true);
    }
    @Override
    public void onPageScrollStateChanged(int i) {
    }
});
}
```

(8) 在 MainActivity 类中创建 initOnClick() 方法,用于底部导航栏被单击时 ViewPager 切换到对应的页面。

```java
private void initOnclick() {
    radio_group.setOnCheckedChangeListener(new RadioGroup.OnCheckedChangeListener() {
        @Override
        public void onCheckedChanged(RadioGroup group, int checkedId) {
            switch (checkedId){
                case R.id.radio_one:
                    positon = 0;
                    view_pager.setCurrentItem(positon);
                    break;
                case R.id.radio_two:
                    positon = 1;
                    view_pager.setCurrentItem(positon);
                    break;
                case R.id.radio_three:
                    positon = 2;
                    view_pager.setCurrentItem(positon);
                    break;
                default:
                    positon = 0;
                    view_pager.setCurrentItem(positon);
                    break;
            }
        }
    });
}
```

(9) 在 onCreate() 方法中调用 initView()、initData()、initMove()、initOnClick() 四个方法。

```java
protected void onCreate(Bundle savedInstanceState) {
    super.onCreate(savedInstanceState);
    setContentView(R.layout.activity_main);
    initView();
```

```
    initData();        //将三张碎片添加到 viwePager 容器里面
    initMove();        //滑屏时按钮随着变化
    initOnclick();     //单击按钮,碎片跟着变化
}
```

4. 程序运行与功能测试

运行程序,能看到底部导航栏和三个页面的占位文字,并且可以滑动页面或单击导航栏按钮切换页面。

9.4.2 菜谱 App 主界面设计

微课视频

【任务描述】

实现从网络 HTTP 接口获取菜谱数据,并完成菜谱 App 主界面的布局设计。

【设计思路】

数据层面,可以使用 showapi 的菜谱接口。HTTP 请求层面,可以使用 OkHttp 开源库。数据解析层面,可以使用 GSON 解析库。数据展示层面,可以使用垂直的 RecyclerView 控件展示数据,使用水平的 RecyclerView 控件展示菜谱类型标签,数据里涉及的图片可以使用 Glide 库进行展示。

【任务实施】

1. 注册 API 账号

本书完成的菜谱 App 以"万维易源"网站提供的 API 数据接口为例,该数据接口针对注册的会员免费。因此,在使用该 API 接口前,需到该网站完成注册 API 账号,注册完成后,前往"个人中心"创建新的应用,以此获得相应的 appid 和调用密钥。然后通过访问 https://www.showapi.com/apiGateway/view/? apiCode=1164&pointCode=1 获取"菜谱大全"接口的相应数据。"菜谱大全"接口调用格式,如图 9.7 所示。

图 9.7 "菜谱大全"接口调用格式

2. 导入依赖

在前面创建的 Cookbook 工程的 app/build.gradle 文件下添加 RecyclerView、OkHttp、Glide 和 GSON 的依赖。

```
implementation 'com.android.support:recyclerview-v7:28.0.0'
implementation 'com.squareup.okhttp3:okhttp:4.4.0'
implementation 'com.github.bumptech.glide:glide:4.4.0'
implementation 'com.google.code.gson:gson:2.8.6'
```

3. 布局界面设计

（1）菜谱主界面布局设计。

在 app 目录结构下找到 res/layout，右击 layout 文件夹，选择 New→Layout resource file，输入新建的布局文件名称"fragment_one"，并修改 Root element 类型为 LinearLayout，单击 OK 按钮，然后在线性布局中添加两个 RecyclerView 控件。

```xml
<?xml version="1.0" encoding="utf-8"?>
<LinearLayout xmlns:android="http://schemas.android.com/apk/res/android"
    android:orientation="vertical"
    android:layout_width="match_parent"
    android:layout_height="match_parent">
    <android.support.v7.widget.RecyclerView
        android:id="@+id/recy_view_tags"
        android:layout_width="match_parent"
        android:layout_height="50dp"
        />
    <android.support.v7.widget.RecyclerView
        android:id="@+id/recy_view1"
        android:layout_width="match_parent"
        android:layout_height="wrap_content"/>
</LinearLayout>
```

（2）菜谱子界面布局设计。

在 app 目录结构下的 res/layout 文件夹，新建 recycleritem.xml，并修改 Root element 类型为 LinearLayout，然后在线性布局中添加 ImageView 控件和两个 TextView 控件。

```xml
<?xml version="1.0" encoding="utf-8"?>
<LinearLayout xmlns:android="http://schemas.android.com/apk/res/android"
    android:orientation="horizontal"
    android:layout_width="match_parent"
    android:layout_height="wrap_content">
    <ImageView
        android:id="@+id/ke_img"
        android:layout_width="80dp"
        android:layout_height="80dp"/>
    <LinearLayout
        android:layout_width="match_parent"
```

```xml
            android:layout_height = "wrap_content"
            android:orientation = "vertical"
            android:layout_gravity = "center">
        <TextView
            android:id = "@+id/ke_name"
            android:layout_width = "match_parent"
            android:layout_height = "wrap_content"
            android:text = "菜谱名称"
            android:textSize = "20sp"
            android:textColor = "#000000"
            />
        <TextView
            android:id = "@+id/ke_des"
            android:layout_width = "wrap_content"
            android:layout_height = "wrap_content"
            android:text = "菜谱描述"/>
    </LinearLayout>
</LinearLayout>
```

(3) 菜谱类型布局设计。

在 app 目录结构下找到 res/drawable,右击 drawable 文件夹,选择 New→Drawable resource file,输入新建的资源文件名称"my_round_rectangle1",单击 OK 按钮,进入文件后将 selector 修改为 shape 属性。按照同样的操作,再创建资源文件 my_round_rectangle2,这两个文件分别用于菜谱类型未选中和被选中的图片资源。其中,my_round_rectangle1.xml 资源文件代码如下:

```xml
<?xml version = "1.0" encoding = "utf-8"?>
<shape xmlns:android = "http://schemas.android.com/apk/res/android">
    <solid android:color = "#cccccc" />
    <stroke
        android:width = "1dp"
        android:color = "#cccccc" />
    <corners
        android:radius = "6dp"
        />
</shape>
```

my_round_rectangle2.xml 资源文件代码如下:

```xml
<?xml version = "1.0" encoding = "utf-8"?>
<shape xmlns:android = "http://schemas.android.com/apk/res/android">
    <solid android:color = "#00BCD4" />
    <stroke
        android:width = "1dp"
        android:color = "#cccccc" />
    <corners
        android:radius = "6dp"
```

```
        />
</shape>
```

在 app 目录结构下的 res/layout 文件夹,新建 layouttagtext.xml,并修改 Root element 类型为 LinearLayout,然后在线性布局中添加一个 TextView 控件。

```xml
<?xml version = "1.0" encoding = "utf-8"?>
<LinearLayout xmlns:android = "http://schemas.android.com/apk/res/android"
    android:orientation = "vertical"
    android:layout_width = "wrap_content"
    android:layout_height = "match_parent">
    <TextView
        android:layout_width = "wrap_content"
        android:layout_height = "match_parent"
        android:gravity = "center"
        android:background = "@drawable/my_round_rectangle1"
        android:paddingLeft = "10dp"
        android:paddingRight = "10dp"
        android:layout_marginTop = "10dp"
        android:layout_marginBottom = "10dp"
        android:layout_marginLeft = "10dp"
        android:layout_marginRight = "10dp"
        android:id = "@ + id/text"/>
</LinearLayout>
```

4. 逻辑代码的实现

(1) 在 OneFragment 类中进行控件的绑定以及添加展示布局。

```java
public class OneFragment extends Fragment {
    private RecyclerView recy_view1;
    private RecyclerView recy_view_tags;
    @Override
    public View onCreateView(LayoutInflater inflater,ViewGroup container,Bundle savedInstanceState) {
        //删除原来的 TextView
        View view = inflater.inflate(R.layout.fragment_one, container, false);
        initView(view);
        return view;
    }
    private void initView(View view) {
        recy_view1 = (RecyclerView) view.findViewById(R.id.recy_view1);
        recy_view_tags = (RecyclerView) view.findViewById(R.id.recy_view_tags);
    }
}
```

(2) 创建数据实体类。

根据输出参数 showapi_res_body 字段的 JSON 数据结构,新建数据实体类 RecipeBean,并给 RecipeBean 及其子类 StepsBean、YlBean 添加 Serializable 接口实现,以便使用 Intent 传递数

据。其中，实体类 RecipeBean 的定义如下：

```java
public class RecipeBean implements Serializable {
    private String cpName;
    private String ct;
    private String des;
    private String id;
    private String largeImg;
    private String smallImg;
    private String tip;
    private String type;
    private String type_v1;
    private String type_v2;
    private String type_v3;
    private List<StepsBean> steps;
    private List<YlBean> yl;
    public String getCpName() {
        return cpName;
    }
    public void setCpName(String cpName) {
        this.cpName = cpName;
    }
    public String getCt() {
        return ct;
    }
    public void setCt(String ct) {
        this.ct = ct;
    }
    public String getDes() {
        return des;
    }
    public void setDes(String des) {
        this.des = des;
    }
    public String getId() {
        return id;
    }
    public void setId(String id) {
        this.id = id;
    }
    public String getLargeImg() {
        return largeImg;
    }
    public void setLargeImg(String largeImg) {
        this.largeImg = largeImg;
    }
    public String getSmallImg() {
        return smallImg;
    }
    public void setSmallImg(String smallImg) {
```

```java
            this.smallImg = smallImg;
        }
        public String getTip() {
            return tip;
        }
        public void setTip(String tip) {
            this.tip = tip;
        }
        public String getType() {
            return type;
        }
        public void setType(String type) {
            this.type = type;
        }
        public String getType_v1() {
            return type_v1;
        }
        public void setType_v1(String type_v1) {
            this.type_v1 = type_v1;
        }
        public String getType_v2() {
            return type_v2;
        }
        public void setType_v2(String type_v2) {
            this.type_v2 = type_v2;
        }
        public String getType_v3() {
            return type_v3;
        }
        public void setType_v3(String type_v3) {
            this.type_v3 = type_v3;
        }
        public List< StepsBean > getSteps() {
            return steps;
        }
        public void setSteps(List< StepsBean > steps) {
            this.steps = steps;
        }
        public List< YlBean > getYl() {
            return yl;
        }
        public void setYl(List< YlBean > yl) {
            this.yl = yl;
        }
}
```

实体类 StepsBean 主要用来存放对应食谱的制作步骤,格式定义如下。

```java
public class StepsBean implements Serializable {
    private String content;
```

```java
    private String imgUrl;
    private int orderNum;
    public String getContent() {
        return content;
    }
    public void setContent(String content) {
        this.content = content;
    }
    public String getImgUrl() {
        return imgUrl;
    }
    public void setImgUrl(String imgUrl) {
        this.imgUrl = imgUrl;
    }
    public int getOrderNum() {
        return orderNum;
    }
    public void setOrderNum(int orderNum) {
        this.orderNum = orderNum;
    }
}
```

实体类 YlBean 主要用来存放对应食谱所需的原材料，格式定义如下。

```java
public class YlBean implements Serializable {
    private String ylName;
    private String ylUnit;
    public String getYlName() {
        return ylName;
    }
    public void setYlName(String ylName) {
        this.ylName = ylName;
    }
    public String getYlUnit() {
        return ylUnit;
    }
    public void setYlUnit(String ylUnit) {
        this.ylUnit = ylUnit;
    }
}
```

（3）创建网络工具类。

新建工具类 NetUtil，用于处理网络 HTTP 请求，同时在 NetUtil 类中新建接口类 OnRecipeQueriedCallback，用于菜谱查询成功或失败时的回调。

```java
public class NetUtil {
    public interface OnRecipeQueriedCallback{
        void onSuccess(List< RecipeBean > recipes);
        void onFailure(Exception e);
    }
}
```

另外，在 NetUtil 类中新建 queryRecipe() 方法，通过使用 OkHttp 异步调用菜谱接口，并传入菜谱类型和回调函数。

```java
public static void queryRecipe(String tag, final OnRecipeQueriedCallback callback){
    OkHttpClient okHttpClient = new OkHttpClient();
    Request request = new Request.Builder().url(String.format("http://route.showapi.com/1164-1?showapi_appid=填写你的APPID&type=%s&showapi_sign=填写你的APPKEY", tag)).build();
    okHttpClient.newCall(request).enqueue(new Callback() {
        @Override
        public void onFailure(Call call, IOException e) {
            e.printStackTrace();
            callback.onFailure(e);
        }
        @Override
        public void onResponse(Call call, Response response) throws IOException {
            try {
                String jsonStr = response.body().string();
                List<RecipeBean> recipes = new Gson().fromJson(new JSONObject(jsonStr).getJSONObject("showapi_res_body").getJSONArray("datas").toString(), new TypeToken<List<RecipeBean>>(){}.getType());
                callback.onSuccess(recipes);
            } catch (Exception e) {
                callback.onFailure(e);
            }
        }
    });
}
```

（4）创建列表适配器。

新建 RecipeAdapter 类，并把该类作为菜谱列表的适配器类。

```java
public class RecipeAdapter extends RecyclerView.Adapter<RecipeAdapter.ViewHolder>{
    private Context mcontext;
    private List<RecipeBean> arrlist2;
    public RecipeAdapter(Context mcontext, List<RecipeBean> arrlist2) {
        this.mcontext = mcontext;
        this.arrlist2 = arrlist2;
    }
    @Override
    public RecipeAdapter.ViewHolder onCreateViewHolder(ViewGroup parent, int i) {
        View view = LayoutInflater.from(mcontext).inflate(R.layout.recycleritem, parent, false);
        ViewHolder myholder = new ViewHolder(view);
        return myholder;
    }
    @Override
    public void onBindViewHolder(final RecipeAdapter.ViewHolder myholder, int i) {
        final RecipeBean courseInfo = arrlist2.get(i);
        myholder.ke_name.setText(courseInfo.getCpName());
```

```java
            myholder.ke_des.setText(courseInfo.getDes());
            Glide.with(mcontext).load(courseInfo.getSmallImg()).into(myholder.ke_img);
            myholder.itemView.setOnClickListener(new View.OnClickListener() {
                @Override
                public void onClick(View v) {
                    //单击跳转到菜谱详情页面
                    Intent intent = new Intent(mcontext, RecipeDetailActivity.class);
                    intent.putExtra("data" ,courseInfo);
                    mcontext.startActivity(intent);
                }
            });
        }
        @Override
        public int getItemCount() {
            return arrlist2.size();
        }
        public class ViewHolder extends RecyclerView.ViewHolder {
            ImageView ke_img;
            TextView ke_name,ke_des;
            public ViewHolder(View itemView) {
                super(itemView);
                ke_img = itemView.findViewById(R.id.ke_img);
                ke_name = itemView.findViewById(R.id.ke_name);
                ke_des = itemView.findViewById(R.id.ke_des);
            }
        }
}
```

(5) 菜谱类型筛选。

新建 TagAdapter 类,用于展示菜谱类型标签,并实现菜谱类型标签被单击时主页面加载对应的菜谱数据,这里使用固定的类型标签数据。

```java
public class TagAdapter extends RecyclerView.Adapter<TagAdapter.ViewHolder> {
    public static final String[] DEFAULT_TAGS = {"蛋类","虾","螃蟹","鱼","贝","烘焙","奶制品","甜品","特色食品",};
    private final OneFragment fragment;
    private String[] tags = DEFAULT_TAGS;
    private int selectedTag = 0;
    public TagAdapter(OneFragment fragment){
        this.fragment = fragment;
    }
    @Override
    public ViewHolder onCreateViewHolder(ViewGroup viewGroup, int i) {
        View view = LayoutInflater.from(viewGroup.getContext()).inflate(R.layout.layouttagtext,viewGroup,false);
        ViewHolder holder = new ViewHolder(view);
        return holder;
    }
    @Override
```

```java
        public void onBindViewHolder(final ViewHolder viewHolder, int i) {
            viewHolder.text.setText(tags[i]);
            viewHolder.itemView.setOnClickListener(new View.OnClickListener() {
                @Override
                public void onClick(View view) {
                    fragment.notifyOnTagClick(tags[viewHolder.getAdapterPosition()]);
                    TagAdapter.this.selectedTag = viewHolder.getAdapterPosition();
                    TagAdapter.this.notifyDataSetChanged();
                }
            });
            if(selectedTag == i){
                viewHolder.text.setBackgroundResource(R.drawable.my_round_rectangle2);
            }else{
                viewHolder.text.setBackgroundResource(R.drawable.my_round_rectangle1);
            }
        }
        @Override
        public int getItemCount() {
            return tags.length;
        }
        public class ViewHolder extends RecyclerView.ViewHolder {
            public TextView text;
            public ViewHolder(View itemView) {
                super(itemView);
                this.text = itemView.findViewById(R.id.text);
            }
        }
    }
```

接着,在 OneFragment 类中创建 currentTag 实体变量,用于记录当前选中的菜谱类型,默认值为 TagAdapter 中 DEFAULT_TAGS 的第一个。

```java
private String currentTag = TagAdapter.DEFAULT_TAGS[0];
```

然后,在 OneFragment 类中创建 notifyOnTagClick()方法,用于动态加载数据。

```java
public void notifyOnTagClick(String tag) {
    currentTag = tag;
    initData();
}
```

最后,在 OneFragment 类的 initView()方法中创建 TagAdapter 的实例,并以此作为水平方向 RecyclerView 控件的适配器。

```java
//initView 方法后追加
LinearLayoutManager layoutManager = new LinearLayoutManager(getContext());
layoutManager.setOrientation(LinearLayoutManager.HORIZONTAL);
recy_view_tags.setLayoutManager(layoutManager);
recy_view_tags.setAdapter(new TagAdapter(this));
```

(6) 查询和展示数据。

在 OneFragment 类中创建 initData( ) 方法,用于数据的查询和展示,并重写 onActivityCreated()方法,并在其中调用 initData()方法,实现在界面创建时加载数据。

```java
private List<RecipeBean> recipeList = new ArrayList<>(); //定义食谱数组
@Override
public void onActivityCreated(Bundle savedInstanceState) {
    super.onActivityCreated(savedInstanceState);
    initData();        //数据的初始化
}
private void initData() {
    //访问网络 JSON 解析 ====》数据
    NetUtil.queryRecipe(currentTag, new NetUtil.OnRecipeQueriedCallback() {
        @Override
        public void onSuccess(List<RecipeBean> recipes) {
            if(getActivity() == null)
                    return;
            recipeList.clear();
            recipeList.addAll(recipes);
            getActivity().runOnUiThread(new Runnable() {
                @Override
                public void run() {
                    RecipeAdapter adapter = new RecipeAdapter(getActivity(), recipeList);
                    //布局管理器让数据显示到 RecyclerView 控件里面
                    StaggeredGridLayoutManager st = new StaggeredGridLayoutManager(1, StaggeredGridLayoutManager.VERTICAL);
                    recy_view1.setLayoutManager(st);
                    recy_view1.setAdapter(adapter);
                }
            });
        }
        @Override
        public void onFailure(Exception e) {
            if(getActivity() == null) return;
            getActivity().runOnUiThread(new Runnable() {
                @Override
                public void run() {
                    Toast.makeText(getActivity(), "网络错误", Toast.LENGTH_SHORT).show();
                }
            });
        }
    });
}
```

5. 程序运行与功能测试

运行上述程序,可以看到菜谱的主界面已经制作完成,单击上方水平展示的菜谱类型,可以切换显示对应的菜谱列表,运行效果如图 9.8 所示。

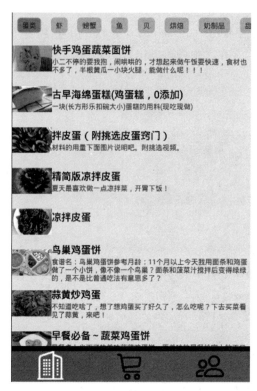

图 9.8 菜谱主界面运行效果

微课视频

9.4.3 菜谱 App 详情界面设计

【任务描述】

动态展示菜谱信息,主要包括食材用料和菜谱做法步骤两部分数据。

【设计思路】

因为食材用料和具体做法在不同的菜谱中内容是不一样的,所以可以考虑使用 RecyclerView 控件实现。主要包括以下几个操作。

(1) 设计"食材用料"的 RecyclerView 适配器。

(2) 设计"做法步骤"的 RecyclerView 适配器。

(3) 菜谱详情主界面设计与实现。

【任务实施】

1. 食材用料适配器的设计

(1) 食材用料子布局界面设计。

在 app 目录结构下的 res/layout 文件夹中,新建 recy_item_yl.xml,并修改 Root element 类型为 LinearLayout,然后在线性布局中添加两个 TextView 控件。

```
<?xml version = "1.0" encoding = "utf-8"?>
<LinearLayout xmlns:android = "http://schemas.android.com/apk/res/android"
```

```
        android:orientation = "vertical"
        android:layout_width = "match_parent"
        android:layout_height = "wrap_content"
        android:layout_gravity = "center">
    < TextView
        android:id = "@ + id/i_ylname"
        android:layout_width = "wrap_content"
        android:layout_height = "wrap_content"
        android:textSize = "25sp"
        android:textColor = " #000000"
        android:layout_gravity = "center"
        android:text = "aa"/>
    < TextView
        android:id = "@ + id/i_ylUnit"
        android:layout_width = "wrap_content"
        android:layout_height = "wrap_content"
        android:textSize = "15sp"
        android:textColor = " #aaaaaa"
        android:text = "bb"
        android:layout_gravity = "center"/>
</LinearLayout >
```

(2) 创建食材用料适配器文件。

接下来需要为食材用料的 RecyclerView 控件准备一个适配器,新建 YlAdapter 类,并让这个适配器继承自 RecyclerView.Adapter,并将泛型指定为 YlAdapter.ViewHolder。其中,ViewHolder 是将要在 YlAdapter 中定义的一个内部类,代码如下。

```
public class YlAdapter extends RecyclerView.Adapter < YlAdapter.ViewHolder > {
    private List < YlBean > mYLlist;
    private Context homecontext;
    public class ViewHolder extends RecyclerView.ViewHolder{
        TextView i_nameyl,i_unityl;
        public ViewHolder(View view){
            super(view);
            i_nameyl = view.findViewById(R.id.i_ylname);
            i_unityl = view.findViewById(R.id.i_ylUnit);
        }
    }
    public YlAdapter(Context context,List < YlBean > YLlist){
        homecontext = context;
        mYLlist = YLlist;
    }
    @Override
    public ViewHolder onCreateViewHolder(ViewGroup parent, int i) {
View view = LayoutInflater.from(parent.getContext()).inflate(R.layout.recy_item_yl,parent,
false);
        ViewHolder holder = new ViewHolder(view);
```

```
            return holder;
        }
        @Override
        public void onBindViewHolder(ViewHolder holder, int position) {
            YlBean ylBean = mYLlist.get(position);
            holder.i_nameyl.setText(ylBean.getYlName());
            holder.i_unityl.setText(ylBean.getYlUnit());
        }
        @Override
        public int getItemCount() {
            return mYLlist.size();
        }
    }
```

其中,ViewHolder 继承自 RecyclerView.ViewHolder,并在其构造函数中传入一个 view 参数,这个参数通常就是 RecyclerView 子项的最外层布局,接着就可以通过 findViewById()方法来获取到布局中的两个 TextView 的实例。

另外,由于 YlAdapter 是继承自 RecyclerView.Adapter 的,因此就必须重写 onCreateViewHolder( )、onBindViewHolder( )、getItemCount( )这 3 个方法,在 onCreateViewHolder()方法中,将 recy_item_yl 布局加载进来,同时创建 ViewHolder 的实例,并把加载出来的布局传入到构造函数当中,最后将 ViewHolder 的实例返回。onBindViewHolder()方法用于对 RecyclerView 子项的数据进行赋值,在每个子项被滚动到屏幕内的时候执行,通过 position 参数得到当前项的 YlBean 实例,然后再将数据设置到 ViewHolder 的两个 TextView 控件中。getItemCount()方法用于告诉 RecyclerView 一共有多少个子项,因此函数中直接返回数据源的长度即可。

2. 做法步骤适配器的设计

(1) 做法步骤子布局界面设计。

在 app 目录结构下的 res/layout 文件夹,新建 recy_item_step.xml,并修改 Root element 类型为 LinearLayout,然后在线性布局中添加两个 TextView 控件和两个 ImageView 控件。原型图如图 9.9 所示。

图 9.9 做法步骤的原型图

```xml
<?xml version = "1.0" encoding = "utf-8"?>
<LinearLayout xmlns:android = "http://schemas.android.com/apk/res/android"
    android:orientation = "horizontal"
    android:layout_width = "match_parent"
    android:layout_height = "170dp">
    <ImageView
        android:id = "@+id/i_imageurlstep"
        android:layout_width = "180dp"
        android:layout_height = "150dp"
        android:src = "@mipmap/ic_launcher_round"
        android:layout_marginRight = "10dp"/>
    <LinearLayout
        android:layout_width = "match_parent"
        android:layout_height = "150dp"
        android:layout_marginTop = "10dp"
        android:orientation = "vertical">
        <LinearLayout
            android:layout_width = "match_parent"
            android:layout_height = "wrap_content"
            android:orientation = "horizontal">
            <ImageView
                android:layout_width = "30dp"
                android:layout_height = "30dp"
                android:src = "@drawable/dot"/>
            <TextView
                android:id = "@+id/i_ordernumstep"
                android:layout_width = "match_parent"
                android:layout_height = "wrap_content"
                android:textSize = "30sp"
                android:layout_marginLeft = "10dp"
                android:textColor = "#f87f00"/>
        </LinearLayout>
        <TextView
            android:id = "@+id/i_contentstep"
            android:layout_width = "match_parent"
            android:layout_height = "wrap_content"
            android:textSize = "20sp"
            android:textColor = "#0000ff"/>
    </LinearLayout>
</LinearLayout>
```

(2) 创建做法步骤适配器文件。

做法步骤适配器的实现过程和食材用料适配器的实现类似,具体实现代码如下。

```java
public class StepAdapter extends RecyclerView.Adapter<StepAdapter.ViewHolder> {
    private List<StepsBean> mStepList;
    private Context homecontext;
    public class ViewHolder extends RecyclerView.ViewHolder{
        ImageView i_imageurlstep;
```

```
            TextView i_ordernumstep,i_contentstep;
            public ViewHolder(View view){
                super(view);
                i_imageurlstep = view.findViewById(R.id.i_imageurlstep);
                i_ordernumstep = view.findViewById(R.id.i_ordernumstep);
                i_contentstep = view.findViewById(R.id.i_contentstep);
            }
        }
        //构造函数
        public StepAdapter(Context context,List<StepsBean> StepList){
            mStepList = StepList;
            homecontext = context;
        }
        @Override
        public ViewHolder onCreateViewHolder(ViewGroup parent, int i) {
            View view = LayoutInflater.from(parent.getContext()).inflate(R.layout.recy_item_step,parent,false);
            ViewHolder holder = new ViewHolder(view);
            return holder;
        }
        @Override
        public void onBindViewHolder(ViewHolder holder, int position) {
            StepsBean stepsBean = mStepList.get(position);
            //这里选用了 Glide 框架,其中,load()中传入的为图片地址,into()中表明将图片指向
            //哪个 ImageView
            Glide.with(homecontext).load(stepsBean.getImgUrl()).into(holder.i_imageurlstep);
            holder.i_ordernumstep.setText("第" + stepsBean.getOrderNum() + "步");
            holder.i_contentstep.setText(stepsBean.getContent());
        }
        @Override
        public int getItemCount() {
            return mStepList.size();
        }
    }
```

3. 菜谱详情主界面的设计与实现

(1) 菜谱详情主界面设计。

在 app 目录结构下的 res/layout 文件夹中,新建 recipe_detail.xml,并修改 Root element 类型为 LinearLayout。原型图如图 9.10 所示。

```xml
<?xml version = "1.0" encoding = "utf-8"?>
<LinearLayout xmlns:android = "http://schemas.android.com/apk/res/android"
    android:orientation = "vertical"
    android:layout_width = "match_parent"
    android:layout_height = "match_parent"
    android:layout_marginTop = "20dp">
    <LinearLayout
        android:layout_width = "match_parent"
```

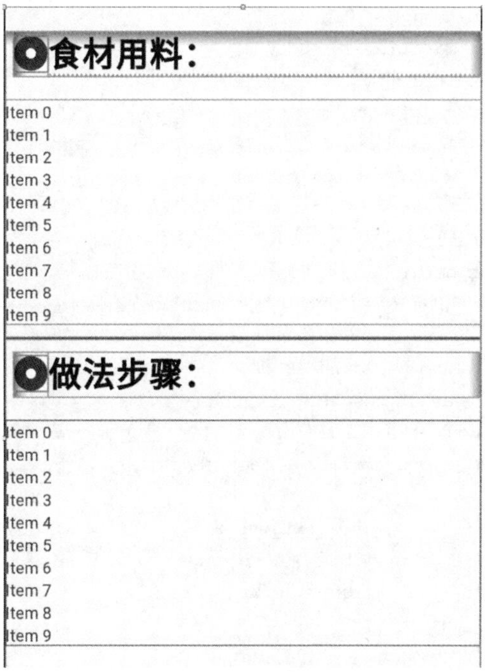

图 9.10 菜谱详情主界面

```
        android:layout_height = "wrap_content"
        android:orientation = "horizontal"
        android:layout_marginLeft = "5dp"
        android:layout_gravity = "center">
    < ImageView
        android:layout_width = "30dp"
        android:layout_height = "30dp"
        android:src = "@drawable/dot"
        android:layout_gravity = "center"/>
    < TextView
        android:layout_width = "0dp"
        android:layout_weight = "1"
        android:layout_height = "wrap_content"
        android:text = "食材用料:"
        android:textColor = "#0000ff"
        android:textSize = "28sp"
        android:textStyle = "bold"/>
</LinearLayout >
< android.support.v7.widget.RecyclerView
    android:id = "@ + id/recy_view_yl"
    android:layout_width = "match_parent"
    android:layout_height = "wrap_content"
    android:layout_marginTop = "20dp"
    android:layout_gravity = "center"/>
< View
    android:layout_width = "match_parent"
```

```xml
            android:layout_height = "2dp"
            android:layout_marginTop = "10dp"
            android:layout_marginBottom = "10dp"
            android:background = "@color/colorAccent"/>
    <LinearLayout
        android:layout_width = "match_parent"
        android:layout_height = "wrap_content"
        android:orientation = "horizontal"
        android:layout_marginLeft = "5dp"
        android:layout_gravity = "center">
        <ImageView
            android:layout_width = "30dp"
            android:layout_height = "30dp"
            android:src = "@drawable/dot"
            android:layout_gravity = "center"/>
        <TextView
            android:id = "@ + id/text_zuofa"
            android:layout_width = "match_parent"
            android:layout_height = "wrap_content"
            android:text = "做法步骤:"
            android:textColor = "#0000ff"
            android:textSize = "28sp"
            android:textStyle = "bold"/>
    </LinearLayout>
    <android.support.v7.widget.RecyclerView
        android:id = "@ + id/recy_view_step"
        android:layout_width = "match_parent"
        android:layout_height = "wrap_content"
        android:layout_marginTop = "20dp"
        android:layout_gravity = "center"/>
</LinearLayout>
```

（2）菜谱详情主界面逻辑代码实现。

新建 RecipeDetailActivity.java 文件，并在其中实现菜谱详情主界面的逻辑。首先声明需要的全局变量。

```java
RecyclerView recyclerView_yl;           //"食谱用料"所对应的列表
RecyclerView recyclerView_step;         //"做法步骤"所对应的列表
RecipeBean bean;                        //存放从上一个界面传递过来的参数
```

在 RecipeDetailActivity.java 中创建 initView() 方法，用于对 UI 界面中的控件进行绑定以及全局变量的初始化。

```java
private void initView() {
    recyclerView_yl = findViewById(R.id.recy_view_yl);
    recyclerView_step = findViewById(R.id.recy_view_step);
    bean = (RecipeBean) getIntent().getSerializableExtra("data");
}
```

上面对 RecyclerView 控件进行了绑定，并将上一页面传过来的数据赋值给了 bean 对象。接着便是解决如何将数据动态更新到 UI 界面上，在 RecipeDetailActivity.java 中新建 loadData() 方法。

```java
private void loadData() {
    List<YlBean> arrlistyl = new ArrayList<>(bean.getYl());
    List<StepsBean> arrliststep = new ArrayList<>(bean.getSteps());
    YlAdapter adapteryl = new YlAdapter(this, arrlistyl);
    StaggeredGridLayoutManager layoutManager = new StaggeredGridLayoutManager(3, StaggeredGridLayoutManager.VERTICAL);
    recyclerView_yl.setLayoutManager(layoutManager);
    recyclerView_yl.setAdapter(adapteryl);
    StepAdapter adapterstep = new StepAdapter(this, arrliststep);
    StaggeredGridLayoutManager layoutManager1 = new StaggeredGridLayoutManager(1, StaggeredGridLayoutManager.VERTICAL);
    recyclerView_step.setLayoutManager(layoutManager1);
    recyclerView_step.setAdapter(adapterstep);
}
```

在该方法中，先创建了两个 List 存放从上个界面传过来的"食材用料"数据和"做法步骤"数据，并通过 ArrayList 构造传参方式赋值给这两个 List。

接着为 recyclerView_yl 列表创建了一个界面布局管理对象，这里使用的是 StaggeredGridLayoutManager 网格管理器来完成网格的效果，在构造参数中的 3 代表网格为 3 列，StaggeredGridLayoutManager.VERTICAL 代表垂直排列。后面同样也给 recyclerView_step 列表创建了界面管理对象。

最后，在 RecipeDetailActivity.java 文件的 onCreate() 方法中，调用 initView()、loadData() 方法，具体代码如下。

```java
@Override
protected void onCreate(Bundle savedInstanceState) {
    super.onCreate(savedInstanceState);
    setContentView(R.layout.recipe_detail);
    initView();
    loadData();
}
```

4. 程序运行与功能测试

完成上述步骤后，运行程序，任意单击一个菜单后便能看到该菜单的详情界面了，具体如图 9.11 所示。

9.4.4 菜谱收藏功能的设计与实现

【任务描述】

用户在使用过程中，虽然能够通过搜索功能快速地找到相应的菜谱，但是如果用户对某个菜谱经常性使用，而每次都需要进行搜索才可以再次找到这个菜谱，对用户来说，在使用

微课视频

图 9.11　菜谱详情界面运行效果图

上有着诸多不便,因此希望能够实现对常用菜谱的收藏与取消收藏的功能,为用户提供方便。

【设计思路】

(1) 在菜谱详情界面添加一个"收藏"按钮。

(2) 创建数据库,保存已收藏的菜谱数据。

(3) 当用户进入菜谱收藏界面后,首先根据菜谱的 id,在已收藏的数据库中进行数据查找,如果数据库中已经存在该菜谱,"收藏"按钮提示文字则修改为"取消收藏",否则,文字修改为"收藏"。

(4) 当用户单击按钮,如果菜谱已经收藏,就将该菜谱从数据库中删除,如果菜谱未收藏,就将该菜谱添加到数据库中。

(5) 创建一个收藏界面,用于展示"已收藏"的菜谱信息。

【任务实施】

1. 在菜谱详情界面添加"收藏"按钮

修改前面创建的菜谱详情界面所对应的布局文件 recipe_detail.xml,在界面中"食材用料"的后面添加一个按钮,文字显示为"收藏"。修改代码如下。

```
 …
<TextView
    android:layout_width = "0dp"
```

```
        android:layout_weight = "1"
        android:layout_height = "wrap_content"
        android:text = "食材用料:"
        android:textColor = "#0000ff"
        android:textSize = "28sp"
        android:textStyle = "bold"/>
< Button
        android:layout_width = "wrap_content"
        android:layout_height = "wrap_content"
        android:id = "@ + id/button_favorite"
        android:text = "收藏"/>
    …
```

添加"收藏"按钮后原型图设计效果如图 9.12 所示。

图 9.12　添加"收藏"按钮后的效果图

2. 创建菜谱收藏数据库

在该数据库中需要保存菜谱的 id、菜谱名称、菜谱图片地址,还有菜谱的描述等信息,另外设置一个编号字段,并设定为主键,同时实现自动增长,因此数据库中需要 5 个字段。这里先创建一个 MyDBHelper 类,继承 SQLiteOpenHelper 类,并重写 OnCreate() 和 onUpgrade() 方法,并在 OnCreate() 方法中执行建表操作。创建数据库的代码如下。

```
public class MyDBHelper extends SQLiteOpenHelper {
    //数据库的名称,数据库的版本号
    private static final String DBNAME = "zscourse1.db";
    private static final int VERSION = 1;
    //构造方法
    public MyDBHelper(Context context) {
        super(context, DBNAME, null, VERSION);
    }
    //创建数据库
    @Override
    public void onCreate(SQLiteDatabase db) {
        db.execSQL("create table tb_recipe(" +
                "id integer primary key autoincrement," +
                "recipeId varchar(50)," +
                "recipeName varchar(20)," +
                "recipeDesc varchar(20)," +
                "recipeSmallImgUrl varchar(100));");
    }
    //升级数据库
    @Override
    public void onUpgrade(SQLiteDatabase db, int oldVersion, int newVersion) {
    }
}
```

创建的表格中，id 代表记录的编号，设为主键，recipeId 代表菜谱的 id，recipeName 代表菜谱的名称，recipeDesc 代表菜谱的描述信息，recipeSmallImgUrl 代表菜谱的封面图片链接地址。

3. 声明 isFavorite 字段，实现"收藏"功能

在菜谱详情界面声明 isFavorite 字段，判断菜谱是否已收藏。当用户进入菜谱详情界面时，从数据库中进行查找，若该菜谱已经存在，则将 isFavorite 字段设为 true，同时按钮的文字修改为"取消收藏"，否则，就将 isFavorite 字段设为 false，按钮的文字修改为"收藏"。在 RecipeDetailActivity.java 文件中，声明 isFavorite 字段，具体修改代码如下。

```
Button button_favorite;                    //"收藏"按钮
boolean isFavorite = false;                //代表该条数据是否已收藏
MyDBHelper dbHelper;                       //SQLite 数据库的声明
```

判断菜谱是否已收藏，可以放在 loadData()方法中，代码如下。

```
Cursor cursor = dbHelper.getReadableDatabase().rawQuery("SELECT id FROM tb_recipe WHERE recipeId = ?", new String[]{bean.getId()});
isFavorite = cursor.getCount() > 0;
cursor.close();
button_favorite.setText(isFavorite?"取消收藏":"收藏");
```

另外，还需要给该按钮设置一个单击事件，当用户单击时，若 isFavorite 字段为 true，就从数据库中删除该菜谱，否则就将该菜谱添加到数据库。在 initView()方法中添加按钮的单击事件，代码修改如下。

```
dbHelper = new MyDBHelper(this);           //创建 MyDBHelper 类的对象
button_favorite = findViewById(R.id.button_favorite);
button_favorite.setOnClickListener(new View.OnClickListener() {
@Override
public void onClick(View view) {
    if(isFavorite){
        dbHelper.getWritableDatabase().execSQL("DELETE FROM tb_recipe WHERE recipeId = ?",
new Object[]{bean.getId()});
    }else{
        dbHelper.getWritableDatabase().execSQL("INSERT INTO tb_recipe(recipeId,recipeName,
recipeDesc,recipeSmallImgUrl) VALUES (?,?,?,?)",new Object[]{bean.getId(),bean.getCpName
(),bean.getDes(),bean.getSmallImg()});
    }
    isFavorite = !isFavorite;
    button_favorite.setText(isFavorite?"取消收藏":"收藏");
}
});
```

此时，已经实现了菜谱的"收藏"和"取消收藏"的功能，再次运行程序，并进入菜谱的详情界面，单击"收藏"后，按钮变为"取消收藏"，单击"取消收藏"，则按钮变为"收藏"。运行界面如图 9.13 和图 9.14 所示。

图 9.13 "收藏"界面

图 9.14 "取消收藏"界面

4. 创建收藏界面,用于展示"已收藏"的菜谱信息

(1) 创建 fragment_two.xml 布局文件,在里面放置一个 TextView 控件与一个 RecyclerView 控件,其中,TextView 用于展示提示信息,RecyclerView 用于展示收藏的菜谱信息。布局代码如下。

```xml
<?xml version = "1.0" encoding = "utf-8"?>
<LinearLayout xmlns:android = "http://schemas.android.com/apk/res/android"
    android:orientation = "vertical"
    android:layout_width = "match_parent"
    android:layout_height = "match_parent">
    <TextView
        android:layout_width = "match_parent"
        android:layout_height = "wrap_content"
        android:text = "您好,您收藏的菜谱信息如下:"
        android:textSize = "25sp"
        android:textColor = "#ff0000"/>
    <android.support.v7.widget.RecyclerView
        android:id = "@+id/recy_view2"
        android:layout_width = "match_parent"
        android:layout_height = "wrap_content"/>
</LinearLayout>
```

(2) 创建 TwoFragment 类,并让它继承 Fragment 类,然后在 onCreateView()方法里引入布局文件 fragment_two.xml。具体代码如下。

```
@Override
public View onCreateView ( LayoutInflater inflater, ViewGroup container, Bundle
savedInstanceState) {
    View view = inflater.inflate(R.layout.fragment_two, container, false);
    initView(view);
    return view;
}
```

(3) 创建"已收藏"菜谱列表的子布局。新建 recy_sc_item.xml 布局文件,并修改 Root element 类型为 LinearLayout。

```xml
<LinearLayout xmlns:android = "http://schemas.android.com/apk/res/android"
    android:orientation = "horizontal"
    android:layout_width = "match_parent"
    android:layout_height = "wrap_content">
    <TextView
        android:id = "@ + id/sc_id"
        android:layout_width = "wrap_content"
        android:layout_height = "wrap_content"
        android:text = "11"
        android:textSize = "50sp"
        android:textColor = "#000000"/>
    <ImageView
        android:id = "@ + id/sc_img"
        android:layout_width = "80dp"
        android:layout_height = "80dp"
        android:src = "@mipmap/ic_launcher"/>
    <LinearLayout
        android:layout_width = "match_parent"
        android:layout_height = "wrap_content"
        android:orientation = "vertical"
        android:layout_gravity = "center">
        <TextView
            android:id = "@ + id/sc_name"
            android:layout_width = "match_parent"
            android:layout_height = "wrap_content"
            android:text = ""
            android:textSize = "20sp"
            android:textColor = "#000000"
            />
        <TextView
            android:id = "@ + id/sc_des"
            android:layout_width = "wrap_content"
            android:layout_height = "wrap_content"
            android:text = ""/>
    </LinearLayout>
</LinearLayout>
```

(4) 为"已收藏"菜谱列表的 RecyclerView 控件准备适配器,新建 CourseSCAdapter

类,并让该适配器继承自 RecyclerView.Adapter,指定泛型为 CourseSCAdapter.ViewHolder,同时重写 onCreateViewHolder()和 onBindViewHolder()方法,代码如下。

```java
public class CourseSCAdapter extends RecyclerView.Adapter<CourseSCAdapter.ViewHolder> {
    private Context ncontext;
    private List<RecipeBean> arrlist4;
    public CourseSCAdapter(Context ncontext, List<RecipeBean> arrlist4) {
        this.ncontext = ncontext;
        this.arrlist4 = arrlist4;
    }
    @Override
    public ViewHolder onCreateViewHolder(ViewGroup parent, int i) {
     View view = LayoutInflater.from(ncontext).inflate(R.layout.recy_sc_item, parent, false);
        ViewHolder nholder = new ViewHolder(view);
        return nholder;
    }
    @Override
    public void onBindViewHolder(final CourseSCAdapter.ViewHolder nholder, int i) {
        final RecipeBean courseInfo = arrlist4.get(i);
        nholder.sc_id.setText(i + 1 + "");
        Glide.with(ncontext).load(courseInfo.getSmallImg()).into(nholder.sc_img);
        nholder.sc_name.setText(courseInfo.getCpName());
nholder.sc_des.setText(courseInfo.getDes());
nholder.itemView.setOnClickListener(new View.OnClickListener() {
@Override
  public void onClick(View v) {
      //数据库里只存储了部分数据,需要使用时查询完整的数据
NetUtil.queryRecipe(courseInfo.getId(), new NetUtil.OnRecipeQueriedCallback() {
        @Override
        public void onSuccess(List<RecipeBean> recipes) {
            if(recipes.size() < 1){
                onFailure(new Exception("data not queried"));
                return;
            }
            //跳转页面的代码
            Intent intent = new Intent(ncontext, RecipeDetailActivity.class);
            intent.putExtra("data",recipes.get(0));
            ncontext.startActivity(intent);
        }
        @Override
        public void onFailure(Exception e) {
            new Handler().post(new Runnable() {
                @Override
                public void run() {
Toast.makeText(ncontext, "数据查询失败", Toast.LENGTH_SHORT).show();
                }
            });
        }
```

```java
            });
        }
    });
}
@Override
public int getItemCount() {
    return arrlist4.size();
}
//创建内部类 ViewHolder
public class ViewHolder extends RecyclerView.ViewHolder {
    TextView sc_id;
    ImageView sc_img;
    TextView sc_name;
    TextView sc_des;
    public ViewHolder(View itemView) {
        super(itemView);
        sc_id = itemView.findViewById(R.id.sc_id);
        sc_img = itemView.findViewById(R.id.sc_img);
        sc_name = itemView.findViewById(R.id.sc_name);
        sc_des = itemView.findViewById(R.id.sc_des);
    }
}
```

（5）修改 TwoFragment 类的 onActivityCreated()方法。当用户进入收藏界面后，就会执行 onActivityCreated()方法，在该方法中调用 initData()方法，目的是获取收藏数据库中的全部数据，然后将数据交给适配器进行数据的展示。代码如下。

```java
@Override
public void onActivityCreated(Bundle savedInstanceState) {
    super.onActivityCreated(savedInstanceState);
    initData();        //数据的初始化
}
private void initData() {
    //获取数据
    arrlist3 = new ArrayList<RecipeBean>();
    Cursor cursor = db.rawQuery("select * from tb_recipe", null);
    while (cursor.moveToNext()) {
        String name = cursor.getString(cursor.getColumnIndex("recipeName"));
        String resId = cursor.getString(cursor.getColumnIndex("recipeId"));
        String desc = cursor.getString(cursor.getColumnIndex("recipeDesc"));
        String smallImg = cursor.getString(cursor.getColumnIndex("recipeSmallImgUrl"));
        RecipeBean bean = new RecipeBean();
        bean.setCpName(name);
        bean.setId(resId);
        bean.setDes(desc);
        bean.setSmallImg(smallImg);
        arrlist3.add(bean);
    }
```

```
    cursor.close();
    //创建自定义适配器的对象
    CourseSCAdapter adapter = new CourseSCAdapter(getActivity(), arrlist3);
    //用网格布局管理器将数据显示出来
    StaggeredGridLayoutManager st = new StaggeredGridLayoutManager(1, Staggered
GridLayoutManager.VERTICAL);
    recy_view2.setLayoutManager(st);
    recy_view2.setAdapter(adapter);
}
```

（6）完善 TwoFragment 类中的相应方法。

```
//定义对象
    private RecyclerView recy_view2;
    private MyDBHelper mhelper;
    private SQLiteDatabase db;
    ArrayList < RecipeBean > arrlist3;
    …
    private void initView(View view) {
        recy_view2 = (RecyclerView) view.findViewById(R.id.recy_view2);
        mhelper = new MyDBHelper(getActivity());
        db = mhelper.getWritableDatabase();
    }
    @Override
    //退出时,回收资源
    public void onResume() {
        super.onResume();
        arrlist3.clear();
        initData();
    }
```

5. 程序运行与功能测试

完成上述步骤后,运行程序,单击主页面下方的"收藏"栏目,运行效果如图 9.15 所示。

9.4.5 菜谱搜索功能的设计与实现

微课视频

【任务描述】

用户可以通过主页的分类标签,快速地找到对应类别的菜谱,但是对于用户来说,可能更多情况还是希望从中找到某一个菜谱,所以还是需要经常在大量的菜谱中进行筛选,费时费力,用户体验差,因此需要考虑在 App 中添加菜谱搜索的功能,可以让用户根据自己的需求,进行菜谱的快速查找,最后将查找到的数据展示到界面上。

【设计思路】

（1）在主页面中添加一个搜索框以及一个"搜索"按钮,完成搜索条件的输入。

（2）在搜索结果界面放置一个 TextView 控件用于展示搜索到的记录数量,然后界面中再放置一个 RecyclerView 控件,用于搜索结果数据的展示。

（3）创建一个子布局,里面放置一个 ImageView 控件用于展示菜谱图片,然后再添加一个

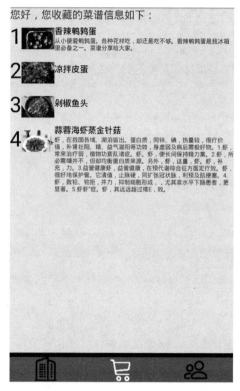

图 9.15 "收藏"栏目运行效果

LinearLayout 布局，里面放置两个 TextView 控件，用于展示菜谱名称和菜谱描述信息。

【任务实施】

1. 添加搜索框与"搜索"按钮

在 App 主页面菜谱分类的上方，设置一个搜索输入框以及"搜索"按钮，方便用户输入搜索条件。修改 fragment_one.xml 布局文件，添加搜索框以及"搜索"按钮，修改代码如下。

```xml
<LinearLayout
    android:layout_width = "match_parent"
    android:layout_height = "wrap_content"
    android:orientation = "horizontal">
    <EditText
        android:id = "@ + id/input_kc"
        android:layout_width = "wrap_content"
        android:layout_height = "wrap_content"
        android:hint = "请输入菜谱类型"
        android:layout_weight = "1"
        android:gravity = "center"/>
    <Button
        android:id = "@ + id/btn_search"
        android:layout_width = "60dp"
        android:layout_height = "wrap_content"
        android:background = "@drawable/search_bar_icon_normal"/>
</LinearLayout>
```

当用户单击"搜索"按钮时,将输入的条件传送给搜索界面进行搜索,因此需要给"搜索"按钮设置单击事件。修改 OneFragment.java 文件,具体修改代码如下。

```java
private EditText input_kc;                              //定义搜索框对象
private Button btn_search;                              //定义"搜索"按钮对象
…
input_kc = view.findViewById(R.id.input_kc);            //搜索框控件绑定
btn_search = view.findViewById(R.id.btn_search);        //"搜索"按钮控件绑定
//给"搜索"按钮设置单击事件
btn_search.setOnClickListener(new View.OnClickListener() {
    @Override
    public void onClick(View v) {
        Intent intent = new Intent(getActivity(), MenuSearchActivity.class);
        String input_name = input_kc.getText().toString();
        intent.putExtra("search", input_name);
        getActivity().startActivity(intent);
    }
});
```

2. 创建搜索结果主界面

创建菜谱搜索结果的显示界面。新建 recy_cx_result.xml 布局文件,并修改 Root element 类型为 LinearLayout。

```xml
<?xml version = "1.0" encoding = "utf-8"?>
< LinearLayout xmlns:android = "http://schemas.android.com/apk/res/android"
    android:orientation = "vertical"
    android:layout_width = "match_parent"
    android:layout_height = "match_parent">
    < TextView
        android:id = "@ + id/txt_result"
        android:layout_width = "match_parent"
        android:layout_height = "wrap_content"
        android:text = "查询结果:"
        android:textSize = "25sp"
        android:textColor = "#ff0000"/>
    < android.support.v7.widget.RecyclerView
        android:id = "@ + id/recy_view3"
        android:layout_width = "match_parent"
        android:layout_height = "wrap_content"/>
</LinearLayout >
```

3. 创建搜索结果列表的子界面

创建搜索结果列表的子布局。新建 recy_cx_item.xml 布局文件,并修改 Root element 类型为 LinearLayout。

```xml
<?xml version = "1.0" encoding = "utf-8"?>
< LinearLayout xmlns:android = "http://schemas.android.com/apk/res/android"
```

```xml
            android:orientation = "horizontal"
            android:layout_width = "match_parent"
            android:layout_height = "wrap_content">
    <ImageView
            android:id = "@ + id/cx_img"
            android:layout_width = "80dp"
            android:layout_height = "80dp"
            android:src = "@mipmap/ic_launcher"/>
    <LinearLayout
            android:layout_width = "match_parent"
            android:layout_height = "wrap_content"
            android:orientation = "vertical"
            android:layout_gravity = "center">
        <TextView
            android:id = "@ + id/cx_name"
            android:layout_width = "match_parent"
            android:layout_height = "wrap_content"
            android:text = "名称"
            android:textSize = "20sp"
            android:textColor = "#000000"
            />
        <TextView
            android:id = "@ + id/cx_des"
            android:layout_width = "wrap_content"
            android:layout_height = "wrap_content"
            android:text = "介绍"/>
    </LinearLayout>
</LinearLayout>
```

4. 为搜索结果菜谱列表的 RecyclerView 控件准备适配器

新建 RecipeSearchAdapter 类，并让该适配器继承 RecyclerView.Adapter，指定泛型 RecipeSearchAdapter.ViewHolder，同时重写 onCreateViewHolder()和 onBindViewHolder()方法，代码如下。

```java
    public class RecipeSearchAdapter extends RecyclerView.Adapter < RecipeSearchAdapter.ViewHolder > {
        private Context wcontext;
        private List < RecipeBean > arrlist6;
        //创建构造方法
        public RecipeSearchAdapter(Context wcontext, List < RecipeBean > arrlist6) {
            this.wcontext = wcontext;
            this.arrlist6 = arrlist6;
    }
    @Override
public RecipeSearchAdapter.ViewHolder onCreateViewHolder(ViewGroup parent, int i) {
     View view = LayoutInflater.from(wcontext).inflate(R.layout.recy_cx_item, parent, false);
     ViewHolder myholder = new ViewHolder(view);
```

```java
        return myholder;
    }
    @Override
    public void onBindViewHolder(RecipeSearchAdapter.ViewHolder myholder, int i) {
        final RecipeBean courseInfo = arrlist6.get(i);
        myholder.cx_name.setText(courseInfo.getCpName());
        myholder.cx_des.setText(courseInfo.getDes());
        Glide.with(wcontext).load(courseInfo.getSmallImg()).into(myholder.cx_img);
        myholder.itemView.setOnClickListener(new View.OnClickListener() {
            @Override
            public void onClick(View v) {
                Intent intent = new Intent(wcontext, RecipeDetailActivity.class);
                intent.putExtra("data",courseInfo);
                wcontext.startActivity(intent);
            }
        });
    }
    @Override
    public int getItemCount() {
        return arrlist6.size();
    }
    public class ViewHolder extends RecyclerView.ViewHolder {
        ImageView cx_img;
        TextView cx_name,cx_des;
        public ViewHolder(View itemView) {
            super(itemView);
            cx_img = itemView.findViewById(R.id.cx_img);
            cx_name = itemView.findViewById(R.id.cx_name);
            cx_des = itemView.findViewById(R.id.cx_des);
        }
    }
}
```

5. 完成搜索结果界面数据的展示

当用户在主页面的输入框中输入搜索条件并单击"搜索"按钮后，程序会将条件传送给搜索界面，然后搜索界面使用封装的网络请求（NetUtil）进行数据请求操作，请求成功后将数据交给适配器进行数据展示，并在顶部的 TextView 控件中显示搜索到的记录数量，如果搜索失败，则显示"未查询到相关菜谱！"。

完善 MenuSearchActivity.java 文件，接收从 OneFragment 页面传送过来的搜索条件，实现代码如下。

```java
public class MenuSearchActivity extends AppCompatActivity {
    private TextView txt_result;          //搜索结果布局中的提示信息控件
    private RecyclerView recy_view3;      //搜索结果布局中的列表控件
    @Override
    protected void onCreate(Bundle savedInstanceState)
    {
```

```
            super.onCreate(savedInstanceState);
            setContentView(R.layout.recy_cx_result);    //加载搜索结果布局文件
            initview();
            loadData();
        }
    public void initview(){
            txt_result = findViewById(R.id.txt_result);
            recy_view3 = findViewById(R.id.recy_view3); }
    private void loadData() {
            String ke_name = getIntent().getStringExtra("search");
            NetUtil.queryRecipe(ke_name, new NetUtil.OnRecipeQueriedCallback() {
                @Override
                public void onSuccess(final List<RecipeBean> recipes) {
                    final List<RecipeBean> arrlist5 = new ArrayList<>(recipes);
                    runOnUiThread(new Runnable() {
                        @Override
      public void run() {
        txt_result.setText("您好,共查询到" + recipes.size() + "条数据,分别如下:");
RecipeSearchAdapter adapter = new RecipeSearchAdapter(MenuSearchActivity.this,arrlist5);
        StaggeredGridLayoutManager st = new StaggeredGridLayoutManager(1,StaggeredGridLayoutManager.
VERTICAL);
            recy_view3.setLayoutManager(st);
            recy_view3.setAdapter(adapter);
                        }
                    });
                }
                @Override
                public void onFailure(Exception e) {
                    runOnUiThread(new Runnable() {
                        @Override
                        public void run() {
                            txt_result.setText("未查询到相关菜谱!"); }
                    });
                }
            });
        }
    }
```

上述代码中,NetUtil 网络请求接收一个参数,该参数就是从 OneFragment 页面传过来的搜索条件,然后发起网络请求,请求成功后返回菜谱列表 recipes,接着将 recipes 赋值给 arrlist5,最后将 arrlist5 交给适配器并展示数据。如果请求失败,说明未搜索到相关的菜谱,则在 TextView 控件中显示"未查询到相关菜谱!"。

6. 程序运行与功能测试

完成上述步骤后,运行程序,在主页面上方的输入框中输入搜索条件,然后单击旁边的"搜索"按钮,即可查看搜索到的相关菜谱信息,运行效果如图 9.16 和图 9.17 所示。

项目9 我的第一道菜——菜谱App的设计与实现 | 451

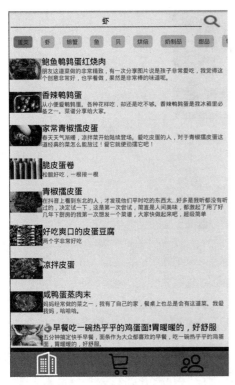

图 9.16 搜索框运行效果图

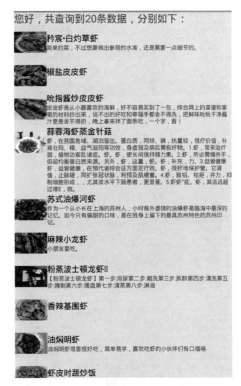

图 9.17 菜谱搜索结果效果图

9.4.6 个人中心的设计与实现

【任务描述】

在个人中心界面,主要用于展示登录用户的相关信息,以及作品介绍的相关内容。

【设计思路】

使用 WebView 控件展示作品介绍的相关内容。

微课视频

【任务实施】

1. 个人中心布局文件的设计

新建 fragment_three.xm 布局文件,并修改 Root element 类型为 ScrollView。界面设计原型图如图 9.18 所示。

实现的 UI 代码如下。

```xml
<?xml version = "1.0" encoding = "utf - 8"?>
< ScrollView xmlns:android = "http://schemas.android.com/apk/res/android"
    android:layout_width = "match_parent"
    android:layout_height = "match_parent">
    < LinearLayout
        android:layout_width = "match_parent"
        android:layout_height = "match_parent"
        android:orientation = "vertical">
        < FrameLayout
            android:layout_width = "match_parent"
```

图 9.18 "个人中心"设计原型图

```
        android:layout_height = "wrap_content"
        android:background = "@drawable/myinfo_login_bg">
        <ImageView
            android:layout_width = "80dp"
            android:layout_height = "80dp"
            android:src = "@drawable/default_icon"
            android:layout_gravity = "center"
            android:layout_marginTop = " - 20dp"/>
        <TextView
            android:layout_width = "wrap_content"
            android:layout_height = "wrap_content"
            android:text = "admin"
            android:textSize = "30sp"
            android:textColor = " # ffffff"
            android:layout_gravity = "center"
            android:layout_marginTop = "40dp"/>
    </FrameLayout>
    <View
        android:layout_width = "match_parent"
        android:layout_height = "5dp"
        android:background = " # C51D1D"
        android:layout_marginTop = "10dp"
        android:layout_marginBottom = "10dp"/>
    <WebView
        android:id = "@ + id/web_view"
        android:layout_width = "match_parent"
        android:layout_height = "wrap_content">
    </WebView>
    </LinearLayout>
</ScrollView>
```

2. "个人中心"碎片的实现

创建 ThreeFragment.java 文件,并让它继承 Fragment 类,完成碎片的创建。然后在该类文件中重写 onCreateView()方法进行界面的绑定。

```java
@Override
public View onCreateView ( LayoutInflater inflater, ViewGroup container, Bundle savedInstanceState) {
    View view = inflater.inflate(R.layout.fragment_three, container, false);
    initView(view);
    return view;
}
```

在该方法中,和适配器绑定界面一样,使用 View 加载布局文件,同时定义了 initView 方法,在该方法中,完成界面中 WebView 控件的绑定。

```java
private void initView(View view) {
    web_view = (WebView) view.findViewById(R.id.web_view);
}
```

重写 onActivityCreated()方法,这个方法在碎片依赖的 Activity 加载成功后会被调用,同时还在这个方法中定义了 initData()方法,用来完成将作品信息显示到 WebView 控件。实现代码如下。

```java
@Override
public void onActivityCreated(Bundle savedInstanceState) {
    super.onActivityCreated(savedInstanceState);
    initData();
}
private void initData() {
    StringBuffer st = new StringBuffer();
    st.append("<div>作品简介:");
    st.append("<ul>");
    st.append("<li>浏览菜谱界面:作品运行后,可以看到菜谱列表,上下滚动即可浏览菜谱.</li>");
    st.append("<li>菜谱详情界面:在菜谱界面中,单击某行,即可进入菜谱详情页面,可以查看菜谱的用料和制作过程.</li>");
    st.append("<li>菜谱收藏页面:在菜谱详情页面中单击收藏按钮,即可收藏,再次单击按钮,即可取消收藏,收藏的菜谱可在收藏页面中显示.</li>");
    st.append("<li>菜谱搜索界面:在菜谱列表界面上方,输入菜谱类型,单击右侧的查询按钮,即可完成查询.</li>");
    st.append("<li>个人中心界面:显示用户头像、姓名、菜谱简介.</li>");
    st.append("<li>作品涵盖的知识点:滑屏 ViewPager、碎片 Fragment、列表 RecyclerView、访问网络 OkHttp、数据库 SQLiteOpenHelper、数据库的添加、删除、查询等功能. </li>");
    st.append("</ul>");
    st.append("</div>");
    web_view.loadDataWithBaseURL(null, st.toString(),"text/html","utf-8",null);
}
```

在 initData()方法中，创建了一个 StringBuffer 对象，用于存放一个 HTML 的布局文件，后面实现的就是 HTML 页面代码的拼接，最后通过 WebView 的 loadDataBaseURL()方法展示上面的 HTML 页面。其中，第一个参数为 baseurl，代表网页指向的 URL，这里仅展示上述静态内容不需要指向某个网站因此选择 null。第二个参数是一个字符型的数据，WebView 接收到这个参数后，会尝试将其解析为网页。第三个和第四个参数便是解析类型和编码方式了，由于解析为网页，所以指定为 text/html。最后一个参数为 historyUrl，代表历史记录的 HTML，同样也要设置为 null。

3. 程序运行与功能测试

完成上述步骤后运行程序，单击"个人中心"栏目，显示效果如图 9.19 所示。

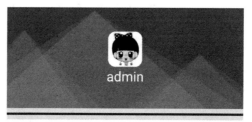

图 9.19　个人中心运行效果

项目小结

本章借助于菜谱综合案例，主要学习了如何使用 ViewPager 来实现手机中页面的滑屏效果，使用 OkHttp 协议来进行网络交互，利用 Java 的回调机制来将服务器响应的数据进行返回，同时还学会了如何对 JSON 格式的数据进行解析。通过这几个知识点的综合运用，对学生 Android 网络编程的能力提出了更高的要求，在作品设计与实现的过程中，不仅扩宽了学生的视野，也让原本在本地端运行的程序迁移到了云端。同时在完成本案例的过程中，也充分运用了前面章节的相关内容,如数据库的创建、数据的增删改查、RecyclerView 列表数据的显示等。希望同学们能够举一反三,达到知识的灵活运用,开发出更优质的作品。

习题

一、填空题

1. 当客户端与服务器端建立连接后,向服务器发送的请求,被称为( )。
2. Android 客户端访问网络发送 HTTP 请求的方式一般有两种:( )和( )。
3. 与服务器交互过程中,最常用的两种数据提交方式是( )和( )。

二、选择题

1. 下列通信方式中,不是 Android 系统提供的是( )。
   A. Socket 通信　　　B. HTTP 通信　　　C. URL 通信　　　D. 以太网通信
2. 关于 HttpURLConnection 访问网络的基本用法,描述错误的是( )。
   A. HttpURLConnection 对象需要设置请求网络的方式
   B. HttpURLConnection 对象需要设置超时时间
   C. 需要通过 new 关键字来创建 HttpURLConnection 对象
   D. 访问网络完毕需要关闭 HTTP 连接
3. 下列选项中,关于 GET 和 POST 请求方式,描述错误的是( )。
   A. 使用 GET 方式访问网络 URL 的长度是有限制的
   B. HTTP 规定 GET 方式请求 URL 的长度不超过 2k
   C. POST 方式对 URL 的长度是没有限制的
   D. GET 请求方式向服务器提交的参数跟在请求 URL 后面

三、简答题

1. 简述使用 HttpURLConnection 访问网络的基本步骤。
2. 简述 Handler 机制 4 个关键对象的作用。
3. 简述使用 gson 解析 JSON 数组的操作步骤。

项目 10

实战案例——移动互联网软件开发竞赛1

【教学导航】

学习目标	（1）培养在企业真实项目环境下移动应用开发的工程实践能力。 （2）培养设计能力和创新能力。 （3）培养团队协作、沟通力、抗压力、职业规范等职场素质。 （4）提高学生移动互联网应用软件开发技能水平。
教学方法	任务驱动法、理论实践一体化、探究学习法、分组讨论法
课时建议	4 课时

10.1 赛项简介

移动互联网应用软件开发赛项分为两个级别：省级和国家级。比赛主题每 4 年左右更换一次，最初比赛内容是智能农业移动应用开发系统，现阶段比赛内容更新为智能交通领域企业真实案例。

10.1.1 竞赛目的

通过竞赛，考察参赛选手实际工程项目的分析理解能力、编码与程序排错能力、文档编写能力、一定的创意创新能力，培养计算机类相关专业学生移动应用软件产品开发意识和用户体验设计能力，激发其对移动互联网应用软件开发领域的学习和研究兴趣，提高其软件编程能力和职业素养。通过相应能力的考察和训练，让参赛选手尽可能地适应未来工作岗位的需要，为移动开发工作岗位和实习就业奠定坚实的基础。推进移动应用开发专业对接最新行业标准和岗位规范，紧贴工程前沿技术和实际生产，调整课程结构，更新课程内容，有效开展实践性教学，促进产教深度融合的人才培养模式改革。

10.1.2 竞赛内容

项目竞赛内容是应用智能交通领域企业真实案例,通过"系统文档""程序排错""功能编码"及"创意设计"4 种赛题形式,考察参赛选手实际工程项目的编码能力、文档编写能力、综合分析能力、技术架构设计能力、创意创新能力、大数据分析能力。本赛项关联就业岗位主要包括移动应用开发工程师、软件开发工程师、UI 工程师、测试工程师等核心岗位。赛项内容涵盖上述岗位的核心知识、技能,考核技术点包括:MVP 设计模式、UI 设计标准、MaterialDesign、四大组件、资源使用、Handler/多线程/定时器、网络请求框架、数据封装和解析、多媒体、手势识别、依赖注入、事件传递、内存泄漏管理、数据存储、业务逻辑、数据挖掘和开源图表库 MPAndroidChart API 等。

10.1.3 竞赛方式

竞赛以团队方式进行,每支参赛队由 1 名领队、3 名选手(其中队长 1 名)、2 名指导教师组成。所有参赛队在现场根据给定的项目任务,相互配合,在设备上完成移动应用开发项目的系统文档、程序排错和功能编码,最后以配置文件、提交的截图、文档和竞赛作品作为最终评分依据。3 名选手自行分配系统设计题、"程序排错"题、"功能编码"题,以保证整个团队并行开发和调试。比赛时间为 4 个小时,参赛选手必须在规定时间内完成比赛内容并提交相关文档。

10.2 比赛器材及技术平台

10.2.1 计算机配置

个人计算机最低软硬件配置要求如下。
(1)操作系统:Windows 10(64 位)或更新版本。
(2)处理器:英特尔酷睿 i3 以上处理器(支持 VT 技术,该技术可加速模拟器的运行速度)。
(3)内存:4GB 或以上(根据参赛经历,建议内存至少为 8GB 以上)。
(4)硬盘:100GB 或以上。
(5)显卡:支持 DirectX 9 256MB 或以上。
(6)显示器:分辨率 1024×768px 或以上。

10.2.2 比赛平台

移动互联网应用软件开发实训系统,包含如下硬件。
(1)1 台移动互联开发平台。
(2)1 套智能交通应用后台服务系统。
(3)1 个智能交通仿真沙盘。
为降低赛项与设备的关联性,赛项关键技能(UI、Activity、Service、Broadcast Receiver

和 Content Provider 四大组件、资源使用、网络编程、Handler/多线程/定时器、多媒体、手势识别、数据存储、大数据分析、业务逻辑)可使用 PC 模拟器进行调试工作,模拟沙盘可增强赛项的观赏性。

10.2.3 软件版本

(1) JDK 8 或以上。
(2) Android Studio 3.0 或以上。
(3) Android 4.0 或以上。
(4) MySQL 5 或以上。
(5) Tomcat 8 或以上。

10.3 考察知识点

本竞赛采用建立试题库的方式,比赛前由裁判长从试题库中随机抽取一套试题作为竞赛题目,竞赛满分为 100 分。比赛成绩评判将根据"系统文档""程序排错""功能编码"及"创意设计"四个部分评分,分值比例分别为 5%、10%、80% 和 5%。

竞赛总得分=系统文档得分+程序排错得分+功能编码得分+创意设计得分

根据竞赛成绩,从高到低排序,按参赛人数的 10% 设一等奖,20% 设二等奖,30% 设三等奖。竞赛中考察的知识点及评分标准如表 10.1 所示。

表 10.1 智能交通赛项考察知识点及评分标准

考试模块	考察点	权重	描述	评分标准
系统文档	流程图	1%	绘制系统指定模块的流程图	结果评分(客观)(每组两名裁判随机抽取独立评分)
	类图/领域模型	1%	绘制系统指定模块的流程图的类图/领域模型	
	时序图	1%	绘制系统指定模块的时序图	
	数据库	1%	绘制系统指定模块的数据库表设计	
	代码规范	1%	展示系统指定模块的部分功能编码	
程序排错	UI 设计	5%	根据界面原型与实际显示之间的差异,定位并修改相应代码,以实现正确功能	结果评分(客观)(每组两名裁判随机抽取独立评分)
	业务逻辑	5%	根据需求描述及对功能的理解,定位并修复系统中业务逻辑存在的错误	
功能编码	UI 设计	5%	根据给定的资源和界面原型,自行设计/编写布局代码,实现与原型相一致的界面布局功能	结果评分(客观)(每组两名裁判随机抽取独立评分)
	四大组件	6%	Activity、Service、Broadcast Receiver 和 Content Provider 的使用	
	自定义控件	3%	实现自定义控件设计	
	资源管理	6%	各种类型资源的使用,例如,布局、图形、字符串和颜色等资源	
	动画效果	3%	动画技术的使用,例如,属性动画、视图动画、过度动画	

续表

考试模块	考察点	权重	描述	评分标准
功能编码	网络请求框架使用	5%	根据给定的网络通信 API，实现网络数据请求	结果评分（客观）（每组两名裁判随机抽取独立评分）
	数据封装和解析	5%	编程实现网络数据的传送和解析	
	Handler/多线程/定时器	5%	利用 Handler、多线程、定时器等技术，实现系统的同步/异步信息处理	
	多媒体	5%	图片、音频和视频等的使用	
	通知	3%	利用 Notification 实现消息提示	
	事件处理和手势识别	6%	手势识别 API 的使用	
	数据存储	7%	SharedPreferences、文件、SQLite 等数据存储方式的使用，以及第三方开源 LitePal 等插件使用	
	WebView	5%	WebView 使用及与原生态数据交互	
	地图导航	5%	利用高德离线地图，实现导航等功能	
	大数据分析	5%	利用服务器提供大数据 API，进行数据分析	
	依赖关系	5%	利用第三方开源图表库进行图形化分析显示	
创意设计	美观性	2%	界面具备可视化，美观简洁易懂，操作符合人体工程学	结果评分（客观）（每组两名裁判随机抽取独立评分）
	主题性	1%	符合智能交通主题，传递的理念积极向上，融于智能理念	
	实用性	1%	构思与设计的完善性与合理性，能为生活提供服务便利	
	技术性	1%	模块的技术含量，以及复杂度	
	创意性	1%	创意新颖程度	
扣分项	违纪扣分		视情节而定	裁判长

10.4 地图导航案例

10.4.1 编码实现离线地图 1

1. 功能说明

实现地图的坐标点定位、测距等功能。

注意以下两点。

（1）请先检查平台内是否有 amap 文件夹。如没有，请将离线地图文件夹中的 amap 文件夹整体复制至联想开发平台根目录内。

（2）离线地图所需的 jar 包和 so 库文件本题库不提供，需考生自己去高德官网下载并集成在项目中：https://lbs.amap.com/api/android-sdk/download。

微课视频

微课视频

2. 题目要求

(1) 完成如图 10.1 所示的布局(细节部分已框出),实现离线地图的初始化加载,地图可放大缩小、旋转、拖动等。

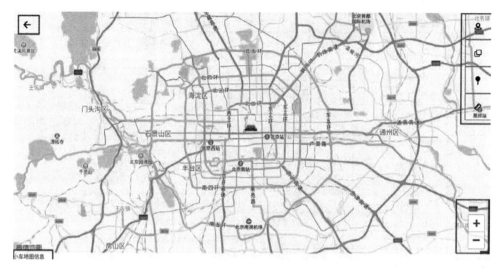

图 10.1　界面原型

(2) 完成如图 10.2 所示的功能。单击右侧的 ♀ 图标,在地图中显示 1、2、3、4 号小车的标记点,并在地图左下角弹出提示信息"1、2、3、4 号小车地图标记已完成"。单击 1 号小车标记点,弹出文本框,文本框中内容显示为 1 号小车目前的位置。同理,单击 2、3、4 号小车标记点,也会弹出每辆小车的位置。

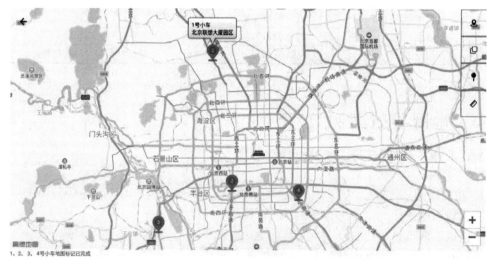

图 10.2　界面原型

3. 操作视频

扫描下方的二维码,即可观看实现步骤及运行效果。

10.4.2　编码实现离线地图 2

1. 功能说明

实现地图的图层切换等功能。

注意以下两点。

（1）请先检查平台内是否有 amap 文件夹。如没有，请将离线地图文件夹中的 amap 文件夹整体复制至联想开发平台根目录内。

（2）离线地图所需的 jar 包和 so 库文件本题库不提供，需考生自己去高德官网下载并集成在项目中：https://lbs.amap.com/api/android-sdk/download。

2. 题目要求

（1）完成如图 10.3 所示的布局（细节部分已框出），实现离线地图的初始化加载，地图可放大缩小、旋转、拖动等。

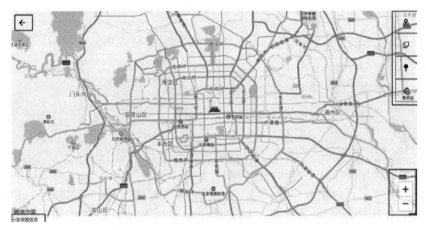

图 10.3　界面原型

（2）完成如图 10.4 所示的功能。单击右侧的 图标进行图层切换，选择对应图层并在左下角提示出切换对应图层，如图 10.5 所示。

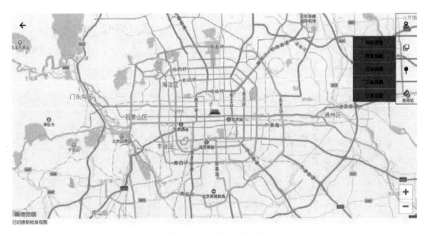

图 10.4　界面原型

图 10.5 界面原型

（3）完成导航视图、夜景视图、标准视图、卫星视图、交通视图的切换。

3. 操作视频

扫描下方的二维码，即可观看实现步骤及运行效果。

项目小结

本项目内容紧紧围绕移动互联网应用软件开发赛项展开，以智能交通为载体，教授学生学习 Android 基于位置的服务、地图导航、手势识别等方面内容，拓宽学生知识学习的境界和视野。培养计算机类相关专业学生移动应用软件产品开发意识和用户体验设计能力，激发其对移动互联网应用软件开发领域的学习和研究兴趣，提高其软件编程能力和职业素养。

项目 11

实战案例——移动互联网软件开发竞赛2

【教学导航】

学习目标	(1) 培养在企业真实项目环境下移动应用开发的工程实践能力。 (2) 培养设计能力和创新能力。 (3) 培养团队协作、沟通力、抗压力、职业规范等职场素质。 (4) 提高学生移动互联网应用软件开发技能水平。
教学方法	任务驱动法、理论实践一体化、探究学习法、分组讨论法
课时建议	6课时

移动互联网应用软件开发赛项,采用智能交通领域企业真实案例,竞赛编程模块分为:红绿灯管理、环境指标、交通单双号管制、车管局账户车辆管理、公交查询、路况查询、生活助手、车辆违章、数据分析、个人中心、车辆ETC账户管理、车辆违章视频浏览播放、意见反馈、城市地铁、高速路况查询、旅行助手、天气信息、定制班车、二维码支付、新闻客户端、IC卡充值、停车场信息管理、小车充值、我的座驾、我的交通、离线地图、实时交通、违章类型分析等。本章选择生活助手、旅行助手、数据分析三个模块进行解析,学习网络请求框架、Handler/多线程/定时器、数据封装和解析、多媒体、依赖注入、事件传递、数据挖掘和开源图表库MPAndroidChart API等相关知识。

11.1 编程案例——生活助手

11.1.1 功能说明

通过生活助手功能,可以查询当地昨天、今天以及未来4天的气象信息,并且为用户提供生活指数以及整点天气实况。

微课视频

11.1.2 题目要求

单击侧边栏/主界面的"生活助手"列表项,进入生活助手界面,如图 11.1 所示。

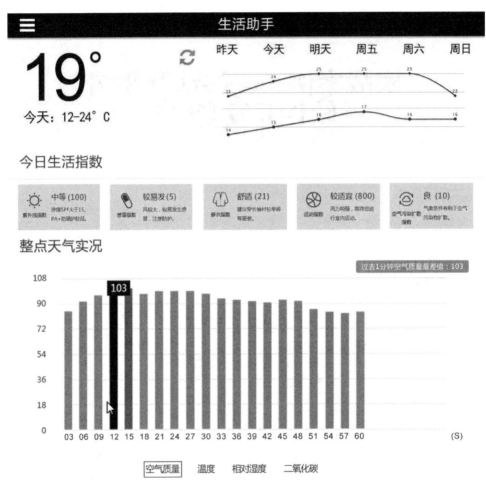

图 11.1 界面原型

(1) 上方天气栏目,左侧显示当天即时温度以及当天温度区间,右侧显示昨天、今天以及未来 4 天的天气数据,单击"刷新"按钮,显示当前天气数据。

(2) 今日生活指数栏目,实时(每隔 3s)显示当前生活信息指数,包括 PM2.5、空气湿度和温度等,如表 11.1~表 11.5 所示。

表 11.1 紫外线指数:根据光照强度值显示

紫外线强度	(0,1000)	[1000,3000]	(3000,∞)
强度描述	弱	中等	强
提示信息	辐射较弱,涂擦 SPF12~15、PA+护肤品	涂擦 SPF 大于 15、PA+防晒护肤品	尽量减少外出,需要涂抹高倍数防晒霜

表 11.2　感冒指数：根据温度值显示

感冒指数	(∞,8)	[8,∞)
强度描述	较易发	少发
提示信息	温度低,风较大,较易发生感冒,注意防护	无明显降温,感冒概率较低

表 11.3　穿衣指数：根据温度值显示

穿衣指数	(∞,12)	[12,21]	(21,∞)
强度描述	冷	舒适	热
提示信息	建议穿长袖衬衫、单裤等服装	建议穿短袖衬衫、单裤等服装	适合穿T恤、短薄外套等夏季服装

表 11.4　运动指数：根据二氧化碳值显示

运动指数	(0,3000)	[300,6000]	(6000,∞)
强度描述	适宜	中	较不宜
提示信息	气候适宜,推荐您进行户外运动	易感人群应适当减少室外活动	空气氧气含量低,请在室内进行休闲运动

表 11.5　空气污染扩散指数：根据 PM2.5 值显示

空气污染扩散指数	(0,30)	[30,100]	(100,∞)
强度描述	优	良	污染
提示信息	空气质量非常好,非常适合户外活动,趁机出去多呼吸新鲜空气	易感人群应适当减少室外活动	空气质量差,不适合户外活动

（3）整点天气实况,显示空气质量(PM2.5)、温度、相对湿度、二氧化碳指标。
① 手势左右滑动,进行指标间切换。
② X 轴：时间轴,最大 60s,周期为 3s,即 3s 更新一次数据,并存储过 去1min 的数据。
③ Y 抽：显示存储的过去1min 的数据值。
④ 在图标右上方显示当前图表过去1min 内数据的最大值或最小值。
（4）空气质量(PM2.5)指标：柱状图显示,如图 11.1 最下部分所示,过去 1min 内,空气质量最高值为 103。
（5）温度指标：折现图显示,如图 11.2 所示,过去 1min 最高气温为 25℃,最低气温为 16℃。
（6）相对湿度指标：折现图显示,如图 11.3 所示,过去 1min 最大相对湿度为 67%。
（7）二氧化碳指标：折现图显示,如图 11.4 所示,过去 1min 最大相对浓度为 97。

11.1.3　操作视频

扫描下方的二维码,即可观看实现步骤及运行效果。

微课视频

微课视频

微课视频

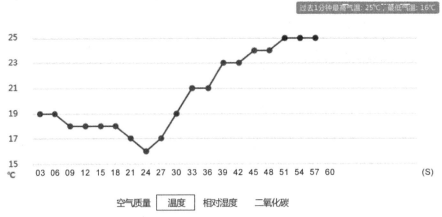

图 11.2 温度折线图界面原型

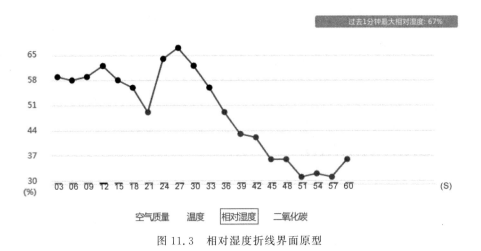

图 11.3 相对湿度折线界面原型

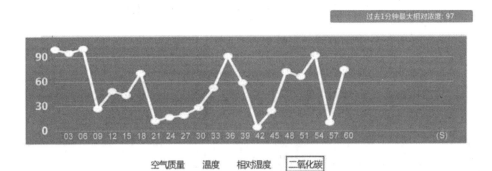

图 11.4 二氧化碳界面原型

11.2 编程案例——旅行助手

11.2.1 功能说明

通过旅行助手功能,实现城市景点的介绍,打造城市名片。

11.2.2 题目要求

打开侧滑菜单的"旅行助手",单击进入界面,效果如图 11.5 所示。

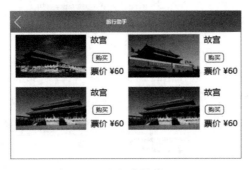

图 11.5 旅行助手界面原型

(1) 完成网格显示城市景点等布局信息。

(2) 在图 11.5 中,单击景点图片进入"详细信息"页面,如图 11.6 所示,单击电话号码可以自动跳转至电话拨打页面。

图 11.6 详细信息

(3) 在图 11.5 中,单击"购买"按钮进入"我的购物车"界面。选择使用日期,如果是当天,显示为灰色框不可用。可以购买第二天以后的门票,如图 11.7 所示,第二天可订则可用(框为黑色),更多日期的箭头可以弹出日期框进行日期的选择。列表显示购买产品的图片、标题、简介、数量和产品价格;数量可通过"+"或"-"进行增减,减至为 0 时"-"置灰色不可用状态,产品价格为单价,"数量×单价"为总金额。"清空购物车"则删除列表里产品信息。

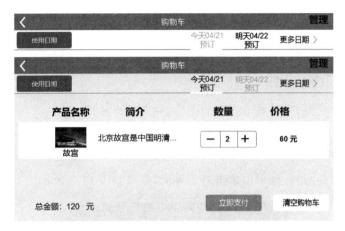

图 11.7　购物车页面

（4）单击更多日期的箭头可以弹出日期选择框进行日期的选择，如图 11.8 所示。

（5）单击"管理"按钮，可删除对应的购物内容，如图 11.9 所示。

图 11.8　日期选择框

图 11.9　单击"管理"按钮显示删除选项

（6）单击"立即支付"按钮跳转到支付界面。展示二维码，5s 改变一次。

（7）长按二维码图片，显示二维码信息，如图 11.10 所示。

图 11.10　二维码支付界面

11.2.3　操作视频

扫描下方的二维码，即可观看实现步骤及运行效果。

微课视频

11.3 编程案例——数据分析

11.3.1 功能说明

长期以来,交通行业内存在数据资源散、开放共享难、分析应用弱等问题,在一定程度上制约了行业发展。因此,需要借助大数据的力量解决日益紧迫的交通问题,分析司机违章行为,减少违章驾驶习惯等。

11.3.2 题目要求

(1) 单击侧滑菜单的"数据分析"项,进入数据分析界面,手势左右滑动,进行数据指标分析图表间的切换。根据相关接口返回数据,进行数据挖掘分析,所有统计图表数值精确到百分数统计的小数 2 位,如 71.33%。利用饼状图显示平台上有违章车辆和无违章车辆的占比统计,如图 11.11 所示,单击右上角"返回"图标,返回主页。

图 11.11 有违章车辆和无违章车辆的占比统计界面原型

(2) 根据相关接口返回数据,进行数据挖掘分析,利用饼状图显示平台上有无"重复违章记录的车辆"的占比统计,如图 11.12 所示。

(3) 根据相关接口返回数据,进行数据挖掘分析,利用水平柱状图显示违章车辆的违章次数占比分布图统计,如图 11.13 所示界面原型仅供参考。

(4) 根据相关接口返回数据,进行数据挖掘分析,利用多层级的堆叠条形图显示平台上年龄群体车辆违章的占比统计,如图 11.14 所示。

图 11.12　有无"重复违章记录的车辆"的占比统计界面原型

图 11.13　违章车辆的违章次数占比分布界面原型

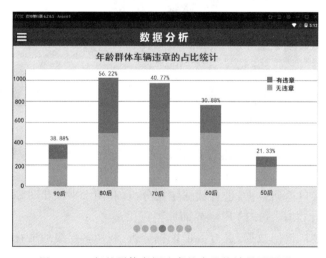

图 11.14　年龄群体车辆违章的占比统计界面原型

（5）根据相关接口返回数据，进行数据挖掘分析，利用多层级的堆叠条形图显示平台上男性和女性有无车辆违章的占比统计，如图11.15所示。

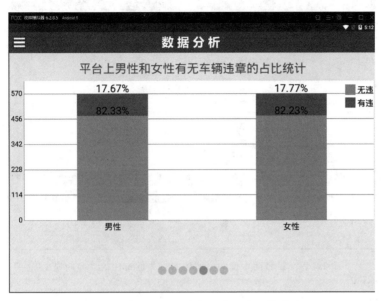

图 11.15　男性和女性有无车辆违章的占比统计界面原型

（6）根据相关接口返回数据，进行数据挖掘分析，利用柱状图显示每日时段内车辆违章的占比统计，图11.16界面原型仅供参考。

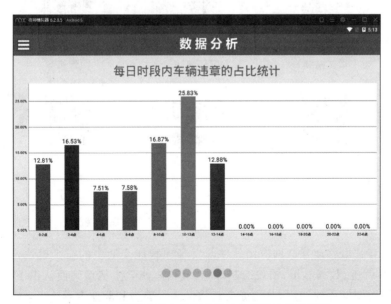

图 11.16　每日时段内车辆违章的占比统计界面原型

（7）根据相关接口返回数据，进行数据挖掘分析，利用水平柱状图排名前十位的交通违法行为的占比统计，图11.17界面原型仅供参考。

图 11.17　排名前十位的交通违法行为的占比统计界面原型

11.3.3　操作视频

扫描下方的二维码,即可观看实现步骤及运行效果。

微课视频　　　　　　微课视频　　　　　　微课视频

微课视频　　　　　　微课视频　　　　　　微课视频

项目小结

　　本项目内容选取移动互联网应用软件开发赛项智能交通中的三个模块,详细介绍了网络访问、JSON 解析、数据显示、图表绘制等相关内容,使学生从点到面认识了 Android 网络编程思路及开源图标库的使用方法,提高其软件编程能力和职业素养。

习题参考答案

项目 1　习题参考答案

1.	2.	3.	4.	5.	6.	7.	8.	9.	10.
C	B	B	A	C	B	×	√	×	×

项目 2　习题参考答案

1.	2.	3.	4.	5.	6.	7.	8.	9.
B	A	C	B	C	A	C	C	A

项目 3　习题参考答案

1.	2.	3.	4.	5.	6.	7.	8.	9.	10.
B	A	C	A	D	C	A	A	A	B

项目 4　习题参考答案

1.	2.	3.	4.	5.	6.	7.	8.	9.	10.
B	A	C	B	A	C	A	A	C	A

项目 5　习题参考答案

1.	2.	3.	4.	5.	6.	7.	8.	9.	10.
B	A	B	D	B	A	B	A	A	D

项目 6　习题参考答案

1.	2.	3.	4.	5.	6.	7.	8.	9.	10.
B	B	A	D	C	B	B	C	A	D

项目 7　习题参考答案

1.	2.	3.	4.	5.	6.	7.	8.	9.	10.
B	A	A	B	D	B	D	A	A	B

项目 8：略

项目 9 习题参考答案

一、填空题

1. HTTP 请求
2. HttpClient、HttpURLConnection
3. GET、POST

二、选择题

1. D
2. C
3. B

三、简答题

1. 使用 HttpURLConnection 访问网络的基本步骤如下：

（1）创建一个 URL 对象：

```
URL url = new URL(http://www.baidu.com);
```

（2）调用 URL 对象的 openConnection() 来获取 HttpURLConnection 对象实例。

```
HttpURLConnection conn = (HttpURLConnection) url.openConnection();
```

（3）设置 HTTP 请求使用的方法：GET 或者 POST。

```
conn.setRequestMethod("GET");
```

（4）设置连接超时，读取超时的毫秒数，以及服务器希望得到的一些消息头。

```
conn.setConnectTimeout(6 * 1000);
conn.setReadTimeout(6 * 1000);
```

（5）调用 getInputStream() 方法获得服务器返回的输入流，然后对输入流进行读取操作。

```
InputStream in = conn.getInputStream();
```

（6）调用 disconnect() 方法将 HTTP 连接关掉。

```
conn.disconnect();
```

2. Handler 机制 4 个关键对象的作用如下：

（1）Message

Message 是在线程之间传递的消息，它可以在内部携带少量的信息，用于在不同线程之间交换数据。Message 的 what 字段可以用来携带一些整型数据，obj 字段可以用来携带一个 Object 对象。

(2) Handler

Handler 顾名思义就是处理者的意思,它主要用于发送消息和处理消息。一般使用 Handler 对象的 sendMessage()方法发送消息,发出的消息经过一系列的辗转处理后,最终会传递到 Handler 对象的 handlerMessage()方法中。

(3) MessageQueue

MessageQueue 是消息队列的意思,它主要用来存放通过 Handler 发送的消息。通过 Handler 发送的消息会存在 MessageQueue 中等待处理。每个线程中只会有一个 MessageQueue 对象。

(4) Looper

Looper 是每个线程中的 MessageQueue 的管家。调用 Looper 的 loop()方法后,就会进入到一个无限循环中。然后每当发现 MessageQueue 中存在一条消息时,就会将它取出,并传递到 Handler 的 HandlerMessage()方法中。此外每个线程也只会有一个 Looper 对象。在主线程中创建 Handler 对象时,系统已经为我们创建了 Looper 对象,所以不用手动创建 Looper 对象,而在子线程中的 Handler 对象,我们需要调用 Looper.loop()方法开启消息循环。

3. 使用 gson 解析数组格式的 JSON 字符串,步骤如下:

```
Gson gson = new Gson();
Type type = new TypeToken<List<String>>() {}.getType();
List<String> jsonList = gson.fromJson(json,type);
listAll.addAll(listData.get(0).getData());
```

图书资源支持

感谢您一直以来对清华版图书的支持和爱护。为了配合本书的使用,本书提供配套的资源,有需求的读者请扫描下方的"书圈"微信公众号二维码,在图书专区下载,也可以拨打电话或发送电子邮件咨询。

如果您在使用本书的过程中遇到了什么问题,或者有相关图书出版计划,也请您发邮件告诉我们,以便我们更好地为您服务。

我们的联系方式:

清华大学出版社计算机与信息分社网站: https://www.shuimushuhui.com/

地　　址: 北京市海淀区双清路学研大厦 A 座 714

邮　　编: 100084

电　　话: 010-83470236　　010-83470237

客服邮箱: 2301891038@qq.com

QQ: 2301891038(请写明您的单位和姓名)

资源下载: 关注公众号"书圈"下载配套资源。

书圈

清华计算机学堂

观看课程直播